T0172906

RUSSIAN TRANSLATIONS SERIES

1. K.Ya. Kondrat'ev et al. (editors): *USSR/USA Bering Sea Experiment*
2. D.V. Nalivkin: *Hurricanes, Storms and Tornadoes*
3. V.M. Novikov (editor): *Handbook of Fishery Technology*, Vol. 1
4. F.G. Martyshev: *Pond Fisheries*
5. R.N. Burukovskii: *Key to Shrimps and Lobsters*
6. V.M. Novikov (editor): *Handbook of Fishery Technology*, Vol. 4
7. V.P. Bykov (editor): *Marine Fishes*
8. N.N. Tsvelev: *Grasses of the Soviet Union*
9. L.V. Metlitskii et al.: *Controlled Atmosphere Storage of Fruits*
10. M.A. Glazovskaya: *Soils of the World* (2 volumes)
11. V.G. Kort & V.S. Samoilenko: *Atlantic Hydrophysical Polygon-70*
12. M.A. Mardzhanishvili: *Seismic Design of Frame-panel Buildings and Their Structural Members*
13. E'.A. Sokolenko (editor): *Water and Salt Regimes of Soils: Modeling and Management*
14. A.P. Bocharov: *A Description of Devices Used in the Study of Wind Erosion of Soils*
15. E.S. Artsybashev: *Forest Fires and Their Control*
16. R.Kh. Makasheva: *The Pea*
17. N.G. Kondrashova: *Shipboard Refrigeration and Fish Processing Equipment*
18. S.M. Uspenskii: *Life in High Latitudes*
19. A.V. Rozova: *Biostratigraphic Zoning and Trilobites of the Upper Cambrian and Lower Ordovician of the Northwestern Siberian Platform*
20. N.I. Barkov: *Ice Shelves of Antarctica*
21. V.P. Averkiev: *Shipboard Fish Scouting and Electronavigational Equipment*
22. D.F. Petrov (Editor-in-Chief): *Apomixis and Its Role in Evolution and Breeding*
23. G.A. Mchedlidze: *General Features of the Paleobiological Evolution of Cetacea*
24. M.G. Ravich et al.: *Geological Structure of Mac. Robertson Land (East Antarctica)*
25. L.A. Timokhov (editor): *Dynamics of Ice Cover*
26. K.Ya. Kondrat'ev: *Changes in Global Climate*
27. P.S. Nartov: *Disk Soil-Working Implements*
28. V.L. Kontrimavichus (Editor-in-Chief): *Beringia in the Cenozoic Era*
29. S.V. Nerpin & A.F. Chudnovskii: *Heat and Mass Transfer in the Plant–Soil–Air System*
30. T.V. Alekseeva et al.: *Highway Machines*
31. N.I. Klenin et al.: *Agricultural Machines*
32. V.K. Rudnev: *Digging of Soils by Earthmovers with Powered Parts*
33. A.N. Zelenin et al.: *Machines for Moving the Earth*
34. *Systematics, Breeding and Seed Production of Potatoes*
35. D.S. Orlov: *Humus Acids of Soils*
36. M.M. Severnev (editor): *Wear of Agricultural Machine Parts*
37. Kh.A. Khachatryan: *Operation of Soil-working Implements in Hilly Regions*
38. L.V. Gyachev: *Theory of Surfaces of Plow Bottoms*
39. S.V. Kardashevskii et al.: *Testing of Agricultural Technological Processes*
40. M.A. Sadovskii (editor): *Physics of the Earthquake Focus*
41. I.M. Dolgin: *Climate of Antarctica*
42. V.V. Egorov et al.: *Classification and Diagnostics of Soils of the USSR*
43. V.A. Moshkin: *Castor*
44. E'.I. Sarukhanyan: *Structure and Variability of the Antarctic Circumpolar Current*
45. V.A. Shapa (Chief Editor): *Biological Plant Protection*
46. A.I. Zakharova: *Estimation of Seismicity Parameters Using a Computer*
47. M.A. Mardzhanishvili & L.M. Mardzhanishvili: *Theoretical and Experimental Analysis of Members of Earthquake-proof Frame-panel Buildings*
48. S.G. Shul'man: *Seismic Pressure of Water on Hydraulic Structures*
49. Yu.A. Ibad-zade: *Movement of Sediments in Open Channels*
50. I.S. Popushoi (Chief Editor): *Biological and Chemical Methods of Plant Protection*
51. K.V. Novozhilov (Chief Editor): *Microbiological Methods for Biological Control of Pests of Agricultural Crops*
52. K.I. Rossinskii (editor): *Dynamics and Thermal Regimes of Rivers*
53. K.V. Gnedin: *Operating Conditions and Hydraulics of Horizontal Settling Tanks*
54. G.A. Zakladnoi & V.F. Ratanova: *Stored-grain Pests and Their Control*
55. Ts.E. Mirtskhulava: *Reliability of Hydro-reclamation Installations*
56. Ia.S. Ageikin: *Off-the-road Mobility of Automobiles*

(continued)

57. A.A. Kmito & Yu.A. Sklyarov: *Pyrheliometry*
58. N.S. Motsonelidze: *Stability and Seismic Resistance of Buttress Dams*
59. Ia.S. Ageikin: *Off-the-road Wheeled and Combined Traction Devices*
60. Iu.N. Fadeev & K.V. Novozhilov: *Integrated Plant Protection*
61. N.A. Izyumova: *Parasitic Fauna of Reservoir Fishes of the USSR and Its Evolution*
62. O.A. Skarlato (Editor-in-Chief): *Investigation of Monogeneans in the USSR*
63. A.I. Ivanov: *Alfalfa*
64. Z.S. Bronshtein: *Fresh-water Ostracoda*
65. M.G. Chukhrii: *An Atlas of the Ultrastructure of Viruses of Lepidopteran Pests of Plants*
66. E.A. Bosoi et al.: *Theory, Construction and Calculations of Agricultural Machines, Vol. 1*
67. G.A. Avsyuk (Editor-in-Chief): *Data of Glaciological Studies*
68. G.A. Mchedlidze: *Fossil Cetacea of the Caucasus*
69. A.M. Akramkhodzhaev: *Geology and Exploration of Oil- and Gas-bearing Ancient Deltas*
70. N.M. Berezina & D.A. Kaushanskii: *Presowing Irradiation of Plant Seeds*
71. G.U. Lindberg & Z.V. Krasyukova: *Fishes of the Sea of Japan and the Adjacent Areas of the Sea of Okhotsk and the Yellow Sea*
72. N.I. Plotnikov & I.I. Roginets: *Hydrogeology of Ore Deposits*
73. A.V. Balushkin: *Morphological Bases of the Systematics and Phylogeny of the Nototheniid Fishes*
74. E.Z. Pozin et al.: *Coal Cutting by Winning Machines*
75. S.S. Shul'man: *Myxosporidia of the USSR*
76. G.N. Gogonenkov: *Seismic Prospecting for Sedimentary Formations*
77. I.M. Batugina & I.M. Petukhov: *Geodynamic Zoning of Mineral Deposits for Planning and Exploitation of Mines*
78. I.I. Abramovich & I.G. Klushin: *Geodynamics and Metallogeny of Folded Belts*
79. M.V. Mina: *Microevolution of Fishes*
80. K.V. Konyaev: *Spectral Analysis of Physical Oceanographic Data*
81. A.I. Tseitlin & A.A. Kusainov: *Role of Internal Friction in Dynamic Analysis of Structures*
82. E.A. Kozlov: *Migration in Seismic Prospecting*
83. E.S. Bosoi et al.: *Theory, Construction and Calculations of Agricultural Machines, Vol. 2*
84. B.B. Kudryashov and A.M. Yakovlev: *Drilling in the Permafrost*
85. T.T. Klubova: *Clayey Reservoirs of Oil and Gas*
86. G.I. Amurskii et al.: *Remote-sensing Methods in Studying Tectonic Fractures in Oil- and Gas-bearing Formations*
87. A.V. Razvalyaev: *Continental Rift Formation and Its Prehistory*
88. V.A. Ivovich and L.N. Pokrovskii: *Dynamic Analysis of Suspended Roof Systems*
89. N.P. Kozlov (Technical Editor): *Earth's Nature from Space*
90. M.M. Grachevskii and A.S. Kravchuk: *Hydrocarbon Potential of Oceanic Reefs of the World*
91. K.V. Mikhailov et al.: *Polymer Concretes and Their Structural Uses*
92. D.S. Orlov: *Soil Chemistry*
93. L.S. Belousova & L.V. Denisova: *Rare Plants of the World*
94. T.I. Frolova et al.: *Magmatism and Transformation of Active Areas of the Earth's Crust*
95. Z.G. Ter-Martirosyan: *Rheological Parameters of Soils and Design of Foundations*
96. S.N. Alekseev et al.: *Durability of Reinforced Concrete in Aggressive Media*
97. F.P. Glushikhin et al.: *Modelling in Geomechanics*
98. I.A. Krupenikov: *History of Soil Science*
99. V.A. Burkov: *General Circulation of the World Ocean*
100. B.V. Preobrazhensky: *Contemporary Reefs*
101. V.P. Petrukhin: *Construction of Structures on Saline Soils*
102. V.V. Bulatov: *Geomechanics of Deep-seated Deposits*
103. A.A. Kuzmenko et al.: *Seismic Effects of Blasting in Rock*
104. L.M. Plotnikov: *Shear Structures in Layered Geological Bodies*
105. M.I. Petrosyan: *Rock Breakage by Blasting*
106. O.L. Kuznetsov and E.M. Simkin: *Transformation and Interaction of Geophysical Fields in the Lithosphere*
107. M.V. Abdulov: *Phase Transformations and Rock Genesis*
108. I.E. Mileikovskii and S.I. Trushin: *Analysis of Thin-walled Structures*

(continued on title facing page)

Historical Geotectonics
PALAEOZOIC

Historical Geotectonics
PALAEOZOIC

V.E. Khain
K.B. Seslavinsky

RUSSIAN TRANSLATIONS SERIES
115

A.A. BALKEMA/ROTTERDAM/BROOKFIELD/1996

Translation of : Istoricheskaya geotektonica-paleozoi, Nedra, Moscow, 1991.
Revised and updated by the author for the English edition
in 1994.

Translator   : Mr. P.M. Rao

Translation Editor : Dr. V.E. Khain

General Editor   : Ms. Margaret Majithia

ISBN (Palaeozoic Volume) 90 5410 226 8
ISBN (Set) 90 5410 224 1

Distributed in USA and Canada by: A.A. Balkema Publishers, Old Post Road,
Brookfield, VT 05036, USA.

Preface

This publication represents the second volume of HISTORICAL GEOTEC-TONICS conceived by V.E. Khain. The first volume, dealing with the Precambrian, was published in 1988 by Nedra Publishers (authors: V.E. Khain and N.A. Bozhko). The third volume covering the Mesozoic and Cenozoic was written by V.E. Khain and A.N. Balukhovsky and published in 1993.

The layout of the present volume conforms to the general plan of the first: descriptions of factual material and palaeotectonic conclusions are grouped independently in the various chapters. The better coverage of the Palaeozoic compared to the Precambrian helped us to examine in greater detail the various aspects of the geotectonic process in the concluding general chapter. From the viewpoint of classification of various tectonic events, it appears more logical to commence the analysis of the tectonic development of the Earth from the Vendian, which many geologists justifiably include in the Palaeozoic, or even from the latter half of the Late Riphean, which has already been covered in the first volume. We have therefore had to revert in part in this second volume to the terminal episodes of the Precambrian.

Concomitant with a study of this volume, we recommend reference to *Atlas of the Lithological-Palaeogeographic Maps of the World. Late Precambrian and Palaeozoic Continents* by A.B. Ronov, V.E. Khain and K.B. Seslavinsky published in 1984 by the Leningrad Division of Nedra Publishers.

The first and second sections of all chapters, except the sixth, have been authored by K.B. Seslavinsky and the third sections by V.E. Khain. Sections 6.1, 6.2, 6.4.1 and 6.4.4 of Chapter 6 were written by V.E. Khain and the other sections by K.B. Seslavinsky. The 'Conclusions' were authored jointly. V.E. Khain is the general editor.

Contents

1

Vendian—Late Cambrian. Development of Phanerozoic Mobile Belts

1.1 REGIONAL REVIEW. CONTINENTAL CRATONS AND THEIR MARGINS

1.1.1 North American Craton

V e n d i a n. At this time, uplift and lowland conditions prevailed within the craton (Fig. 1-1). Marine basins were present only in the marginal zones of this craton. Thus from the Riphean the Precordilleran zone of shelf-type basins continued to develop in the west. Predominantly terrigenous and, less frequently, carbonate formations were deposited in these basins to a thickness of up to 3 km or more in a westerly direction. In the Canadian Rocky Mountains such deposits were represented by the Hamil and Horstiff Creek formations of the Upper Windermere series; in eastern Alaska by carbonate-terrigenous portions of the Tinder group; and in California by Stirling quartzites, Johnny sandstones and dolomites and Nunday dolomites (Cloud et al., 1974; Young, 1977).

Carbonates predominated on the northern slope of the Canadian Shield, in the shelf basins of the Fore-Innuitian zone, but the thickness of these formations is small, not exceeding 700 to 800 m. In eastern Greenland subsidence of the cratonic margin was compensated by accumulation of the Hagen Fjord group of shelf-type terrigenous and carbonate rocks. This group also contains a complex of intermediate volcanic rocks (Harland, 1979). The thickness of the Vendian formations in these regions reaches 5 km.

E a r l y   C a m b r i a n. Subsidence in the marginal zones of the craton intensified in this stage and the area of marine basins enlarged somewhat. This is true primarily of the Fore-Innuitian zone within which sand deposits of small thickness formed over an extensive area under shallow-water conditions. Northward, sand lithofacies were often replaced by carbonates and their thickness here rises sharply to 2 km or more (Trettin and Balkwill, 1979). Accumulation of shelf carbonates predominated in eastern Greenland also.

In the Precordilleran zone the role of coarse clastic and carbonate rocks rose in the composition of the formations, suggesting an increase in tectonic

2

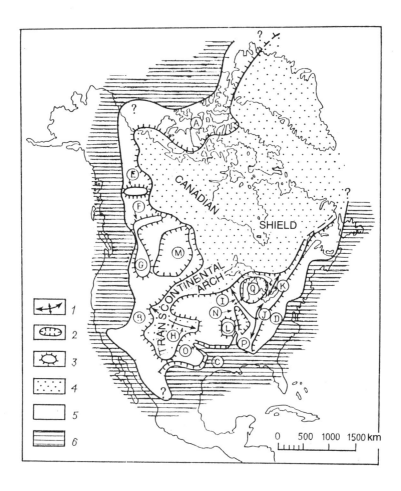

Fig. 1-1. Main physical-geographic and structural features of North America in the Early Palaeozoic (from Frasier and Schwimmer, 1987).

1—structural arch; 2—basin; 3—uplifts; 4—land; 5—shallow sea; 6—ocean basin.

A—Innuitian shelf; B—Cordilleran shelf; C—Wichita trough; D—Appalachian shelf; E—Tatleena arch; F—Peace River arch; G—Montana dome; H—Cambridge-Central Kansas arch; I—Kankakee arch; J—Cincinnati arch; K—Findlay arch; L—Ozark dome; M—Williston basin; N—Illinois basin; O—Southern Oklahoma aulacogen; P—Reelfoot rift; Q—Michigan basin.

differentiation of these shelf basins. This conclusion is also confirmed by data on the outpouring of plateau basalts which originated within Utah and Nevada states (Stewart and Poole, 1974). The thickness of amygdaloidal basalts constitutes tens or even a few hundred metres. All of this clearly points to marginal-continental rifting conditions.

Subsidence and manifestation of felsic and mafic (bimodal) volcanism commenced in the Early Cambrian in the Fore-Appalachian zone. In the southern segment of this zone sedimentary complexes are represented by coarse clastic rocks of littoral and continental facies and in the northern segment, in the Avalonian block, by marine sandstones and shales. Basalts and rhyolites of Macdonalds Brooke, whose formation is evidently associated with rifting in the initial stage of opening of the Iapetus palaeo-ocean, are also developed here (Murphy et al., 1985).

M i d d l e C a m b r i a n. As before, uplift and lowland conditions prevailed over much of the cratonic territory but subsidence intensified and marine basins enlarged even more in the Fore-Appalachian and Precordilleran zones. Transgression was manifest most intensely in the Precordilleran zone. The boundary of the sea here shifted repeatedly throughout this epoch and predominantly sandy sediments were deposited in the broad littoral belt. Farther westward, these sediments were replaced by much thinner clay and carbonate formations, their thickness gradually rising to 1 km (Rowell et al., 1979). In the northern part of this zone lay a basin with high water salinity; saline and gypsum-bearing formations of the Salina River and Samgvin accumulated here at the end of the Middle Cambrian.

A similar picture of distribution and thickness of facies is noticed in the Fore-Appalachian zone but eruptions of basalts and rhyolites continued here in the Avalonian block. The composition of these outpourings suggests conditions of intraplate extension and continental rifting (Greenough et al., 1985), i.e., continuation of opening of the Iapetus palaeo-ocean.

Within the Canadian Arctic archipelago and north-western Greenland (Fore-Innuitian zone), sand formations were succeeded by carbonates and terrigenous-carbonates. Their thickness is not significant, being only 800 m in south-western Ellesmere Island. Regression encompassed northern and north-eastern Greenland.

L a t e C a m b r i a n. Subsidence in the craton intensified even more and the sea covered almost a third of its territory. Transgression initially encompassed extensive areas of the Midcontinent leading to connection of the Fore-Appalachian and Precordilleran basins. The quiescent tectonic regime and the warm climate determined the predominance of limestones and dolomites in the composition of the formations, especially in the southern Midcontinent and on the northern slope of the craton. However, along the periphery of sedimentary basins, sandy clay formations accumulated almost exclusively. The thickness of the Upper Cambrian over extensive areas is insignificant at a few tens of metres; the Michigan-Illinois basin, in which the thickness of carbonates exceeds 1 km in its southern segment, constitutes an exception (Mellen, 1977).

Much of the thickness (over 2 km) of carbonate and terrigenous rocks in the shelf zone accumulated in the Rocky Mountains of Canada while greywackes accumulated and eruptions of basalts and andesitic basalts formed possibly under island arc conditions within the Pelly Mountains.

Thus the magnitude of subsidence and area of sea increased steadily throughout the Vendian and Cambrian in the North American Craton. The small thickness of formations in interior basins and comparatively poor development of terrigenous rocks point to the fairly quiescent general conditions and insignificant contrast of tectonic movements. Land areas represented peneplains and an undulating relief could have existed only at the beginning of the Cambrian.

A high tectonic activity has been observed in the marginal zones of the craton, which developed in a regime of passive margins that survived the stage of rifting. The tempo of subsidence here was rather high and bimodal volcanism manifested locally on a small scale. Miogeosynclinal development of eastern Greenland ended with stabilisation and transition to a platform regime without epigeosynclinal orogenesis.

1.1.2 East European Craton

V e n d i a n. Commencement of cratonic development is *sensu stricto* associated with the Early Vendian. In this stage zones of subsidence extended far beyond the limits of aulacogens that actively subsided in the Late Riphean. In the second half of the Early Vendian, much of the craton was encompassed by an extensive ice sheet (Chumakov, 1984) while basaltic volcanism (Volyn series) was manifest in its south-western part. Volcanic products contain plateau basalts, andesitic dacites, trachyrhyolites and their tuffs up to a thickness of 500 m (in: *Palaeogeography and Lithology of the Vendian and Cambrian of the Western Part of the East European Craton*, 1980). Glaciation and bursts of volcanism were synchronous in the structural reorganisation of the craton, i.e., transition to the true platform stage of its development.

Subsidence extended in the Late Vendian to the entire Baltic-Moscow syneclise; the basin covering it joined the Urals and Vistula-Dnestr pericratonic troughs and probably the Caspian basin through a strait. Marine sediments in all these interior basins are represented by terrigenous rocks (Valday series) up to 300 to 500 m in thickness, the thickness being considerably higher in the marginal zones of the craton. Thus in the Timan section of the Uralian zone, the Churoch series is about 2 km thick while bands of turbidites are present among Vendian complexes of 4 km thickness (Sylvytsa series) and 2 km (Asha series) in the Central and Southern Urals respectively. A thick flysch complex (up to 3.5 km) was formed in the Vendian in the northern part of the platform in the zone of eastern Finnmarksvidda (Finnmarken). Turbidites of the Stapugieds suite, constituting the above complex, point to the existence of a continental slope and rise condition in this region.

E a r l y   C a m b r i a n. The boundaries of the regions of subsidence underwent repeated modifications throughout this epoch (in: *Palaeogeography and Lithology...*, 1980). The total area of the sea, compared to the Vendian period, became somewhat less since extensive regression took place in the Ryazan-Saratov strait and the basins in the south-eastern and north-eastern parts of the craton (Fig. 1–2). These uplifts were partly compensated by a transgression in the western Baltic-Moscow basin and within Scandinavia. Clay deposits predominated throughout the territory except for Scandinavia where sandy facies prevailed. The thickness of the Lower Cambrian rocks usually does not exceed tens of metres and reaches hundreds of metres only in the Fore-Timan zone, evidently as a result of the Timan uplift. At the beginning of the epoch, this zone was probably connected by a strait to the Urals basin. In the south-western part of the craton considerable thickness has been established within the Moesian block.

M i d d l e   C a m b r i a n. The general tendency towards regression continued. The Uralian marginal zone dried up for a long time and the Baltic-Moscow basin contracted in size. The basin became a partially closed, shallow muddy sea with a very low rate of subsidence changing over to uplift in the Maya epoch (Ivanov, 1979). The thickness of the Middle Cambrian clay sediments and of littoral sand deposits nowhere exceeded a few tens of metres. Clay formations of small thickness predominated in the Scandinavian marine basin. Intense subsidence took place only in the south-western part of the craton, in the Moesian block, and coarse clastic deposits a few hundred metres thick formed there.

L a t e   C a m b r i a n. Uplift and regression in the craton attained maximum proportions. Upper Cambrian formations have nowhere been proved and probably the entire territory represented a low plain and, at places, an undulating hilly terrane. Subsidence with marine terrigenous sedimentation survived only in the north-west (Scandinavia) and in a narrow band along the south-western margin of the craton (including the Moesian block).

In this stage as a whole, the role of uplift, although of an extremely quiescent nature, steadily rose in the East European Craton. The area of interior as well as marginal marine basins contracted. The weak contrast of tectonic movements led to low plains conditions in land regions. Sediments, predominantly thin clay formations, entered the basin from these low plains. A very high tectonic activity has been identified only in the northern part of the craton, in eastern Finnmarksvidda (Finnmarken), and in the south-western marginal zone, including the Moesian block.

1.1.3 Siberian Craton

V e n d i a n. The Vendian was preceded by a general uplift and erosion over much of the craton, succeeded later by subsidence and transgression. During the Vendian, the sea gradually covered almost the

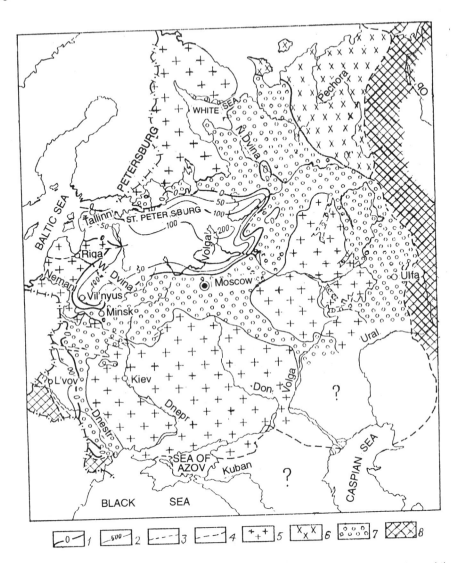

Fig. 1-2. Schematic distribution and thickness variations of marine silt-clay formations of the transgressive stage of Early Palaeozoic development of the East European Craton, i.e., Early Cambrian (after N.S. Igolkina In: *Geological Structure of the USSR and Pattern of Location of Mineral Deposits*. Vol. 1, *Russian Craton*, 1985):

1—boundaries of formations depicted by zero isopachs; 2—isopachs of formations, m; 3—faults; 4—boundaries of East European Craton; 5, 6 —outcrops of basement rocks (5—pre-Riphean and 6—Riphean); 7—zone of distribution of older cover formations; 8—geosynclinal formations of this stage.

entire territory of the Siberian Craton. The general conditions favoured sedimentation of carbonates and thus the Vendian sedimentary rocks are represented predominantly by shallow-water dolomites and limestones up to a few hundred metres thick (in: *Phanerozoic* of *Siberia*. Vol. 1, *Vendian. Palaeozoic*, 1984). Extensive basins of high salinity prevailed in the region of Anabar massif, Irkutsk amphitheatre and Aldan Shield.

E a r l y   C a m b r i a n. Downwarping continued over much of the craton and carbonate deposition prevailed. A maximum amplitude of subsidence was characteristic of troughs arising in continuation of the re-entrant angles of geosynclinal systems—Khantay, Irkineev and Urin—and zones of subsidence developed on the margins of the craton along the boundaries of mobile belts—Fore-Yenisei, Fore-Sayan, Fore-Baikal, Berezovo, Yudoma and Lena—surrounding the craton and also of inner troughs—Central and Sukhana (Fig. 1-3). The thickness of the Lower Cambrian attains hundreds

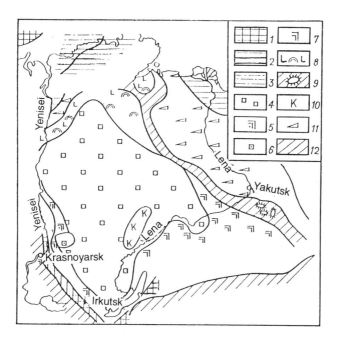

Fig. 1-3. Palaeotectonic scheme of the Siberian Craton for end of the Early Cambrian to commencement of the Middle Cambrian (after N.S. Malich. In: *Geological Structure of the USSR and Pattern of Location of Mineral Deposits*, vol. 4, *Siberian Craton*, 1987).

1—zones of uplift and erosion; 2 to 11—formations (2—clayey limestone, 3—sandy clay, 4—carbonate-halite, 5—sand-clay-dolomite, 6—clay (sand)-dolomite-halite, 7—limestone-dolomite, 8—limestone-siliceous dolomite, 9—limestone-dolomite of reef genesis, 10—halite and potassium-bearing salts, 11—bituminous-siliceous carbonate); 12—geosynclinal regions.

of metres in these troughs. Southern, south-western and central parts of the craton represented a gigantic, partially closed saline basin with thickness of formations containing not only halite, but also potassium salt, reaching 2 km in the Irkutsk amphitheatre. It was bound in the east by a band of barrier reefs extending in a north-north-westerly direction, crossing the middle reaches of the Lena River. Eastward, in relatively deep troughs—Sukhana and Yudoma—and on Taimyr peninsula, under conditions of subsidence uncompensated by sedimentation, bituminous calcareous clay formations of small thickness were formed ('Siberian domanik').

M i d d l e   C a m b r i a n. The pattern of tectonic movements in the craton was inherited from the preceding epoch and a major portion of the deposits again represented by carbonates. Towards the end of the epoch, the Sukhana and Yudoma troughs, formerly uncompensated by sedimentation, had filled up. The thickness of the Middle Cambrian in them exceeded 1.5 km. The area of uplift enlarged on the Aldan Shield and regression manifested.

L a t e   C a m b r i a n. The total area of the craton enlarged because of the Patom Upland in which the orogenic regime was replaced by a cratonic one. In its eastern regions uplift intensified and regression enlarged, encompassing the entire Aldan Shield and adjoining regions. As a result, the contact between the main Siberian basin and Yudoma trough was interrupted. Intense subsidence and accumulation of carbonate and terrigenous formations exceeding 1 km in thickness continued in the Yudoma trough. On the whole, shallow-water conditions, carbonate deposition and a quiescent tectonic regime prevailed in the basins of the craton. Terrigenous rocks played a significant role only in the western and south-western marginal zones adjoining the Yenisei-Sayan orogenic belt whose uplift was the source of clastic material. A basin with high salinity still existed in the south-western part of the craton but its size was extremely reduced.

Thus in the Vendian and Cambrian the characteristic features of tectonic development of the Siberian Craton were: nearly continuous quiescent subsidence and formation of the platform cover represented by carbonate formations. Subsidence was most intense in the western marginal troughs of the craton and in Yudoma trough. At the end of this stage, subsidence in the eastern regions was replaced by uplift, resulting in gradual enlargement of the land area.

1.1.4 Sino-Korea and South China (Yangtze) Cratons

V e n d i a n. The Sino-Korea Craton may include the Qaidam and Tarim blocks. At the transition from the Riphean to the Vendian, the South China Craton experienced fold deformations and later a general uplift, block movements and granitoid intrusions; it was subsequently covered by an ice sheet, traces of which (Nantou tillites) are encountered everywhere. Later, quiescent subsidence predominated and, in shallow seas, carbonate rocks of the Dougshantou and Denin suites accumulated to a total thickness of up to 1 km or more (Liu Hung-yun et al., 1973). In the south-east, however, subsidence was very marked and a thick (up to 5 km) pile of terrigenous formations formed here under the condition of continental slope and rise. This complex is comparable to the Adelaide one in Australia. It has been suggested that the South China block adjoined Australia and was separated from the North China block at the end of the Precambrian (Jin-Lu et al., 1985).

The Sino-Korea Craton with the Qaidam and Tarim blocks experienced only partial subsidence in the Vendian and seas covered half their area. Terrigenous sediments and limestones were formed in the basins with cherts in the form of concretions, lenses and thin bands (Siamalin and Jinerui suites). Similar Vendian limestones with cherts are known in the Bureya and Khanka massifs, which suggests their former structural entity with the Sino-Korea Craton.

E a r l y   C a m b r i a n. The general pattern of tectonic development continued with no significant modifications. The South China Craton continued to subside but some differential tectonic movements occurred and separate basins with a high rate of subsidence and submarine uplifts were formed. The composition of the formations varied and terrigenous sediments began to predominate. The thickness of the latter went up to 1 km or more in basins (Chang, 1980). Sandy shale complexes continued to prevail even in the south-eastern marginal zone. On the Sino-Korea Craton some increase in the role of terrigenous rocks in the composition of Lower Cambrian formations might suggest complications of topography in the land zones.

M i d d l e   C a m b r i a n. Subsidence of the Sino-Korea Craton extended perceptibly and transgression attained maximum proportions at this stage. Carbonate sedimentation began to predominate again in the marine basins. As before, the South China Craton, especially its south-eastern marginal zone, was characterised by very high mobility and the thickness of Middle Cambrian formations in this marginal zone exceeded 1.5 km.

L a t e   C a m b r i a n. Subsidence compensated by carbonate sedimentation continued in all the basins of the craton. The thickness of the Upper Cambrian formations is usually 200 to 400 m except in the extreme south where it reaches 1 km. Along the south-eastern margin of the South China Craton, terrigenous rocks were formed under conditions of a marginal sea and the formation of an andesitic volcanic island arc can similarly be assumed (Chi-ching, 1978).

On the whole, development of the Sino-Korea and South China Cratons in the Vendian and Cambrian was characterised by a predominance of subsidence and shallow marine basins with carbonate sedimentation. Probably development of the south-eastern part of the South China Craton (Cathaysia) proceeded independently in a passive marginal regime. The tectonic activity here in the area of the former (Riphean) Cathaysian mobile belt was comparatively high and andesite (island arc) volcanism manifest.

1.1.5 South American Craton

V e n d i a n. Uplift predominated throughout most of the territory. Development of subsidence and marine terrigenous sedimentation may be presumed only for the western marginal zone adjoining the Andes in the region of Colombia and Ecuador where Vendian metamorphic schists are known and north of the Sierra Pampeanas massif in Argentina where the lower portion of sandstones of the Puncoviscana suite was deposited. But the main episodes were the powerful tectonic (tectonothermal) processes in the east and south-west of the craton which topped the Riphean geosynclinal development of these territories. These episodes also included intense uplift with mountain-building, outpouring of rhyolites and accumulation of molasse beds: Marica, Itagi, Bom-Jardin, Camarinja, Estancia, Giuca, Correros etc. In several regions, for example in the northern Andes of Argentina, granites intruded and processes of regional metamorphism developed (Carlier et al., 1982; Coira et al., 1982).

E a r l y   C a m b r i a n. As before, uplift prevailed and narrow pericratonic zones of subsidence developed only in the Peri-Andean zone within Colombia (sandy shale formations), Bolivia (lower portion of the terrigenous-carbonate formation of Limbo) and Argentina (Puncoviscana sandstones). The activity of tectonic processes decreased significantly, in the east and south-east of the craton, as did the area of their manifestation, and volcanic activity ceased altogether.

M i d d l e   C a m b r i a n. Uplift continued to predominate throughout most of the territory of the craton. In the Peri-Andean zone, on the other hand, the area of marginal seas enlarged somewhat. A basin with sedimentation of clay rocks existed in the north-western part of the Peri-Andean zone while accumulation of the carbonate-terrigenous (with gypsum) Limbo complex continued in the central sector. Terrigenous rocks, less frequently limestones, were formed in the south, in the Precordilleran and Buenos Aires ranges (Argentina). The contours of these basins are tentative and information on the thickness of sediments is not yet available.

Tectonothermal activity in the zones fringing the craton from the east and south-west ceased in the Middle Cambrian and development of these zones no longer differed from the rest of the territory of the craton. The high activity of tectonic movements was preserved only in the Arequipa and Pampeanas massifs which continued to rise. Granitoids intruded in these areas and rapid subsidence in the small basin between the aforesaid massifs was compensated by the accumulation of a thick (over 5 km) terrigenous Meson series complex (Aceñolaza, 1973).

L a t e   C a m b r i a n. On the whole, uplift still prevailed in the craton but the area of subsidence in the Peri-Andean zone enlarged perceptibly and covered nearly all its extent from Venezuela to southern Argentina. The Pampeanas massif experienced subsidence while a deepwater trough arose between the Arequipa massif and the craton, evidently as a result of the movements of continental blocks, and initiated development of a new mobile belt, whose history is discussed later.

Thus the Vendian-Cambrian stage of development of the South American Craton is characterised by a prevalence of uplifts, gradual extinction of post-Riphean tectonic activity along its eastern, southern and south-western fringes and increased subsidence in the Andean marginal zone, evidently associated with the commencement of a new cycle of development of the Andean mobile belt.

1.1.6 African Craton

V e n d i a n. We include the territory of the Arabian peninsula, the Near and Middle East, in addition to Africa and Madagascar Island, in the frame of this craton. Tectonic development of this large craton was complex and varied. In its West African segment predominance of subsidence prevailed from the Riphean while formation of the Taoudenni basin and its filling with terrigenous formations proceeded throughout the Vendian (Nenashev and Petrovsky, 1981). Carbonates, tillites and red beds developed in the littoral facies. This basin included the former Gourma geosynclinal zone where subsidence was maximal and the thickness of sediments reached 2 km. In the Central African segment, on the contrary, a moderate uplift prevailed. An exception was the Congo basin, which was filled with marine and continental sandstone formations up to 1 km thick. In the western part of the South African segment, subsidence of a small marginal zone continued in the Vendian and was accompanied by accumulation of continental coarse clastic formations (molasse) as well as (at the end of the Vendian) marine carbonate and terrigenous sediments, including the namama series.

The Arabian segment of the craton mainly experienced subsidence and was almost wholly covered by shallow sea in which terrigenous sedimentation prevailed. Transgression encompassed these territories in the second half of the Vendian and was preceded by an orogenic regime with powerful uplift and formation of a large volcanic belt within which subaerial eruptions of andesites and rhyolites occurred. The thickness of the volcanics reaches hundreds of metres. An extensive basin with high salinity formed in the region of manifestation of volcanism, giving rise to an evaporite complex comprising carbonates, clay rocks, sulphates and rock salt of the Hormuz formation.

The extensive zones separating the above segments of the craton represented regions of manifestation of tectonothermal reworking called

'Pan-african reactivation'. These processes encompassed extensive expanses of the ancient craton in some cases (for example in the Mozambique belt) and thus were epiplatformal. In other cases, in belts such as the Libya-Nigerian, Damara-Kibaran, Western Congolides, Mauritania-Senegalese, and especially the Red Sea and in Gariep and Malmsberry zones in the extreme south-west, tectonothermal processes succeeded the preceding geosynclinal development and were epigeosynclinal. In all the aforesaid belts molasse formed rather extensively, pointing to powerful uplift and development of a rugged topography. Apart from uplift, reactivation was manifest in widespread magmatic and metamorphic processes, i.e., intrusion of granitoids, radiometric rejuvenation of ancient metamorphic complexes, formation of migmatite and pegmatite fields, mineralisation and sometimes eruptions of rhyolites. In the Atakora zone subaqueous eruption of basalts and andesites occurred at the beginning of the Vendian.

E a r l y C a m b r i a n. Subsidence continued in this epoch only in the Peri-Atlas marginal zone and in the Arabian segment of the platform. The rest of the territory experienced uplift and erosion. In the Peri-Atlas zone subsidence was compensated by accumulation of sand beds to a thickness of 800 m under marine and continental conditions. Within the Arabian segment, marine basins represented shallow lagoons with a low rate of subsidence and a very wide distribution of sand deposits.

The intensity of tectonothermal processes decreased significantly, especially in the southern part of the craton. Ruggedness of relief lessened and molasse is almost unknown. High endogenic activity was maintained in the Mozambique, Libya-Nigerian and Western Congolides belts, as evidenced by manifestation of metamorphism and intrusion of granitoids.

M i d d l e C a m b r i a n. Intense downwarping continued in the Peri-Atlas marginal zone where the thickness of marine and continental terrigenous formations increased in the north to 1 km or more (Aliev et al., 1979; Buggisch et al., 1979). The Arabian segment, as before, represented the second region with a predominance of subsidence. Subcontinental sandstones of the Lalun formation and its analogues formed over much of the territory to a thickness of tens of metres. In the Middle Eastern territory the thickness of the Middle Cambrian exceeded a hundred metres and is represented by only marine carbonate-terrigenous complexes.

Middle Cambrian formations are not known in other regions of the craton; evidently uplift prevailed in these regions. A comparatively high tectonothermal activity was maintained in the Middle Cambrian only in the Libya-Nigerian belt. Continental sandstones, evidently representing molasse in its northern portion, and small manifestations of rhyolite volcanism were noticed in this part of the belt.

L a t e C a m b r i a n. No Late Cambrian formations are known within the vast territory of the craton. It has been assumed that the entire craton was uplifted and underwent erosion. In the development of the Peri-Atlas marginal zone, inversion occurred and subsidence was replaced by uplift in the Anti-Atlas region and slowed down elsewhere. The maximum thickness of rocks of the Upper Cambrian does not exceed 150 m. Subsidence of the basins of the Arabian segment continued at the same tempo as before and carbonate-terrigenous sediments of 300 to 400 m thickness were formed (900 m in the Telbessa region).

In the Libya-Nigerian belt tectonothermal activity was residual in character and died out towards the end of the Cambrian. Subsidence of a small sedimentary basin continued in the north and molasse-like marine and continental sandstones accumulated in the basin.

The main features in the development of the African Craton from the commencement of the Vendian to the end of the Cambrian are gradual reduction of the tectonic activity encompassing many of its regions in the Vendian, with a transition to gentle elevation. Development of the Peri-Atlas marginal zone followed a scenario characteristic of a passive margin while the depressions in the Arabian segment represented interior basins.

1.1.7 Indian Craton

V e n d i a n. Much of the craton experienced uplift. Subsidence was confined to its northern margin and the size of the marine basins there decreased gradually. Carbonates and terrigenous formations predominated in the composition of the formations while gypsum and anhydrites formed additionally in the littoral parts of these basins. In the Salt Range, salt-bearing formations of the Punjab series accumulated to a thickness exceeding 2 km. The south-eastern segment of the craton (including Sri Lanka Island) experienced tectonothermal reactivation in the Vendian, similar to the African and South American cratons. Traces of this reactivation in the south-eastern part of the craton are revealed in the radiometric 'rejuvenation' of the age of rocks, in manifestation of metamorphism and the formation of pegmatite fields.

E a r l y   C a m b r i a n. Subsidence and sedimentation extended only along the northern margin of the craton and the boundary of the sea continued to gradually recede northward while marine facies in the marginal part of the basin were replaced by continental facies from time to time. Marine terrigenous and more rarely carbonate deposits of 500 to 800 m thickness were formed in the north (Garzanti et al., 1986; Jain et al., 1980). In the Late Cambrian the north-western part of the Himalayan territory underwent intense uplift and a mountain relief arose (Garzanti et al., 1986). Elsewhere in the subcontinent moderate uplift prevailed.

M i d d l e   and   L a t e   C a m b r i a n. The overall palaeotectonic conditions in the craton underwent no change: subsidence of its northern margin continued with the formation of carbonate and terrigenous rocks while the rest of the subcontinent experienced uplift and erosion.

As a result, the general pattern of development of the craton was characterised by the very same features as in the case of the South American and African cratons: tectonic activity, regression and predominance of weak uplift.

1.1.8 Australian Craton

V e n d i a n. Subsidence of the Kimberley and Amadeus basins continued from the Riphean in the central part of the craton and consequently their areal extent underwent successive contraction. These basins were interconnected and filled with sandy clay deposits constituting the Albert-Edward and Louisa-Downs formations in the north and Arambera in the south. Their thickness usually reaches hundreds of metres except in the central Amadeus aulacogen where it exceeds 2 km. Basaltic volcanism manifested in the northern margin of the aulacogen. This basin deepened in a south-easterly direction, became more mobile and joined with the zone usually identified as the Adelaide miogeosyncline, in which the thickness of Vendian formations reaches 5 km (Preiss and Krylov, 1980). This zone developed into a regime typical of a passive continental margin. Uplift prevailed in the remaining territory of the craton.

E a r l y C a m b r i a n. An important event was the extensive outpouring of Antrim plateau basalt in the northern part of the craton and in Officer basin. The thickness of these basalts reaches 1 km or more. Volcanic eruptions occurred in the background of reactivation of differential tectonic movements and continental coarse clastic deposits formed in many regions. Commencement of subsidence and transgression in the Georgina basin with predominantly carbonate sedimentation to a thickness of 100 m is associated with this epoch (Fig. 1-4). Intense subsidence of the Amadeus aulacogen continued, in which terrigenous sediments of the lower Pertuarta group accumulated to a thickness of up to 1 km (Plumb, 1979). West and southward the role of coarse clastic and continental facies increased, pointing to the formation of uplift in the adjoining region.

Significant changes occurred in the development of the south-eastern margin of the craton, i.e., the Adelaide miogeosyncline. Intense uplift already predominated in the northern part under conditions of an orogenic regime. In the south, along the Kanmantoo trough, terrigenous complexes including flysch of the continental slope and rise were formed to a thickness of up to 5 km and within Brocken Hill uplift (microcontinent) andesite and rhyolite complexes of the island-arc type were formed. These events were undoubtedly associated with development of the Tasman mobile belt adjoining from the south-east.

M i d d l e C a m b r i a n. The area of subsidence and of marine basins enlarged significantly and the Georgina basin joined with the Bonaparte and Daly River basins. Fairly intense subsidence of the Amadeus

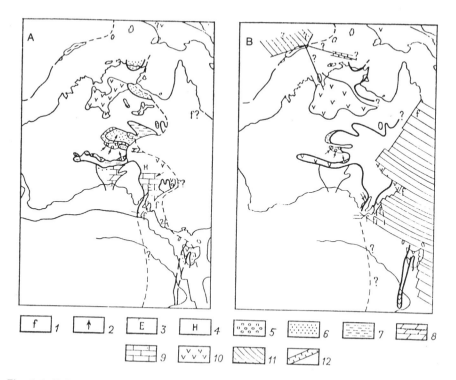

Fig. 1-4. Palaeogeography of Australia in the Early Cambrian (575 to 545 m.y.) (from Veevers, 1986).

A—general scheme; B—interpreted for end of the Early Cambrian. Depiction of spreading of the sea-floor is tentative.

1—flysch; 2—direction of transport of sedimentary material; 3—Ediacaran fauna; 4—halites; 5—conglomerates; 6—sandstones; 7—shales; 8—marls; 9—limestones; 10—basalts; 11—oceanic spreading; 12—graben.

aulacogen and its filling with Pertuarta group sediments continued. Regions with an elevated relief existed west of the aulacogen and continental terrigenous complexes were replaced in the east by marine and later by carbonate-terrigenous complexes. In this same direction, their thickness increased up to 2 km. In the Georgina basin the rate of subsidence, judging from the thickness of formations (over 1.2 km), was also significant.

In the south-eastern margin of the craton, in the Adelaide miogeosyncline, differential tectonic movements intensified in the Middle Cambrian and islands rose actively, while terrigenous sedimentation, mainly coarse clastic, prevailed in the composition of trough formations.

L a t e   C a m b r i a n. The scale of subsidence in the northern part of the craton diminished and almost total regression occurred in the Bonaparte basin. Conditions in the central and southern parts of the craton remained as before. The Amadeus aulacogen underwent active subsidence and was filled with sediments (thickness up to 2 km) with the same lateral zoning as in the Middle Cambrian. The area of the Georgina basin shrank in the north and a relatively high tempo of subsidence and predominance of carbonate formations prevailed in it.

In the south-eastern margin of the craton (Adelaide), inversion of movements was completed, uplift prevailed, folding was manifest and total transition to conditions of an orogenic regime occurred. Coarse clastic continental and marine formations of the molasse type were formed in small residual basins.

Thus uplift prevailed during the Vendian and Cambrian within the Australian Craton. Judging from the large volume of terrigenous rocks, the rate of uplift was high and hence an elevated relief was formed. Subsidence and transgression intensified towards the Middle Cambrian but later this tendency reversed. A comparatively high tectonic activity of the craton was manifest in extensive outpouring of plateau basalts at the beginning of the Cambrian. Evolution of the south-eastern passive margin of the craton (Adelaide miogeosyncline) consisted of the change of rapid subsidence to conditions of differential movements and later a general uplift and transition to an orogenic regime.

1.1.9 East Antarctic Craton

Uplift evidently predominated over much of the territory of the craton in the V e n d i a n. Development of a marine basin with carbonate and terrigenous sediments has been assumed in the west (Shackleton Range). According to radiometric determinations, the marginal zones of the craton were affected by tectonothermal reworking.

Throughout the C a m b r i a n period the East Antarctic Craton continued to experience uplift and served as a source of sediments for the marine basin, which covered the Transantarctic range and West Antarctic regions. Only the marginal zone of the craton adjoining the Transantarctic range was affected by Cambrian transgression. This is particularly relevant to the Shackleton Range where Cambrian deposits rest directly on the pre-Upper Proterozoic crystalline basement.

1.2 REGIONAL REVIEW. MOBILE BELTS

1.2.1 North Atlantic Belt

1.2.1.1 *Appalachian system*. V e n d i a n. Tectonic development of the system was accompanied by a new formation of its southern apophysis, i.e., the Southern Oklahoma zone. In the region of the Arbuckle uplift and Ouachita Mountains, following the rifting processes which divided this segment of the North American Craton, spilite and chert of the Nevajos series and terrigenous rocks of the Timman series formed (King, 1975; Thomas, 1977). Pillow lavas, silicites and terrigenous rocks of the Ashland group, exceeding 3 km in thickness, accumulated in the Southern Appalachians under similar conditions (Pichter and Diecchio, 1986; Hurst, 1973; Wehr and Glover, 1985). Evidently this zone extended into the Northern Appalachians but Vendian formations of yet another type are known here. In the southern part of Newfoundland Island they include the Upper Avalon series and consist of felsic, intermediate and mafic subaerial and submarine volcanics (island arc complex) as well as marine terrigenous formations and molasse (Rast et al., 1983). This helps delineate an island arc system here which subsequently entered the scope of the Avalonian Platform.

E a r l y   C a m b r i a n. In the Southern Oklahoma zone the rift trough was filled with terrigenous sediments and intermediate and felsic volcanics to a thickness exceeding 1 km. In the Southern Appalachians eruptions of submarine lavas ceased and probably pelagic sediments of small thickness accumulated. Similar conditions should be assumed in the Northern Appalachians where shales about 200 m thick fall in the Lower Cambrian (Osberg, 1978). However, within southern Newfoundland Island and Nova Scotia (Avalonian Platform) the Lower Cambrian is represented by sandstones and shales in the shelf facies and also basalts and rhyolites that evidently erupted under conditions of rifting (Greenough et al., 1985; Murphy et al., 1985). This was assumed to be associated with commencement of the opening of the Iapetus palaeo-ocean.

M i d d l e   C a m b r i a n. In the north, in the region of Newfoundland Island, a spilite and chert complex about 3 km in thickness began to form under a deepwater condition while pelagic clay sediments continued to accumulate in the remaining portion (Thomas, 1977). In the zone adjoining the North American Craton, within the continental rise, a thick terrigenous complex including turbidites was formed (Strong and Walker, 1981). One more thick terrigenous sand complex (over 3.5 km) formed in the region of Nova Scotia peninsula. The continental slope and rise of the Avalonian Platform probably lie here.

Development of the Southern Oklahoma zone was completed in this epoch by a general uplift. Subaerial eruption of rhyolites to a thickness exceeding 1 km occurred in this territory and conditions of an orogenic regime prevailed.

L a t e   C a m b r i a n. Processes of reactivation of geosynclinal development (evidently spreading) enlarged and extended from Newfoundland region in the south to the Northern and possibly Southern Appalachians (Osberg, 1978; Rast et al., 1983; and Ruitenberg et al., 1977). Submarine-volcanic, siliceous and clay formations formed under deepwater conditions throughout the belt. The true width of this zone is not known but it is usually regarded as a relict of the Iapetus palaeo-ocean. The thickness of the volcanosedimentary complexes here is significant (up to 5 km, for example, in the Northern Appalachians). In the Avalonian block, representing a submarine uplift, sandy clay formations accumulated in shelf facies and a flysch exceeding 3 km in thickness on its slopes.

The oldest formations in the southern Ouachita-Mexican segment pertain to the Ordovician (Morris, 1974; Thomas, 1977). It may be assumed, however, that deepwater conditions with pelagic sedimentation prevailed here in the Late Cambrian; the region of the Uisatchal-Peregripa anticlinorium in north-eastern Mexico evidently represented the southern branch of this system (De Cserna et al., 1977; King, 1975).

Thus during the Vendian-Cambrian stage the main features of the Appalachian system were rifting (Vendian and Early Cambrian) succeeded by spreading and gradual extension of the latter from the north-eastern to the south-western regions of the system. On the whole, conditions of deepwater sedimentation and submarine volcanism prevailed. Only a small intracratonic Southern Oklahoma zone, an apophysis of the main geosynclinal system forming a triple junction with the zone, underwent rapid evolution with a change in geosynclinal regime to an orogenic one towards the end of the Cambrian period.

1.2.1.2 *British-Scandinavian system.* V e n d i a n. The Hebridian massif represented land in northern Britain. A very thick (exceeding 7 km) complex of sandstones and shales of the Dalradian series was formed in the south (Harris et al., 1980), evidently on the continental slope and rise. Even more southward this complex is succeeded by deepwater shales. Submarine basalts are known in its upper portion in northern Ireland (Max and Long, 1985). In the region of central England a zone of continental slope has been established north of the English Midland microcontinent. A pile of sandstones and lutites (including turbidites of the Strutton series) accumulated here (Wright, 1977). Conditions changed in this zone at the end of the Vendian and coarse clastic sandstones of the Ventor series and later lavas of andesites and rhyolites of the Uricon series formed here under shallow-water conditions. The total thickness of the Vendian formations exceeded 7 km.

The region of Scandinavia and Spitsbergen Island represents the northern portion of this system. In the region of eastern Finnmarksvidda (Finnmarken), a flysch complex up to 3.5 km in thickness was formed in the Vendian. Turbidites of the Stapugieds suite constituting it point to continental slope and rise conditions (Siedlecka, 1973), evidently of the East European continent. On Spitsbergen Island sediments are represented by terrigenous rocks of about 2 km in thickness, which is rather small for Vendian formations.

Early Cambrian. The Hebridian massif underwent subsidence and shelf limestones about 200 m thick were deposited within it. Accumulation of terrigenous deposits continued in the south, their maximum thickness (over 3 km) being known in northern Scotland where a zone of continental slope has been reconstructed. Farther south lay a deepwater zone in which lutites accumulated to a small thickness. In central England a greywacke suite 1 to 1.5 km in thickness (including the upper portion of the Mona complex) was formed under conditions of a shelf basin, its supply source being the Midland microcontinent located more southward (Barber and Max, 1979; Bassett, 1984). In the zone of eastern Finnmarksvidda (Finnmarken), formation of turbidites was succeeded by quiescent conditions and shallow-water sedimentation of small thickness. In the region of Spitsbergen tectonic activity, already weak in the Vendian, decreased even more in the Early Cambrian and the tectonic regime acquired platform features. About 500 m of carbonate and terrigenous sediments of the Sofecatmen series of the shelf type accumulated here.

Middle Cambrian. The overall tectonic zoning, regime of development and composition of formations continued throughout the entire British-Scandinavian system from the Early Cambrian almost without change (Bassett, 1984; Leggett, 1980). A flysch complex exceeding 2 km in thickness is known in southern Ireland and points to the position of the slope of the southern shelf zone of this belt. In its axial portion extending (within present-day co-ordinates) from Britain to the north-western shelf of Scandinavia, the formation of ophiolite complexes, i.e., oceanic crust, is possible.

Late Cambrian. In the zone of the Hebridian massif, quiescent subsidence and accumulation of shallow-water limestones continued. Continental slope conditions prevailed in the south and, farther south, lay a deepwater zone with thin lutites. Terrigenous formations of southern Ireland exceeding 2 km in thickness were formed in the continental slope zone of Midland (microcontinent) while the shallow-water terrigenous complexes of Wales and central England were formed in the shelf of this microcontinent, the bulk of which continued to rise (Bassett, 1984). In the zone of eastern Finnmarksvidda (Finnmarken) intense tectonic activity and manifestation of early impulses of fold deformations have been noted.

Thus the general tectonic zoning of the system formed in the Vendian was preserved without significant change throughout the Cambrian. Zones of continental slopes and continental shelves were situated farther north and south of the axial deepwater zone. The primary width of the deepwater zone in which the formation of ophiolites (oceanic crust) is assumed is not clear and it is regarded as a relict of the Iapetus palaeo-ocean (Anderton, 1982). In such a situation the margins of this palaeo-ocean, noticed here throughout the Vendian and Cambrian, developed in a regime of passive margins.

1.2.1.3 *Innuitian system*. V e n d i a n. Commencement of development of the Innuitian system is assumed to be the result of rifting processes (Surlyk and Hurst, 1983; Young, 1977). This is confirmed by data on basalt eruption in the northern regions of Axel-Heiberg and Ellesmere islands. An ensialic origin is most probable for much of this system but with certain thinning and at places even complete rupture of the continental crust since a deep basin existed here; furthermore, ophiolites are known. The width of this system in the Vendian has not been precised.

E a r l y C a m b r i a n. The eastern segment of the system (northern Greenland) was a narrow trough filled with turbidites (Surlyk and Hurst, 1983). The central zone of the system was slightly wider and subsidence here was not compensated by sedimentation. The western segment of the system probably included the regions of north-eastern Alaska where lutites, cherts, basalts, and carbonates of Neroukrik formations accumulated under deepwater conditions.

M i d d l e C a m b r i a n. This epoch recorded enlargement of the deepwater basin (Surlyk and Hurst, 1983) in which clay formations of small thickness were deposited everywhere. In the region north of Ellesmere Island, uplift evidently prevailed and an island terrane representing the northern flank of the basin was formed (Trettin et al., 1979).

L a t e C a m b r i a n. The axial zone of the system maintained the same deepwater conditions with the accumulation of lutitic and siliceous sediments. In northern Ellesmere Island the land experienced subsidence and comparatively shallow sand deposits were formed here.

Thus the Innuitian system was formed in the Vendian as a result of rifting and was probably a narrow ocean basin with a relatively quiescent type of tectonic development throughout the Cambrian (Surlyk and Hurst, 1983; Trettin and Balkwill, 1979).

1.2.2 Mediterranean Belt

1.2.2.1 *West European region*. V e n d i a n. Submarine-volcanic, siliceous and terrigenous complexes formed in regions south of the Armorican massif to south-west of France and Sardinia Island point to a high tectonic activity in this part of the belt. Northward, in the zone of the Central and Bohemian massifs, more quiescent conditions of tectonic development and terrigenous sedimentation of the shelf type predominated and coarse clastic and turbidite sedimentation on the slopes of the massif, as in the northern Bohemian massif, represented an island terrane (Chaloupsky, 1978). In the Galicia zone Vendian rocks are represented by metalutites and metadiabases, pointing to deepwater conditions and comparatively high tectonic activity (Bonchev, 1985). A similar environment evidently prevailed in the Caucasian zone (Belov, 1981).

In the territory of the Iberian peninsula upheaval of two ancient massifs occurred and thick (over 2 km) terrigenous formations were formed in the West Asturia zone (Julivert et al., 1980). The western and southern parts of the peninsula, viz., Ossa-Morena and Southern Portugal zones, were represented by very thick (up to 4 km) terrigenous formations and volcanics of intermediate and basic composition by the Vendian period.

E a r l y   C a m b r i a n. Significant changes occurred at this time in the development of much of this region. Upheaval of all the ancient massifs representing island terranes occurred and subaerial eruptions of felsic and intermediate lavas and tuffs manifested within their areas as, for example, in the southern Armorican and northern Bohemian massifs. At the beginning of this epoch (or even at the end of the Vendian), folding (Ligerian phase of the Armorican massif) occurred in the massifs and granitoids intruded. Manifestation of volcanism ceased in southern France but deepwater conditions prevailed and hence sediments here are represented by lutites of small thickness (Feist, 1978). Similar deepwater facies are known in regions north of the Central and Bohemian massifs; greywacke complexes up to 2.5 km in thickness are known on the north-western slope of the latter (central Czechia massif) (Havliček, 1971). The nature of tectonic development is not clearly known in the eastern part of the zone, within Crimea, Precaucasus and northern Caucasuş. In the ancient massifs in the zone of the Iberian peninsula, uplift was succeeded by subsidence and accumulation of terrigenous and carbonate-terrigenous shelf formations up to 1 km thick (Guy Tamain, 1978; Julivert et al., 1980). In the West Asturia and Leonese zones subsidence was accompanied by sedimentation of thick (over 2 km) terrigenous formations and additionally by submarine basaltic outpouring in the zone of Ossa-Morena.

M i d d l e   C a m b r i a n. Uplift in all the ancient massifs intensified and, as a result, the Central European massif, uniting the Armorican, Central, Bohemian and some other massifs, was formed (Franke, 1978). This uplift evidently represents quiescent upheaval and was not accompanied by manifestation of folding. Along the northern boundary of the massif, greywackes were evidently deposited under a continental slope condition. The margin of the East European Craton with the continental slope turned southward has been traced even more northward (Cwojdzinski, 1979). Deepwater conditions continued to prevail in south-western France with lutite sedimentation of small thickness but in the south, in the region of Montagne Noire, and eastward, deepwater conditions gave way to shelf conditions, judging from lithology (Floch et al., 1977). Uplift and folding affected the central part of the Balkan peninsula (Bonchev, 1985). In the ancient massifs of the Iberian peninsula also uplift intensified and the size of island terranes enlarged. Deepwater clay deposits formed in the West Asturia and Leonese zones, terrigenous-carbonate and terrigenous complexes were deposited in the south, in the zone of Ossa-Morena, and eruption of andesites occurred (Guy Tamain, 1978; Julivert et al., 1980).

L a t e   C a m b r i a n. The ancient massifs of the region maintained the tendency to rise while island terranes surrounding the deepwater troughs held their position with uncompensated lutitic sedimentation and probably with submarine basic volcanism north-east of the Bohemian massif (Cwojdzinski, 1979; Franke, 1978; Havlicek, 1971). Terrigenous and partly carbonate sediments accumulated and continental felsic volcanics erupted in the narrow southern shelf of the Armorican massif.

Commencing from this epoch, the evolution of structures situated in the territory of the present-day Southern Alps and Carpathians can be well traced (Belov, 1981). It is possible that processes of destruction of massifs and formation of deepwater troughs with an oceanic crust between them occurred at this time.

In the Central Iberian zone the area of uplift increased and subaerial felsic volcanism manifested at some places. In the north, in the zone of West Asturia, the deepwater basin was filled with flysch to a thickness exceeding 2.5 km while shelf-type sand deposits accumulated in the Cantabrian zone. In the southern part of the Iberian peninsula, in the zones of Ossa-Morena, Southern Portugal and Cordillera Betica, sandy shale formations and andesites fall in the Late Cambrian but the palaeotectonic environment of their formation is not clearly known.

Within Dobrogea, Crimea, Precaucasus and Turan, Cambrian deposits are not known to date and shelf conditions with deposition of sediments of small thickness are most likely. Rifting conditions from commencement of formation of a spilite and chert complex (Karabek suite) including low-potassium tholeiites are assumed within the region of Front Range in Caucasus Major (Belov, 1981).

Thus the Vendian-Cambrian stage in the development of the western part of the Mediterranean belt is characterised by the accretion of continental crust, gradual consolidation of massifs along the margins and formation of mobile zones with deepwater conditions between them under conditions of extension.

1.2.2.2 *Qilianshan system.* Development of this structure has been traced from the Vendian when the Qaidam block separated from the North China Craton as a result of rifting (Xiang Dingpu, 1982). In the V e n d i a n evidently a deep basin occurred between them and the Vendian formations were represented by pelagic breccia-like and siliceous limestones, siliceous shales and basalts. In the E a r l y   C a m b r i a n the structure of the system became complicated, its dimensions enlarged and deepwater conditions prevailed as before. In the composition of the formations, clay deposits and mafic submarine volcanics played the main role. In the north-eastern zone of the belt andesites too were extensively developed; the thickness of the Lower Cambrian here attained 3 km. In the M i d d l e   C a m b r i a n, in this obviously island arc zone, volcanic eruptions continued and siliceous complexes formed, their thickness here reaching 3 km. In the south-western zone, manifestation of volcanism ceased and deepwater conditions prevailed but the sediments became very coarse and sandy and their thickness rose. In the L a t e   C a m b r i a n the general conditions changed everywhere. In the north-eastern zone volcanism almost ceased (excluding small eruptions with an intermediate composition) and pelagic limestones and clays were deposited. The south-western zone represented a continental slope on which accumulation of flysch proceeded.

Thus the main feature of development of the Qilianshan system in the Vendian-Cambrian stage was the gradual filling of the deep basin situated between two platform blocks with sediments and products of submarine volcanic activity. By the end of the Cambrian, volcanism had died out and transition to a mature stage of development is noticed.

1.2.3 Urals-Okhotsk Belt

1.2.3.1 *Uralian system.* V e n d i a n. In the northern part of the Uralian system the preceding intense subsidence was succeeded in many regions by uplift and the area of marine basins shrank notably within the Timan-Pechora region. Vendian formations were represented by a sandy-conglomerate complex which included the Churoch series suite of Poludov ridge exceeding 2 km in thickness and beds of sandstones and submarine andesites, possibly of the island-arc type, of the Northern Urals reaching up to 3 km in thickness (in: *Unified and Correlated Stratigraphic Sequence of the Urals*, 1980). Conditions were different in the Polar Urals and in southern Novaya Zemlya where a submarine volcanic and deepwater sequence of siliceous shales continued to form at the beginning of the Vendian (Gryaznov et al., 1986; and in: *Novaya Zemlya in Early Stages of Geological Development*, 1984). Conditions here later underwent a change and limestone-sand-clay and sometimes flysch-like complexes began to predominate in the composition of the deposits; submarine outpourings of basalts and andesite basalts were replaced by subaerial eruptions of intermediate and felsic composition while uplift and intrusion of gabbro-diorite-granodiorite plutons occurred at the end of the Vendian.

In the central and southern parts of the Uralian system, in the shelf zone adjoining the East European Craton and continental slope, terrigenous complexes of diverse composition and with bands of turbidites accumulated, following a minor pre-Vendian erosion. In the Central Uralian zone these are represented by the Serebryanka and Sylvytsa series with a total thickness of about 4 km and in the Southern Urals by the Asha series with a thickness exceeding 2 km.

E a r l y   C a m b r i a n. Upheaval intensified in the northern part of the belt and extensive regions of present-day Timan ridge and Pechora lowland represented an elevated orogenic zone dissected by faults into blocks which experienced upward movement to various degrees, albeit at a small rate and, as a result, folded structures of moderate height were formed. Within the Polar Urals and southern Novaya Zemlya, at the beginning of the epoch, marine basins were still preserved but the general tendency of development here, too, underwent a change with uplift intensifying and these residual basins became intermontane (possibly riftogenic) troughs (Gryaznov et al., 1986). The Central Uralian zone was evidently the region of uplift throughout the Early Cambrian and fold development and intrusion of granitoids were manifest here but on a rather small scale.

M i d d l e   C a m b r i a n. For the Timan-Pechora region and the Urals this was an epoch of tectonic quietude and probably of weak upheaval. The only exception was the region of Novaya Zemlya where turbidites were formed in the north and clay deposits in the south (in: *Geology and Stratigraphy of Novaya Zemlya*, 1979; Kheraskov, 1975).

L a t e   C a m b r i a n. The end of the epoch witnessed the commencement of a new stage of development of the system when its marginal segments underwent rifting and weak subsidence with the accumulation of conglomerate-sand beds. In the Polar Urals, extension led to plutonic magmatism and eruption of diabase-picrite and trachybasalt lavas.

On the whole, the Vendian-Cambrian stage for the Uralian system represented the period in which the Riphean stage of high tectonic activity and manifestation, although weak, of an orogenic regime came to an end. Exceptions were the regions of the Polar Urals and Novaya Zemlya where a somewhat higher intensity of tectonic and magmatic processes has been identified.

1.2.3.2 *Kazakhstan-Tien-Shan region.* V e n d i a n. This period is characterised by uplift of the Ulytau-Northern Tien Shan, Aktau-Junggar, Kokchetav, Syr-Darya and other ancient massifs and accumulation of thick sedimentary and volcanic formations in the basins adjoining them. In the Ishim-Karatau and Chatkal-Naryn zones, tuffaceous-terrigenous and jasperoid-basaltoid complexes of variable composition and 3 to 4 km in thickness formed. In the Yerementau-Chu-Ili and Chingiz-Tarbagatai zones the thickness of formations belonging to the Vendian is significantly smaller and carbonaceous shale and quartz-terrigenous sediments accumulated (Kheraskova, 1986). Complexes of a different composition were deposited within the massifs: quartz-arkose, terrigenous-siliceous-tuffaceous molassoid and sometimes volcanics of high alkalinity.

Early Cambrian. The general environment in this epoch was inherited from the preceding epoch. Shallow-water carbonate and terrigenous deposits a few hundred metres in thickness were deposited in depressions and troughs separating the island terranes. In the Ishim-Karatau and Chatkal-Naryn zones volcanic manifestation ceased and siliceous-carbonate beds of reduced thickness (200 to 400 m) accumulated, evidently under deepwater conditions. Similar formations were established in the Yerementau-Chu-Ili, Chingiz-Tarbagatai and Northern Tien Shan zones where siliceous, clayey and carbonate complexes of tens and several hundred metres thickness belong to the Lower Cambrian (Akhmedzhanov et al., 1979; Borisenok et al., 1979; Kheraskova, 1979). The siliceous-carbonate complexes that formed at the verge of the Vendian-Cambrian within the Kazakhstan-Tien Shan and Altay-Sayan regions were characterised by phosphate mineralisation.

Middle Cambrian. Sedimentation of siliceous and shale complexes continued in comparatively deep troughs. Submarine volcanism manifest mainly in several regions of the Yerementau-Chu-Ili and Chingiz-Tarbagatai zones and the thickness of the Middle Cambrian rose here to 2 to 2.5 km. At the end of the epoch, folding processes involved the Chingiz-Tarbagatai zone and granitoids intruded here. Kazakhstan-Northern Tien Shan and Khanty-mansy massifs maintained the tendency towards uplift and terrigenous beds about 1 km thick accumulated in the area of the Aktau-Junggar massif under a shallow-water condition. Commencing from this epoch, development of the Junggar-Balkhash basin can be traced with sedimentation restricted to siliceous-clay formations of small thickness formed under deepwater conditions.

Late Cambrian. No significant transformation occurred in the palaeotectonic development of the region. An increase in the role of andesites in the composition of submarine eruptions and commencement of formation of submarine andesite complexes of the island-arc type in the north-western and south-eastern parts of the Kazakhstan-Northern Tien Shan massif are of interest. In the Chingiz-Tarbagatai zone the Upper Cambrian is represented by terrigenous-carbonate formations of small thickness. Sedimentation of pelagic clay and less frequently siliceous sediments within the Junggar-Balkhash zone and Tien Shan continued and massifs with a platform regime of development prevailed south of Tien Shan throughout the Cambrian (Akhmedzhanov et al., 1979).

On the whole, Vendian-Cambrian development of the Kazakhstan-Tien Shan region is characterised by predominance of deepwater conditions of sedimentation and basic submarine volcanism in troughs situated between ancient massifs and in a wider basin between the Tuva-Mongolia and Aktau-Junggar massifs. Within the massifs themselves, shallow-water terrigenous

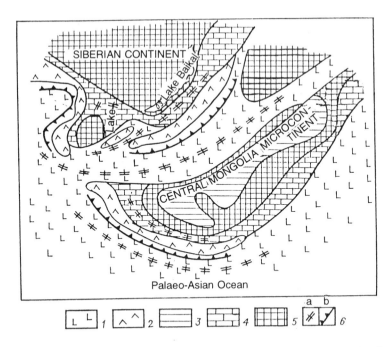

Fig. 1–5. Palaeotectonic reconstruction of Sayan-Baikal-Mongolia region for the Vendian-Early Cambrian period (after Gordienko, 1987).

1—predominantly oceanic including ophiolite complexes; 2—predominantly island arc volcanic complexes; 3—flyschoid complexes of interior seas in the Central Mongolia microcontinent; 4—siliceous-terrigenous, carbonate-terrigenous and carbonate complexes of marginal seas and shelf and continental slope of passive continental margins; 5—blocks of Precambrian continental crust; 6—suggested disposition of (a) spreading and (b) subduction zones including those covered by shelf deposits.

and, less frequently, carbonate complexes were deposited and andesite volcanism manifested at the end of the Cambrian.

1.2.3.3 *Altay-Sayan region.* V e n d i a n. Complexes of various shales (including siliceous), submarine volcanics of basic composition and terrigenous rocks of Vendian age were extensively developed in many parts of the Altay-Sayan region (Fig. 1–5). The thickness of the formations deposited predominantly in deepwater conditions constitutes 3 to 4 km (Belousov et al., 1978; Dergunov et al., 1980; Palei, 1981; Kheraskova and Dergunov, 1986). At the site of the Tuva-Mongolia massif, a shelf zone was situated in which comparatively shallow-water carbonate-terrigenous complexes began to accumulate. In the Selenga-Upper Vitim zone of southern Transbaikalia, intense subsidence was accompanied by the formation of massive terrigenous and volcanic (andesite) complexes, including the Bitu-Dzhida and Oldynda suites.

E a r l y C a m b r i a n. The general palaeotectonic environment over much of the region remained as before: ophiolite association with basic submarine-volcanic, siliceous and terrigenous complexes with a total thickness of 2 to 3 km and up to 5 km or more were formed in many regions. A flysch complex exceeding 4 km in thickness accumulated in the eastern region, along the northern boundary of the Selenga-Upper Vitim deepwater zone, evidently on the continental rise (Butov, 1985). Within the Tuva-Mongolia and Kerulen-Argun massifs, accumulation of terrigenous, carbonate and at places intermediate and felsic volcanics occurred under shallow-water conditions of shelf basins on a crust of the continental type and granite formation was manifest on a small scale in Tuva.

M i d d l e C a m b r i a n. Significant changes occurred over much of the territory in the nature of overall development and composition of sedimentary and volcanic formations. The distribution of siliceous sediments decreased and andesite volcanism of the island-arc type began to play a major role. Many regions of the Salair Range, Western Sayan and almost the entire territory of the Tuva-Mongolia massif underwent uplift. Folding and metamorphic processes developed here, granitoids intruded, eruption of subaerial and submarine volcanics of felsic and intermediate composition occurred and molasse formed (Belousov et al., 1978; Dergunov et al., 1980; Kheraskova and Dergunov, 1986). All of this suggests a transition to an orogenic regime and the joining of these regions to the Yenisei-Patom orogenic region.

In the Selenga-Upper Vitim zone the composition of volcanic eruptions also changed, andesites and their tuffs predominated and granitoids intruded; further, towards commencement of the Middle Cambrian, this deepwater trough was filled and comparatively shallow-water conditions became established (Belichenko, 1977; Butov, 1985).

L a t e C a m b r i a n. Further enlargement of uplift area resulted in the maintenance of marine sedimentation conditions only in the western area of the region, within Western Sayan and the western Salair Range, with turbidites predominating in the composition of sediments (Kheraskov, 1975; Kheraskova, 1986). The sea abandoned the Selenga-Upper Vitim zone where an orogenic regime was established. Volcanic eruptions ceased.

Thus throughout the Vendian-Early Cambrian the Altay-Sayan region represented an arena of deepwater sedimentation and submarine eruptions of mafic lavas (excluding several massifs and their marginal zones); in the Middle Cambrian uplift commenced and island arcs formed; in the Late Cambrian volcanism died out and uplift prevailed except in the western part of the region where accumulation of thick sedimentary complexes of continental slope and rise continued (Altay mountains series).

1.2.3.4 *Yenisei-Patom system*. The Yenisei-Patom system, which developed in an orogenic regime, stretches between the Siberian Craton and the region of Altay-Sayan (Postel'nikov, 1980). In the V e n d i a n the Taimyr-Severnaya Zemlya orogen represented the northern continuation of this system (Ustritsky, 1985). The Vendian development of this orogen was hboxcharacterised by a combination of large massifs of rugged land and intermontane continental and marine basins in which coarse clastic rocks (molasse) formed deposits up to 6 km in thickness. In the E a r l y C a m b r i a n, the Taimyr-Severnaya Zemlya orogen became a constituent of the Siberian Craton. The rest of the system continued to experience active uplift, restraining the basins of the Siberian Craton from the south. The area of intermontane troughs decreased but in Transbaikalia subsidence and transgression of the sea sometimes covered significant areas and carbonates predominated among the formations here as well as in the adjoining platform basins.

In the M i d d l e   C a m b r i a n the area of the folded system rose because the Altay-Sayan region entered an orogenic stage of development and uplift continued here in general. In the Transbaikalia uplift was comparatively weak and barely influenced the carbonate composition of sediments in the adjoining marine basins. In the western part of the system the activity of movements was high and molasse (including marine) up to 3 km in thickness formed here; subaerial volcanic complexes of intermediate and felsic composition also formed here.

In the L a t e   C a m b r i a n the system underwent some modification of its boundaries since the Patom Upland became a part of the Siberian Craton towards the beginning of this epoch and an epigeosynclinal orogenic regime prevailed in the Selenga-Upper Vitim zone which joined this system. Uplift in that part of the system which adjoined the Altay-Sayan region was particularly active. Thick marine terrigenous complexes and intermediate volcanics accumulated in the submontane basins.

1.2.4 Western Pacific Belt

1.2.4.1 *Cathaysian system.* The most active stage of development of the Cathaysian system was completed in the Vendian, the main events of this development having occurred in the Riphean (Igolkina et al., 1981; Ren Jishun et al., 1984). Vendian formations are represented here by greywackes and siliceous shales of up to 5 km in thickness. It is possible to reconstruct the Early Cambrian palaeotectonic conditions in greater detail (Fig. 1–6). A zone of the outer shelf and continental slope adjoined the South China (Yangtze) Craton and clay deposits and dolomites accumulated in this zone. Farther south-east, evidently an island arc arose at the end of the Middle Cambrian. Here volcanics of differentiated basalt-andesite and

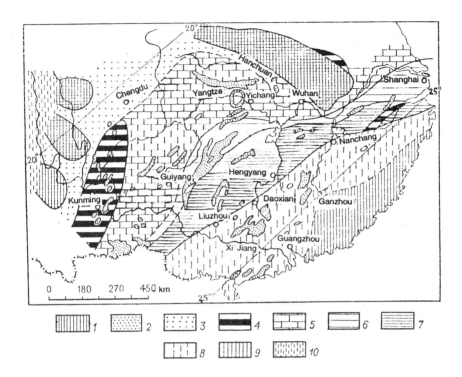

Fig. 1-6. Palaeogeography and Early Cambrian facies of South China (after Il'in, 1986).

1—zone of erosion; 2—surface outcrops of Early Cambrian formations in the modern structure; 3 to 5—inner (littoral) part of Yangtze Sea shelf (3—sandstones, shales, dolomites with shallow-water benthic fauna; 4—predominantly phosphate facies of 'bedded' phosphorites; 5—shallow-water dolomites and limestones); 6, 7—outer part of shelf and continental slope of Jiannan Sea (6—black shales, 'coal' and dolomites; 7—black shales and nodular phosphorites); 8, 9—western Palaeo-Pacific (8—volcanics of differentiated basalt-andesite and tholeiite formations and flysch; 9—andesite-greywacke-chert formation); 10—Palaeo-Asian ocean.

tholeiite complexes and flysch were formed while an andesite-greywacke-chert complex formed more towards the east. Such a zoning was evidently maintained up to the end of the Cambrian but tectonomagmatic events of the Yunnan phase manifested in the Late Cambrian epoch in the region of the island arc and this phase formed the Uishan and Yunkai uplifts. Subsequently these uplifts experienced intense migmatisation and granite formation.

1.2.4.2 *East Australian region.* V e n d i a n. Development of the Adelaide system representing a marginal-continental structure was completed and intense sedimentation of thick terrigenous beds took place under the condition of shelf and continental slope (Preiss and Krylov, 1980; Murray and Kirkegaard, 1978). The Vendian sandy shale Wilpena series here is up to 5 km in thickness.

E a r l y   C a m b r i a n. The northern part of the Adelaide system began to experience uplift and underwent orogeny (Murray and Kirkegaard, 1978). Terrigenous (including flysch) complexes up to 5 km thick accumulated in the Kanmantoo trough in the south while andesite and rhyolite complexes of the island-arc type formed within the Brocken Hill uplift (Borch, 1980; Scheibner, 1978). More eastward the Lachlan system began to develop in this epoch. Metamorphosed complexes, i.e., schists and submarine eruptions of mafic composition, also formed in the Early Cambrian (Gilligan and Scheibner, 1978; Vandenberg, 1978). Precambrian rocks are not known here and the system is presumed to have originated on an oceanic crust. On Tasmania Island, in the Dundas trough, the thickness of such complexes including ophiolites reaches 6 km. But their formation here is associated evidently with rifting and extension of ancient continental blocks (Corbett et al., 1974). Similar complexes are known in the South Island of New Zealand.

M i d d l e   C a m b r i a n. In the Adelaide system uplift intensified and terrigenous, mainly coarse clastic, sedimentation predominated. Development of the Lachlan system continued eastward. Andesites, apart from basalts, appeared in substantial amounts in the composition of the products of submarine eruptions; an island arc formed in which submarine and subaerial andesite volcanics predominated. A greywacke complex about 3 km thick accumulated in the rear part of this arc, in the Ballarat back-arc basin. The Dundas trough was filled with terrigenous formations exceeding 2 km in thickness. In the South Island of New Zealand conditions continued as before.

L a t e   C a m b r i a n. Transition to an orogenic regime concluded in the Adelaide system where uplift predominated, folding was manifest and the Anabama granite pluton intruded. However, marine basins with coarse clastic sedimentation were still preserved along the periphery of the central uplift. Evolution of the Thomson system can be traced in the north from this period. Although information about this system is extremely fragmentary, accumulation of submarine-volcanic and siliceous complexes evidently occurred here (Murray and Kirkegaard, 1978). Formation of the more eastern Molong-Canberra volcanic arc commenced more southward within the Lachlan system. A deepwater zone with lutite sedimentation existed in the west and a narrow continental slope with flysch sedimentation even more westward. On Tasmania Island, Dundas trough continued to be filled with terrigenous formations (including flysch) and, towards the end of the epoch, with continental sandstones and conglomerates.

Thus a distinct tendency for the active regime of development to migrate eastward parallel to the accretion of the Australian continent is noticed in the development of the East Australian region.

1.2.4.3 *West Antarctic region.* In the V e n d i a n this region underwent an uplift under the condition of epigeosynclinal orogeny. The mobile system of the future Transantarctic range, following the Beardmore orogeny manifest at the very end of the Proterozoic, experienced intermittent attenuation of uplift in the C a m b r i a n and was even drawn into subsidence with an accumulation of deposits reaching 5 km in thickness between Beardmore and Bird glaciers. In the E a r l y C a m b r i a n eruptions of felsic volcanics took place concomitant with formation of shallow-marine terrigenous-carbonate sediments; in the M i d d l e and L a t e C a m b r i a n the nature of the sediments hardly changed but volcanism ceased. At the boundary of the Cambrian and Ordovician, the Transantarctic system experienced its concluding orogeny, known as the Ross orogeny, which was accompanied by intrusion of granitoids of the Granite Harbour complex.

West of the Transantarctic range, in the Ellsworth Mountains, the exposed section commences with the Middle Cambrian and contains an admixture of fragments of volcanic material together with terrigenous rocks of deltaic origin. Marine sediments, initially black shales with bands of limestone and later clastic material of fluvial-marine origin with basalt sheets and finally limestones appear upward in the Middle and Upper Cambrian sequence.

Cambrian development in northern Victoria Land differed significantly. Here, in Bowers Hills, the existence of a volcanic arc with eruptions of lavas (from basalts and andesites to rhyolites) has been recorded for the M i d d l e C a m b r i a n. Along the arc extended a trough which was filled with clastic sediments having bands of limestones in the M i d d l e and L a t e C a m b r i a n; the sequence is topped with conglomerates and the total thickness of the Cambrian in this region reaches 6 km.

In the adjoining Robertson Bay western terrane and in Mary Byrd Land, the thickness of the Cambrian (and the lower part of the Ordovician) also attains several kilometres but is lithologically represented by turbidites of submarine debris cones; the sequence is topped with sandstones and con-glomerates containing limestone fragments with C_3-O_1 fauna. All of this reflects the beginning of Ross orogeny accompanied by the intrusion of I-type granites known as the Admiralty complex.

1.2.5 East Pacific Belt

1.2.5.1 *Cordilleran system.* V e n d i a n. The nature of tectonic devel-opment of the system for this period is not yet fully known. Rifting, which had already commenced in the Late Riphean, evidently concluded within the Rocky Mountains, Utah and California states (Bond et al., 1985). This is supported by sporadic development of cross-bedded arkose formations and basalt sheets of the Hamil group alternating in a westerly direction with deepwater lutites and pillow lavas.

E a r l y C a m b r i a n. In the southern, Californian segment of the system, comparatively less thick deepwater siliceous-clay and basalt complexes accumulated (King, 1978; Rowell et al., 1979). Eastward these formations are succeeded by turbidites, pointing to the proximity of the continental slope.

M i d d l e C a m b r i a n. Over much of the western shelf of the North American Craton the accumulation of shallow-water sandstones was replaced by the formation of carbonate-terrigenous clay and siliceous complexes of the pelagic type (in: *Stratigraphic Atlas of North and Central America*, 1975). This change in general environment, i.e., subsidence of the shelf leading to conditions of a deepwater zone, was accompanied at places by eruptions of basalts; the thickness of the Middle Cambrian in such regions exceeds 2 km. So for the Middle Cambrian the zone of predominantly deepwater sedimentation can be traced over the entire stretch of the system from Alaska, where pelagic clay formations are known, to California within which carbonate-clay formations have developed (Churkin and McKee, 1974; Rowell et al., 1979; Stewart and Poole, 1974). The thickness of formations in this zone usually does not exceed tens of or a few hundred metres.

L a t e C a m b r i a n. The overall tectonic conditions of development of the system remained as before. In the deepwater zone clay sediments of small thickness continued to accumulate. Within the Pelly Mountains, greywackes and basalts were formed; their composition enables us to assume the existence here of a more differentiated environment, including submarine uplifts (in: *Stratigraphic Atlas of North* and *Central America*, 1975).

On the whole, a quiescent spreading condition prevailed in the Cordilleran system until the end of the Cambrian.

1.2.5.2 *Andean system.* V e n d i a n. Information is as yet inadequate about the Vendian history of the system. Vendian metamorphic schists are known in the Colombia-Ecuador segment. The lower part of the Puncoviscana sandstones suite was formed in this period in the Peru-Bolivian segment north of the Pampeanas massif (Jezek et al., 1985). In the Patagonian zone a complex of submarine basites and siliceous shales evidently began to accumulate from the Vendian. In the Arequipa, Pampeanas and Patagonia-Deseado massifs of the belt, uplift predominated, granites intruded and orogenic regime conditions prevailed (Coira et al., 1982).

E a r l y C a m b r i a n. In the eastern zone of the Colombia-Ecuador segment (eastern Cordillera) the Lower Cambrian is represented by terrigenous deposition and possibly the metamorphic rocks of Cajamarca (including ophiolites) in the central zone (Central Cordillera) should also be regarded as of the same age. Formations of mafites and siliceous-clay rocks continued in the Patagonian zone. As before, massifs experienced uplift under conditions of an orogenic regime. The northern margin of the Pampeanas massif subsided and sandstones were deposited there. However, this region experienced a rapid uplift and folding towards the end of this epoch (Jezek et al., 1985). Lower Cambrian shallow-water terrigenous and partly carbonate formations are known on the south-western margin of the massif.

Middle Cambrian. The Colombia-Ecuador segment maintained separation into an eastern shelf zone with terrigenous sedimentation and a western, evidently deepwater zone in which submarine eruptions occurred and siliceous and clay complexes were formed. No information is available on changes in the environment compared to the Early Cambrian in the Patagonian zone and in the ancient massifs. Only the intrusion of granitoids into the Pampeanas massif (Borrello, 1969) and the commencement of a new stage of subsidence in the region between the Pampeanas and Arequipa massifs may be mentioned for this period. The rapid subsidence was compensated by the accumulation of a thick (over 5 km) terrigenous complex of the Meson series (Aceñolaza, 1973).

Late Cambrian. Destruction and extension of continental blocks in the central part of the Andean system commenced with the formation of a deepwater zone between the Arequipa massif and the craton (southern Bolivia) where a sandy-clay complex exceeding 2 km in thickness was formed (Dalmayrac et al., 1980). This trough opened in the north-west in the direction of the open ocean and evidently joined with the deepwater zone of the Colombia-Ecuador segment. There is no evidence of volcanic manifestation here and evidently no rupture occurred either, only a thinning of the continental crust. Development of the Peru-Bolivian segment of the system began with these events. Much of the Arequipa and the entire Pampeanas massif was affected by the subsidence but only shelf-type sand complexes deposited on them.

Thus the history of development of the various segments of the Andean system in the Vendian-Cambrian differed. Two zones prevailed in the north—deepwater and shelf—which occupied a marginal position between the continent and the ocean. In the first zone submarine eruptions occurred and siliceous-clay sediments formed; in the second shallow-water terrigenous rocks accumulated. In the central segment active events and formation of a mobile belt occurred at the end of the Cambrian. In the south, in the Patagonian zone, a permanent active development with manifestation of submarine volcanism is assumed.

1.3 PALAEOTECTONIC ANALYSIS

Development of the Earth's crust in the Vendian-Cambrian comprised consolidation of the Gondwana supercontinent which emerged in this period as a single entity, and destruction of Laurasia dissected by the northern Atlantic (Iapetus) and Ural-Okhotsk (Palaeo-Asian) mobile belts (palaeo-oceans) and separated from Gondwana by the Mediterranean belt (Proto-Tethys) (Fig. 1–7).

1.3.1. Continental Cratons

1.3.1.1 *Gondwana.* A fairly large number of mobile zones (monogeosyn-clines) formed in the Late Riphean within the future Gondwana. These zones arose partly on thinned and reworked continental and partly on the axial part of the newly formed oceanic crust. The width of these monogeosyn-clines scarcely exceeded a few hundred kilometres, as demonstrated, for example, by A. Kröner for the Damara system in South Africa (Khain and Bozhko, 1988). Most such mobile systems extended along both sides of the southern Atlantic in South America and Africa and also along the east coast of Africa, i.e., Mozambique belt (Kaz'min, 1988), and probably the north-western margin of India. Their apophyses, for example, the Damara-Katanga system in Africa or Araguaia-Paraguay system in South America, penetrated deep into the continental massifs. The closure of all these mobile zones occurred at the beginning or during the Vendian and from then onwards Gondwana existed as a single supercontinent. Moreover, fold defor-mations, regional metamorphism and granitoid magmatism encompassed at the beginning of the Vendian (at some places, later) an extensive area of Riphean Proto-Tethys temporarily joining to Gondwana, its future northern margin, a territory extending in Europe up to southern England, southern Germany and in Asia, up to the Caucasus, southern Tien Shan, northern Tibet and the northern margin of the South China Platform[1]. In Soviet lit-erature this region is known as the Peri-Gondwana Epibaikalian Platform (Belov, 1981; Khain and Bozhko, 1988); the corresponding folding in West-ern Europe is called Cadomian, in Africa Pan-African and in South Amer-ica Brazilian. In the Balkans, Carpathians, the Caucasus and partly in Iran, commencement of cratonisation extended up to the Middle Cambrian, which is supported in particular by the finds of Archaeocyatha and other organic remains in metamorphic complexes. The same is true of the Adelaide system in Australia.

The belated conclusion of the geosynclinal development in the Gond-wana region made for the predominance of an orogenic regime over con-siderable expanses of almost all the continents constituting it. This resulted in the accumulation of molassoid or even molassic clastic formations in dif-ferent regions.

---

[1] Here and later, in modern co-ordinates.

At the end of the Vendian, the northern periphery of Gondwana in Africa, Europe, Asia and especially Australia gradually began to be covered by shallow sea that developed maximally in the Middle Cambrian. In the region of the Persian and Oman gulfs a vast salt basin was formed in the Vendian to the Early Cambrian. This basin also covered the Salt Range of India. Transgression occurred from north and north-east, from the side of the residual ocean space of the Proto-Tethys. In some regions bands with marine fauna, for example in the Antarctica, are intercalated with molasse. In the Middle to the Late Cambrian the sea advanced westward from the Iapetus ocean and eastward from the residual Proto-Tethys.

The Pacific margin of Gondwana, i.e., south-eastern China, eastern Australia and western Antarctica, represented an active margin. The same is true of the European margin from southern England to northern Caucasus. Rifts (aulacogens) in this period were no longer characteristic of Gondwana; the Amadeus aulacogen in central Australia represents the most striking exception. In Australia (Kimberley plateau), plateau basalts erupted in the Cambrian (Antrim).

1.3.1.2 *Northern row of cratons.* Evolution of the northern cratons proceeded in a significantly different manner than that of those in Gondwana. Foremost, they were distinctly isolated from each other by the opening of intermediary ocean basins. The East European and Siberian cratons entered the platform stage of development in the Late Vendian, Sino-Korea in the Early Cambrian and North American in the Late Cambrian. In the latter case this delay was associated in part with the Southern Oklahoma aulacogen already developed south of the cratons in the Early-Middle Cambrian; this aulacogen closed only at around 525 m.y. with the formation of a felsic volcanoplutonic complex. Another aulacogen, Reelfoot, in triple junction with the Southern Oklahoma aulacogen developed during the Early Palaeozoic extending in a submeridional direction.

The tendency of vertical movements differed for different cratons during the Vendian-Cambrian. While the North American Craton experienced growing subsidence, the East European Craton contrarily gradually rose. The Siberian and Sino-Korea cratons remained relatively stable. Regression became perceptible only in the Late Cambrian on the Siberian Craton which experienced maximum subsidence until then. In the Early-Middle Cambrian an extensive salt basin which had formed in the south-western part of this craton was separated by a barrier reef from the more open and deepwater marine zone in the north-east and east. With this exception, the deposits in all the northern cratons were characteristically of the shallow-water type. In the western cratons—North American and East European—sandy clay beds predominated (in North America they had been increasingly replaced by carbonates by the end of the Cambrian). In the eastern cratons—Siberian

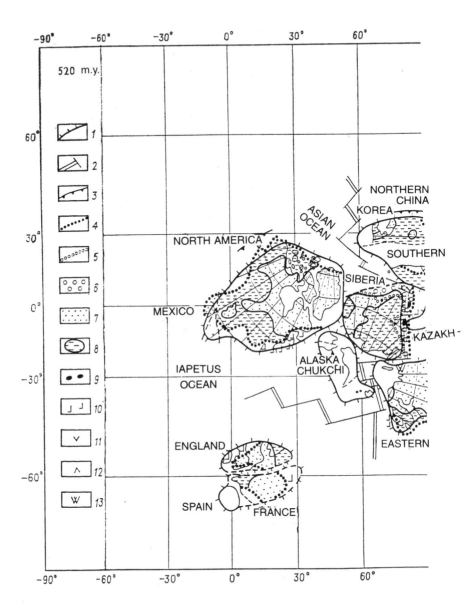

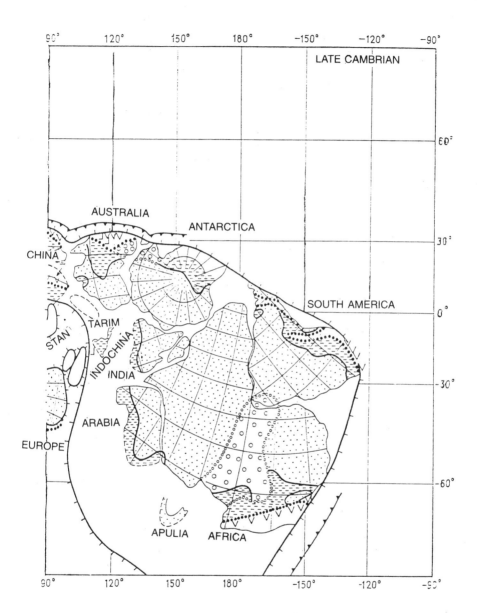

and Sino-Korea—carbonate accumulations predominated. These differences were due not so much to climatic conditions since all the 'northern' cratons were situated in the lower latitudes, as to the land in North America and Eastern Europe occupying a considerably larger area than in Siberia, northern China and the Korean peninsula.

1.3.2 Mobile Belts

1.3.2.1 *North Atlantic belt-Iapetus ocean.* The Vendian-Cambrian represented the period of opening of the Palaeo-Iapetus ocean. It arose in the northern part (Greenland, Spitsbergen and Scandinavia) on a Grenville basement, i.e., as a result of rifting within the Early-Middle Riphean belt of tectonothermal reworking. In the southern part (British Isles, Newfoundland and Appalachians) the rift formed at the junction of Grenville and Baikalian cratonisation regions (Fig. 1–8). Deposits of the continental rifting phase are represented in the Appalachians by the Ocoee supergroup and its analogues; in Newfoundland by the Fleur-de-Lys supergroup and Gander group; in eastern Greenland by Hagen Fjord and Eleanore Bay; in Spitsbergen by Hecla-Hoeck; in Scandinavia by Ost-Finnmark of northern and 'sparagmite' of southern Norway; and in Scotland by Torridonian. Commencement of rifting preceded a more restricted extension accompanied by formation of swarms of dolerite dykes known in Newfoundland, the Appalachians, Greenland and Scandinavia, i.e., almost everywhere. Radiometric ages of this dyke

---

Fig. 1-7. Palaeotectonic reconstruction for the Late Cambrian (Mercator projection with centre at 0°N lat. and 90°E long.).

1—boundaries of continental crust regions (Zonenshain et al., 1987); 2—axes of spreading and transform faults (Zonenshain et al., 1987); 3—subduction zones (Zonenshain et al., 1987, modified and supplemented); 4—boundaries of continental crust regions in areas where our data do not agree with those of Zonenshain and coauthors (1987); 5—boundaries of major regions with orogenic regime of development; 6—zones of uplift in continents with a high relief; 7—zones of uplift in continents with a low relief; 8—marine sedimentary basins in continents and their boundaries; 9—regions of manifestation of granitisation; 10 to 13—magmatic complexes (10—ophiolite; 11—island arc; 12—marginal-continental volcanic belts; 13—bimodal)

The reconstruction of many continents in this scheme of Zonenshain and coauthors (1987), in the opinion of the present authors, is not satisfactory and contradicts palaeotectonic data, primarily of the Arctic. Other reconstructions in which the Arctic represents a part of Hyperborea and adjoins the Innuitian system of the North American continent are more correct. There is no place in the south of the Siberian continent for major orogenic systems, i.e., Altay-Sayan, which fell in the spreading zone and northern Mongolia. For this epoch it is probably incorrect to depict the western Antarctic region almost in the present-day boundaries. The Patagonian mobile zone is not reflected in South America. There is no agreement with the scheme of extension conditions (and spreading) in the Polar Urals. The location of Cathaysia is unsatisfactory as the marginal-continental sedimentary complexes and volcanogenic island arc formations fall within the continent.

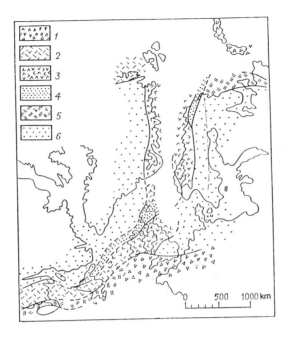

Fig. 1-8. Reconstruction of Caledonian orogen before continental rifting (after D. Roberts and G. Hail. In: Frasier and Schwimmer, 1987). Areas of distribution of tectonic deformations of different ages.

1—Carboniferous; 2—Devonian; 3—Silurian; 4—Late Cambrian-Early Ordovician; 5—end of Proterozoic; 6—Grenville and much earlier orogeny.

complex fall around 600 m.y. Accumulation of synrift deposits was accompanied by manifestation of bimodal volcanism, also known in almost all the regions of the North Atlantic province although distributed unevenly. The clastic, significantly rudaceous, arkose or greywacke composition of these deposits suggests an elevated relief of rift flanks, i.e., their development proceeded against a background of general uplift of the belt. The presence of tillites and tilloids predominantly in two levels, the Upper Riphean and Lower Vendian, is significant. In facies composition filling of the Proto-Iapetus rift basins includes continental (alluvial and deltaic) as well as shallow-marine and at places even deepwater slope (turbidites) deposits apart from glacial and glacial-marine formations.

Auxiliary, blind branches set off from the main trunk of the rift system in several sections and became extinct subsequently. These are: Southern Oklahoma rift (aulacogen or, more correctly, intracratonic geosyncline), Reelfoot, St. Laurent and Labrador rifts in North America and Osterdahl rift in

Scandinavia. The Proto-Iapetus rift system between Spitsbergen and Nord-kap joined with the Barents-Timan Riphean geosynclinal system, advancing considerably further in its evolution and entering in the Vendian the orogenic stage of this evolution. It is possible that the Greenland-Spitsbergen segment of the Proto-Iapetus was initially very closely associated with the Barents-Timan Baikalides but, unlike the latter, did not undergo compression and deformation before the Vendian or in the Cambrian and, on the contrary, entered the Cambrian in a phase of quasi-platformal development.

Transition from rifting to the spreading phase of the Iapetus ocean ought essentially to be dated according to the maximum radiometric age of ophio-lites. But this age is almost always Ordovician, proximate to the Late Ordovi-cian age of their obduction and is interpreted as pertaining to the ophio-lites of marginal sea origin and not to the oceanic crust proper. Therefore the more indirect features are of vital importance, i.e., replacement of syn-rift by shelf epirift deposits in the passive margin (Devlin and Bond, 1988; Eisbacher, 1985) and discontinuity in the subsidence curve of this margin (Bond et al., 1985). Based on these features, commencement of spreading should be placed at the end of the Vendian to the very beginning of the Cambrian. Correspondingly, during the Cambrian accumulation proceeded in the eugeosynclinal zone of the Appalachians under deepwater conditions of mafic volcanics and siliceous-clay deposits. Naturally, it could not have been wholly simultaneous throughout the extent of the future ocean anymore than its closure was. In the Greenland-Spitsbergen segment ophiolites are generally absent and it is not yet known whether destruction of continental crust proceeded here to its total disruption. It is more likely that it did not occur here and the rift maintained its ensialic character.

I n n u i t i a n   s y s t e m. In the modern structural plan of the Earth, the Innuitian fold system of northern Greenland and the Canadian Arctic archipelago adjoins the eastern Greenland-Spitsbergen segment of the North Atlantic belt. This fold system serves as a connecting link between the North Atlantic and the East Pacific mobile belts.

In the Vendian and the first half of the Early Cambrian, the Innuitian sys-tem underwent the rift stage of its development, judging from the accumula-tion of coarse clastic deposits in the axial zone of the belt in Hazen trough (Fig. 1–9). In the second half of the Early Cambrian, these formations were replaced by thin deepwater siliceous-clay formations whose accumulation continued in the Middle and Late Cambrian and also in the first half of the Ordovician. Evidently this transition coincided with the change from the rift stage to the spreading stage, also marked by ophiolites. These are intruded by Middle Ordovician granites and are probably Cambrian-Early Ordovician in age. In the southern shelf and within the microcontinent of the far north of Ellesmere and Axel-Heiberg islands, deepwater deposits of the Hazen trough are replaced by carbonate deposits in the south and sandy deposits in·the north. The western continuation of the Innuitian system is traced in northern Alaska and Wrangel Island region.

1.3.2.2 *Mediterranean belt.* In the Vendian the palaeo-ocean Iapetus was bound in the south by Cadomian orogeny that included the marginal volcanoplutonic belt of southern England and, in the Cambrian, by the Epicadomian (Epibaikalian) platform margin of Gondwana with the Avalonian, Armorican and English Midland massifs. In the Vendian to Early Cambrian a basin with transitional and oceanic crust encompassing the Carpathians, Balkans, the Caucasus and a part of Iran still extended between this zone and the passive margin of the East European continent. A zone with folding dated Early to Late Vendian extended in the north-east

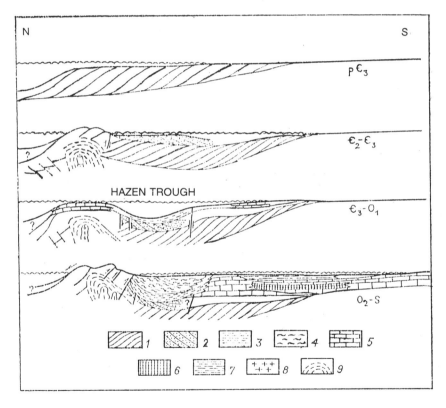

Fig. 1-9. Plate-tectonic model of the Early Palaeozoic history of the Innuitian system (after K. Drummond. In: Frasier and Schwimmer, 1987).

1—progradational shelf wedge; 2—deltaic clastic formations; 3—deepwater formations in a stagnant basin; 4—flysch; 5—carbonates; 6—evaporites; 7—shales; 8—granitoids; 9—metamorphic aureoles.

in the region from the buried Malopolsky massif and farther away under the Outer Carpathians towards Dobrogea (in: *Geology of Poland*, 1977). Farther north-east, it was replaced by the Keltsy zone with thick Upper Vendian, Lower and Middle Cambrian formations which experienced folding in the Late Cambrian to the Early Ordovician ('Sandomrr phase'). Apart from this, other zones of relatively intense subsidence in the Vendian-Cambrian are noticed in western and central Europe. These are: Brabant (Belgium), Leonese-West Asturia and Ossa-Morena (Iberian peninsula) and Montagne Noire (south of the French massif Central). All these zones represent ensialic troughs with no traces of magmatism. They are made up of shallow-marine sandy clay deposits. Later, commencing with the end of the Cambrian, these deposits underwent folding of moderate intensity and were almost not exposed to regional metamorphism. In the rest of the area of western and central Europe, however, the Cambrian was characterised by weak uplift or weak subsidence. In the northern part of the Armorican massif a marginal volcanoplutonic belt was formed in the Late Vendian and Cambrian. J. Cogné (in: Cogné and Wright, 1980) interpreted this belt as the result of subduction from the north, from the side of the western English Channel where a zone of positive magnetic anomalies may point to the existence of an ophiolite belt.

In the eastern part of the Mediterranean belt, characteristics of a geosynclinal regime are not known east of the Caucasus and Iran right up to central Kuen Lun in China. In central Afghanistan and in the northern Pamirs, Vendian-Cambrian formations are platformal in character (Gondwana prominence). We find Vendian-Early Cambrian ophiolite association and much later island arc series only in Qilianshan and northern Qinling, which were probably associated with the Ural-Okhotsk belt through the gap between the Epibaikalian Tarim massif and ancient Sino-Korea Platform (O.A. Seidalin, pers. comm.). It is possible, however, that a basin with oceanic crust could have already been formed in the period under consideration in the extreme northern zone of the northern Pamirs and in their continuation in Kuen Lun (Pai, 1992). Southward, in Tibet and the Sinoburman massif, Vendian and Cambrian are either absent or are represented again by platform-type formations. The Viet-Lao system in north-eastern Indochina constitutes an exception but may regarded as an apophysis of the Pacific belt.

1.3.2.3 *Ural-Okhotsk belt.* Tectonic processes in this belt, which separated the Siberian continent with its Verkhoyansk-Kolyma continuation from eastern Europe and Sino-Korea as early as in the Riphean (Middle ?), were contradictory in nature. On the one hand the fold structures which built them arose in the periphery of the East European and Siberian cratons in this belt. These structures are: Timan-Pechora platform basement and Polar Urals, Taimyr and Yenisei range with an intermediate strip of the left bank of Yenisei and, farther away, Eastern Sayan (pro parte) and Peri-Baikalian region. Considering the Baikalian and Early Precambrian consolidation blocks in the central part of the basement of the West Siberian Platform and in central Kazakhstan and northern Tien Shan, it is highly probable that at the end of the Riphean to the commencement of the Vendian the old cratons bordering the Ural-Okhotsk belt were united by immature Riphean continental crust in the same manner as occurred in the Mediterranean belt. In the Late Vendian and Cambrian the central part of this Epibaikalian Platform began to undergo destruction while its more peripheral parts developed initially in an orogenic and later in a platform regime (Fig. 1-10).

Opening of the Palaeo-Asian ocean in the Vendian-Early Cambrian has been best documented by ophiolites in the Altay-Sayan region—from Kuznetsk Alatau to Eastern Sayan and, more southward, in Western Mongolia (Fig. 1-11). The poorly exposed ophiolites of Transbaikalia and the Amur-Okhotsk segment as well as western Chingiz-Tarbagatai are also probably of the same age. Vendian-Early Cambrian ophiolites are known in Tien Shan, south of the Kazakhstan-Northern Tien Shan microcontinent and north of Tarim. The easternmost belt of the Uralian system traced through Tyumen region could be regarded as the northern continuation of this ophiolite belt. The eastern continuation is probably the ophiolite belt of Qilianshan whose association with the main belt is possible through Beishan between Tarim and the Alashan salient of the Sino-Korea Craton.

The situation changed sharply in the Middle and Late Cambrian (Fig. 1-12). The eastern continental margin within eastern Sayan and western Mongolia experienced accretion due to collision of island arcs and transformed into a margin of Andean type underlain by a seismofocal zone gently inclined eastward. In the central part of the ocean basin, Chingiz-Tarbagatai volcanic arc was formed and divided this basin into two distinct parts: Junggar-Balkhash in the west and Altay-Sayan in the east. After spreading ceased, the latter began to fill increasingly rapidly with clastic material entering mainly from the east, from the mountain framework of the Siberian continent. This is the well-known Gorny-Altay series of Late Cambrian-Ordovician and its analogue in Western Sayan. Spreading continued in the west even in the Ordovician.

1.3.2.4 *West Pacific belt.* In the north the western framework of the Palaeo-Pacific is represented by the passive margin of the Siberian continent traced southward through the Aya zone on the coast of the Sea of Okhotsk to the Shevli zone of the Amur-Okhotsk segment of the Ural-Okhotsk belt, i.e., to the merger with the Palaeo-Asian ocean. The Vendian-Cambrian is represented here mainly by carbonate formation. More southward, the Pacific margin extends along the Hinggan-Bureya and Khanka massifs. In this belt the Lower, partly Middle Cambrian is also made up of shelf carbonates

44

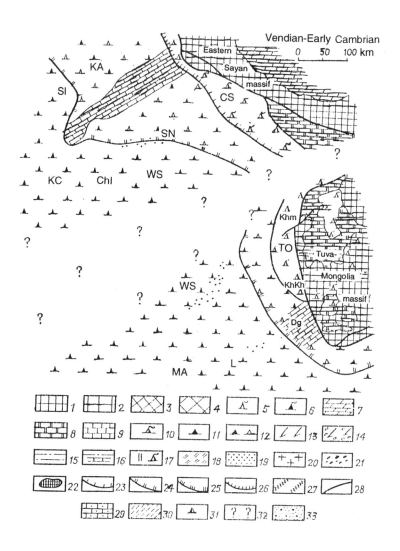

which are replaced in the upper Middle Cambrian in the Khanka massif by terrigenous sediments pointing to the commencement of uplift resulting in the absence of upper Cambrian. In the Hinggan-Bureya massif uplift commenced earlier, even in the Middle Cambrian. In the Japanese islands, the much older western part which originally adjoined the continent, the sequence of non-metamorphosed deposits commences with the Silurian; consequently, uplift can be assumed in the Vendian-Cambrian here. The same is true of the southern Korean peninsula.

Even farther south, in the eastern prolongation of Qinling-Dabieshan, the northern branch of the Mediterranean belt should have merged with the West Pacific belt, i.e., Palaeo-Tethys with Palaeo-Pacific.

We again find a well-developed continental margin in south-eastern China. It is manifest in the transition from clastic Vendian and shelf-carbonate Cambrian deposits of the Yangtze Platform (South China) to the more deepwater flyschoid shale-greywacke, at places including mafic and intermediate volcanics and continental slope and rise deposits of the Cathaysian system. Similar types of formations are developed more southward in the

---

Fig. 1-10. Palaeotectonic scheme of the Caledonides of Central Asia for Vendian-Early Cambrian (after A.B. Dergunov. In: Mossakovskii and Dergunov, 1983).

1 to 4—zones of predominant erosion: 1—with continental crust formed in the Riphean; 2 to 4—with continental crust formed by the Devonian in which a granitic-metamorphic layer was formed at the base (2—by Late Cambrian, 3—Middle Ordovician, 4—Devonian); 5, 6—zones of subaerial volcanism (5—with porphyric, 6—basaltic formations); 7 to 9—zones of carbonate sedimentation: 7—in intracontinental basin (Siberian Platform cover), 8—on the shelf, 9—in marginal sea; 10—zones of island arc volcanism with basaltic-andesite formation; 11 and 12—palaeo-ocean basin (11—central part with diabase-spilite formation, 12—marginal part with spilite-keratophyre formation); 13 and 14—relict palaeo-ocean basin with terrigenous turbidite-flyschoid formation (13—distal, 14—proximal parts); 15, 16—marine residual troughs (15—with terrigenous, 16—with terrigenous-carbonate formations); 17, 18—destructive troughs—monogeosynclines (17—with andesitic-basalt formation, 18—with sandy silt); 19—intermontane basins with molasse; 20—zones of granite formation; 21—zone of olistostrome accumulation; 22—outliers of tectonic nappes; 23 to 26—subduction zones (23—Vendian, 24—Early-Middle Cambrian, 25—Middle-Late Cambrian, 26—Devonian); 27—assumed destruction zones; 28—tectonic boundaries; 29 to 31—formations and palaeotectonic zones of adjoining Hercynian basin (29—carbonate-terrigenous shelf; 30—black shale of deepwater trench, 31—volcanics of axial part of basin); 32—assumed extension of palaeoocean basin; 33—terrigenous coarse clastic deposits in the cover of the Siberian Platform.

Structural zones (assumed positions): Sl—Salair, KA—Kuznetsk-Alatau, CS—Cambrian Sayan, RA—Rudnyi Altay, Chr—Charysh, Tl—Talytsa, ACh—Anuy-Chu, Kt—Katun, KC—Kholzun-Chu, Chl—Chulyshman, WS—Western Sayan, NS—Northern Sayan, TO—Tannu-Ola, Khm—Khamsary.

Structural zones of western Mongolia. MA—Mongol-Altay, L—Lake, Dg—Degandel', KhKh—Khan-Khukhey.

Superposed basins. Mn—Minusa, UL—Uimen-Lebed', Tv—Tuva.

Destructive troughs. SA—Southern Altay, DYu—Delun-Yustyd, Kb—Kobda.

46

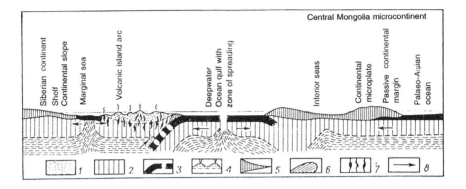

Fig. 1-11. Schematic reconstruction of Sayan-Baikal-Mongolia region in the period of formation of Vendian-Early Cambrian island arcs and marginal seas (after Gordienko, 1987).

1—asthenosphere; 2—lithosphere; 3—blocked oceanic plates; 4—volcanic island arcs; 5—continents and microcontinents; 6—accretionary prism of deposits in deepwater trench; 7—currents of fluids and magmatic solutions; 8—direction of stresses and motion in plates.

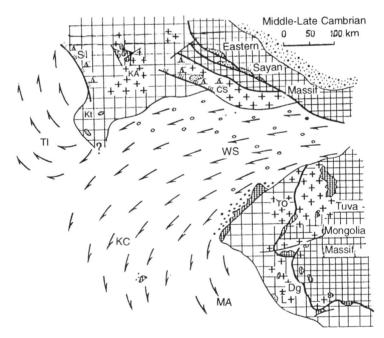

Fig. 1-12. Palaeotectonic scheme of the Caledonides of Central Asia in the Middle-Late Cambrian (after A.B. Dergunov. In: Mossakovsky and Dergunov, 1983).

Legend same as in Fig. 1-10.

Viet-Lao geosynclinal system which could be considered in some sense a branch of the Cathaysian system of south-eastern China. Ophiolites are also known in North Vietnam, most likely of Vendian-Early Cambrian age. In the south-east, the Viet-Lao system is cut off by the South China Sea coast in the immediate proximity of the approaching southern Kontum salient of the Indosinian massif (microcontinent) formed by Early Precambrian rocks.

After a large gap associated with the much later opening of the Tethys, the margin of the Palaeo-Pacific is again recognisable in southern Australia. Its shelf part is represented predominantly by carbonate formations of the Lower Vendian-Lower Middle Cambrian Adelaide system succeeded by continental molasse of the upper part of the Middle Cambrian. The developed more eastern and south-eastern greywackes and shales of Bancania and Kanmantoo zones correspond evidently to the continental slope and rise and also to deposits of a small marginal sea. It was bound in the east by the Mount Wright volcanic arc overlying the Wonnaminta microcontinent. On Tasmania Island, the Dundas trough (rift) (Scheibner, 1985) represents the probable southern continuation of the Bancania-Kanmantoo trough (Scheibner, 1987). The Dundas trough splits the uplift (island arc ?) formed before the Vendian. Ophiolites together with spilites are known at the base of the thick siliceous-clayey-greywacke filling of the Dundas trough. Dundas trough was bound in the east by the Mount Read volcanic arc, which represents a probable continuation of the arc of Mount Wright on the mainland. Ophiolites emerge at places east of the latter in Victoria state among deepwater siliceous-clay deposits of the Lower Palaeozoic. These ophiolites are most likely of Vendian-Early Cambrian age. Ophiolites of proximate age have also been established on the South Island of New Zealand; here, clay-greywacke deposits of Vendian-Middle Cambrian age associated with ophiolites are replaced towards the east by more shallow-water clastic and carbonate deposits. In this epoch New Zealand should have closely adjoined southern Australia and Tasmania and, together with the latter, the Antarctica where continuation of the Tasmanian Dundas trough has been established.

Recently, Middle Cambrian age, 530 $\pm$ 6 m.y. (ion microprobe U-Pb method on zircon), was recorded for plagiogranite from the ophiolite complex of the Great Serpentinite belt west of the New England fold system and roughly 1000 km east of the ophiolites in Victoria state.

Ophiolites of Victoria, New South Wales and New Zealand may be regarded as relics of the Palaeo-Pacific floor or of its marginal sea. In the Vendian-Early Cambrian the Australian margin developed as a passive margin; from the Middle Cambrian it was transformed at places into an active margin with the formation of a seismofocal zone above which the Mount Wright-Mount Read volcanic arc has been formed. The intense folding of Bancania-Kanmantoo deposits and their analogues in New Zealand, their metamorphism and development of Anabama granite pluton, which is in fact unique in the Adelaide system, should be associated with the activity of the aforesaid volcanic arc zone, which reached culmination at the end of the Cambrian. Nappes translated westward with the participation of obducted ophiolites were formed in New Zealand in this epoch of diastrophism, called Delamerian in Australia.

Further continuation of the Pacific margin of Gondwana extends through the Antarctica along Victoria Land and the east coast of the Ross Sea towards the east coast of the Weddell Sea through the Transantarctic range. In the Vendian this margin was of active, evidently Andean type. Roughly 650 m.y. ago, 'Beardmore orogeny' manifested in the form of intense folding of a flysch (turbidites!)-shale-greywacke formation of the same name, regional metamorphism to biotitic subfacies and intrusion and unconformable superposition of this formation by a felsic volcanoplutonic complex (Stump et al., 1986). These relationships were established in La Horse hills; at other places (Nimrod glacier and Pensacola Mts.) a shelf carbonate-terrigenous Cambrian formation was deposited unconformably directly on Riphean geosynclinal strata. The shelf formation was dislocated in turn and intruded by the anatectic granite complex of Granite Harbour in the Late Cambrian to the Early Ordovician. This Ross orogeny is contemporaneous with the Delamerian diastrophism of Australia.

In the extreme north-eastern part of the mainland, on Victoria Land, Bowers graben extended with a north-north-western strike from the Pacific Ocean to the Ross Sea between the Riphean Rosside metamorphic complex of the end of the Transantarctic range and the Robertson Bay group of Lower Palaeozoic age. Its filling consists of volcanosedimentary beds below and clastic-carbonate beds above corresponding to Middle and Middle-Late Cambrian age and supposedly Upper Cambrian continental molasse. Bowers graben is usually compared to the Tasmanian Dundas graben (Stump et al., 1983). The clay-turbidite Robertson Bay formation with olistoliths of Upper Cambrian-Lower Ordovician age located westward may represent continental slope deposits. As pointed out by Stump and coauthors, granites west of Bowers graben belong to type S, i.e., anatectic, and those east to type I, i.e., magmatic, suggesting replacement of a continental crust by an oceanic one.

Ellsworth Mountains, whose present position relative to the Baikalian-Salairian system of the Rossides is regarded as due to the most recent horizontal displacement, also represent an element of submarine margin of Antarctica. These mountains are disposed more westward and are characterised by beds striking perpendicular to the Rossides. A volcanic suite of calc-alkaline composition lies at the base of the sedimentary sequence commencing with terrigenous-carbonate Middle Cambrian beds and including the carbonate Upper Cambrian; it represents a possible analogue of beds described from the Transantarctic range.

The above discussion reveals that the Antarctic margin of Gondwana belonged to the Andean type in the Vendian, was transformed into a passive margin in the Early Cambrian and again reactivated in the Middle Cambrian with the formation of Bowers volcanic arc complicated by a rift bearing the same name. Cambrian development concluded with Ross orogeny.

1.3.2.5 *East Pacific belt.* Encircling the Antarctica, the Pacific margin of Gondwana continues into South America along the Andean fold system, although at present the connection between them has been disturbed due to formation of the Scotia Sea basin in the Oligocene and later. For understanding the geodynamic set-up in the Palaeozoic, the Andes section falling between Chile and Argentina (between 29 and 33°S lat.) is most revelatory (Ramos, 1988).

A typical passive continental margin with a carbonate shelf and turbidite (turbidites also of carbonate composition) continental slope with olistoliths existed in the Precordillera in the Cambrian (and Ordovician). A deepwater basin on oceanic or suboceanic crust has been assumed more westward: its relicts are tholeiitic basalts and ultramafics appearing in the Middle and Late Ordovician olistostromes. It has been suggested that a microcontinent existed west of this deepwater basin, representing most probably a marginal sea. This microcontinent corresponds to the Cordillera Frontal in contemporary structure and the Palaeo-Pacific per se should be situated even farther away.

This fairly simple picture is complicated by the presence of the Sierra Pampeanas massif east of the Precordillera. This massif comprises Late Proterozoic metaterrigenous and metavolcanic rocks and granitoids. It is evidently the southern continuation of the Baikalian fold complex of the Eastern Cordillera of Peru and Bolivia where the age of folding and metamorphism has been determined roughly at 600 ± 50 m.y. (Dalmayrac et al., 1980). But even younger granitoids intruded into this complex, these granitoids being essentially Late Ordovician-Silurian and Early Carboniferous. The presence of Cambrian granitoids has been suggested but this would make necessary the assumption of a second convergence of the Sierra Pampeanas and Precordillera (Ramos, 1988). Most probably, however, the development of this region proceeded in the same manner as in Antarctica, i.e., after Vendian diastrophism a pause occurred in the Cambrian, with the formation of a passive margin, and later tectonomagmatic activity renewed at the end of the Cambrian to the Early Ordovician (see Section 2.3.2.6).

North of the above-described region in the Andes, the Cambrian has not been established with certitude but the sandy-shale formations in the base of the Ordovician in the Eastern Cordillera of Bolivia and Colombia have been tentatively placed in it. A subaerial volcanic suite of felsic composition with ignimbrites of northern Peru belongs most probably to the Vendian (once again, a similarity with the Antarctica!). On analogy with the Ordovician (see Section 2.3.2.6), it may be suggested that a deepwater basin extended in the Cambrian along the Andes. The basin was separated from the west from the Palaeo-Pacific by microcontinents—'borderlands'. One of them was the Chilean and the other the Arequipa massif on the coast of Peru. In the north the Early Palaeozoic geosynclinal system at present is cut off by the Caribbean coast. Its northern continuation re-emerges in Belize, Honduras and Mexico; it is directly associated with the North American Appalachian—Ouachita—Marathon orogen, i.e., with the North Atlantic belt, the Iapetus ocean.

In Mexico the Palaeozoic structural-formation zones extend in a northeasterly direction and are cut off in the west by the Pacific coast. In the innermost zone, comparable to the Piedmont zone of the Appalachians, the Palaeozoic formations have been metamorphosed and granitised; this zone reaches the Monterey region in the north-east. Roughly between Masatlan on the Pacific coast and Monterey, a zone of deepwater deposits of the Marathon-Ouachita type is traced; these are evidently continental slope deposits. In the extreme north-west, in Sonora, the Lower Palaeozoic is represented by shelf carbonates; this evidently represents the extreme southwestern salient of the North American Craton extending into California and Arizona.

On the other side of this salient, in southern California (Walker et al., 1986), a strip of Vendian-Cambrian formations of the western submarine margin of the North American Craton commences and can be traced almost without interruption right up to Alaska (Churkin and Eberlein, 1977; Young, 1982). This strip has been well studied and the tectonic nature of the formations constituting it is now evident. Terminal Riphean and Vendian formations, i.e., the Windermere supergroup and its analogues, represent synrift deposits. They include glacial (at the bottom), black shale, carbonate and clastic sediments as well as basalt sheets—continental tholeiites (in the lower parts). There are signs of synsedimentational development of faults, evidently of the listric type (Eisbacher, 1985). On transition to the Cambrian, the conditions change sharply—facies diversity, characteristic of the rift stage of evolution of the continental margin, is replaced by uniform carbonate sedimentation (Fig. 1-13) corresponding to commencement of opening of the ocean basin (Bond et al., 1985). This basin should have been connected with the Andean one, i.e., with continuation of the North Atlantic in the bypass of the California-Sonoran salient of the North American continent.

The Vendian-Cambrian history of the Earth's crust reveals a similar sequence of events almost everywhere. Intense accretion of new sedimentary material with its metamorphism and granitisation occurred in the Vendian in the active margins of Gondwana and some continents of the Laurasian

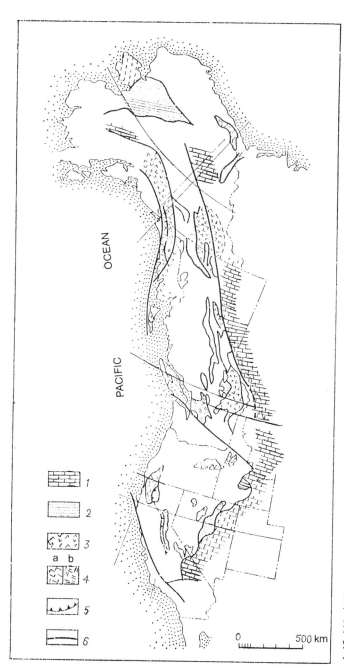

Fig. 1-13. Lithofacies of 'island belts' (or 'borderland terranes') of westernmost North America (after Churkin and Eberlein, 1977, with simplification). *Miogeosyncline:* 1—mainly carbonates, quartzites and shales (shelf formations). *Eugeosyncline:* 2—mixed volcanic and plutonic rocks of Late Palaeozoic age, including Permian and Triassic (volcanic arc formations); 3—cherts, shales, basalts and ultramafic rocks (oceanic formations partly representing two periods of ophiolite formations: Cambrian-Middle Devonian and Carboniferous-Triassic); 4—mixed volcanic and plutonic rocks and volcanoclastic deposits (active and remnant volcanic arc formations); a—juvenile, mainly Carboniferous-Triassic rocks without proven ancient basement; b—ancient, Precambrian-Devonian, at places extending into the Permian and Triassic; 5—overthrusts; 6—other faults.

group—Eastern Europe and Siberia. This epoch of tectonomagmatic activity was called Baikalian, Cadomian (Western Europe), Brazilian and Pan-African. In the other future passive margins this was the period of rift formation while the development of rifts in the central parts of continental cratons had ceased by the Late Vendian or Cambrian and was replaced by the period of commencement of platform cover formation. At the beginning of the Cambrian, even in margins that were active before, a regime characteristic of the post-rift stage of passive margins had been established. Arkose or quartz sandstones with non-skeletal invertebrates of the Ediacaran type have been deposited in the uppermost Vendian, in many regions far removed from each other. At the end of the Vendian to commencement of the Cambrian, opening of the Palaeo-Atlantic and Palaeo-Asian oceans, probably secondary, took place, as documented by ophiolites. Ophiolites of the Cambrian, possibly Late Vendian, are known on both sides of the southern half of the Pacific (Australia, Tasmania, New Zealand and Chile-Argentinian Andes). Atlantic type spreading extended into much of the area of the palaeo-oceans almost throughout the Cambrian. At this time, in the broad and warm shelf seas of passive margins of continents, carbonate beds, largely biogenic, accumulated. It may be assumed that the change of palaeogeographic environment at the end of the Vendian with the formation of extensive shallow-water basins, levelling and general warming of climate promoted the appearance of the skeletal fauna of invertebrates. Their shells and armour were essential for protection against tidal and wave movements of water masses in the shallow seas. The formation of deep ocean basins provided necessary conditions for upwelling, leading in turn to the first large-scale phosphate accumulation. According to the theory of A.V. Kazakov (1939), extinct organisms which prospered in the Late Vendian (Ediacaran of M. Glaessner in: Cloud and Glaessner, 1982) should have supplied phosphorus in the deep waters.

In the second half of the Cambrian, first on the Altay-Sayan flank of the Ural-Okhotsk belt and last at the commencement of the Ordovician in the northern North Atlantic belt (Newfoundland, Scotland and northern Scandinavia) and also in the Antarctica, revival of tectonomagmatic activity took place, deep seismofocal zones were formed and transition to active margins of the Andean type occurred. This epoch of tectonomagmatic activity is known as Salairian (Southern Siberia), Delamerian (Australia), Grampian (Scotland) and Sardian (Sardinia); in Africa, the corresponding events, like the much earlier ones, are included in the concept of Pan-African orogeny or reactivation. Not only in Africa, but also in many other regions, it repeatedly affected regions that had already experienced the action of the much earlier (Vendian) epoch of tectonomagmatic activity.

2

Ordovician. Mature Stage of Caledonian and Initial Stage of Development of Hercynian Mobile Belts

2.1 REGIONAL REVIEW. CONTINENTAL CRATONS AND THEIR MARGINS

2.1.1 North American Craton

E a r l y   O r d o v i c i a n. Compared to the Late Cambrian, the dimensions of Williston basin slightly decreased and later transgression covered northern and eastern Greenland while other basins generally maintained their dimensions (Henriksen, 1978; Ross, 1976; *Stratigraphic Atlas of North and Central America,* 1975; Surlyk et al., 1980). Unlike in the preceding epochs, the relief and erosion rate were probably low and, therefore, there was almost no development of terrigenous lithofacies in the marginal parts of marine basins (except for Williston basin) and carbonate sedimentation generally prevailed. The thickness of the Lower Ordovician was usually small and did not exceed 100 to 200 m but rose to 1 km or more in the pericratonic zones. This is true of the Peri-Appalachian as well as Peri-Cordilleran zones and especially of the Peri-Innuitian zone where the thickness of carbonate deposits exceeds 2 km. At the end of the Early Ordovician (in the Arenigian), gypsum-bearing evaporite deposits were formed in the north-eastern Michigan basin and in the Peri-Innuitian zone in the northern part of the craton (McGill, 1979).

M i d d l e   O r d o v i c i a n. The overall area of marine basins increased as a result of transgression into south-eastern and northern Canada. Erosion processes somewhat intensified in this epoch and the role of terrigenous rocks increased in the sedimentary basins of central and southern parts of the craton, foremost in the Williston basin, where sandstones predominated. But carbonates prevailed in the Michigan basin and northern part of the craton. The thickness of sedimentary Middle Ordovician formations usually constitutes tens of metres. The tempo of subsidence in the pericratonic zones remained high. In the Peri-Appalachian and Peri-Cordilleran zones the thickness gradually rose to 1 to 1.5 km. The thickness of the Middle Ordovician formations of the Peri-Innuitian zone and eastern Greenland exceeded 1 km. In the Caradocian age a zone with highly saline water existed on the northern margin of the continent and gypsum and anhydrite were deposited here in small amounts. A similar zone is known in the south-western Michigan basin.

L a t e   O r d o v i c i a n. Subsidence and transgression covered much of the territory of the craton. The sea area in this epoch was maximal for the entire Palaeozoic. Land regions had the form of low peneplains with almost no supply of clastic material while carbonate sedimentation prevailed almost everywhere in the marine basins. The thickness of the formations was very small and usually did not exceed a few tens of metres. In the central parts of Williston and Hudson basins it increased to 350 m, in the marginal zones of the continent to 600 to 700 m and in the Peri-Appalachian zone to 1000 m.

At the beginning of the Late Ordovician (Late Caradocian), deposition of carbonates in the Williston and Hudson basins proceeded under conditions of high water salinity and bands of gypsum and anhydrite were formed here. A typical complex of bituminous micritic limestones and siliceous rocks of small thickness was formed in the southern part of the continent. The Hudson basin zone was evidently distinguished by an unstable tectonic regime. This represents a unique region in which sand facies predominated in the composition of deposits; their thickness did not exceed 100 m and most of the facies were of continental genesis.

Eastern Greenland experienced folding and uplift in this epoch and an orogenic regime is manifest here from this period.

On the whole, tectonic development of the North American Craton in the Ordovician was extremely quiescent. Subsidence and transgression grew successively but subsidence rates were very small, judging from the thickness of deposits. Nearly all the land regions were low peneplains with weak erosional processes, excluding the western regions in the Middle Ordovician. The subsidence rate was significantly higher in the marginal zones of the craton and a much thicker sedimentary (mainly carbonate) cover was built up successively here.

2.1.2 East European Craton

E a r l y   O r d o v i c i a n. Subsidence and marine conditions gradually encompassed the Baltic-Moscow basin territory (Ivanov, 1979). Quartz and glauconitic sands, graptolite muds and detrital limestones, their thickness exceeding 100 m, are most characteristic of deposits in this extensive but shallow marine basin (Fig. 2-1). A very similar environment prevailed in the north-western Scandinavian part of the basin but the deposits there differ in composition. In littoral zones they are more clayey and fine grained, in the central part carbonaceous, while the role of sandstones increases and thickness rises to 300–400 m in the marginal zone of the craton. In the

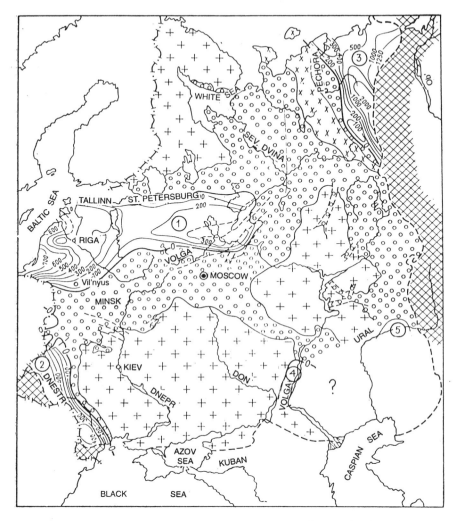

Fig. 2-1. Scheme showing spread and changes in thickness of inundation formations in the Early Palaeozoic stage of development of the Russian Platform, Ordovician-Silurian (after N.S. Igolkina. In: *Geological Structure of the USSR and Pattern of Location of Mineral Deposits.* Vol. 1, *Russian Craton,* 1985). See Fig. 1-2 for legend.

Structures of Early Palaeozoic stage (circled numbers): 1—Baltic-Moscow syneclise; 2—Dnestr pericratonic trough; 3—Pechora syneclise; 4—western slope of Peri-Caspian basin (Linev depression); 5—Sol'-lletsk salient.

south-western (Peri-Carpathian) pericratonic zone of the continent and in the Moesian block, the conditions were more dynamic and sand deposits predominate (Spassov et al., 1978).

Special conditions for rapid uncompensated subsidence prevailed in the Rügen-Pomorie trough, eastern margin of the Mediterranean mobile belt and the middle European segment of the latter where a complex of black shale formations up to 800 m in thickness formed.

A narrow zone of gentle subsidence stretched along the eastern border of the craton in the form of a continuous band from north to south. From west to east in this zone coarse clastic alluvial formations of 1 to 2 km thickness were replaced by shelf rhythmic sandy-clay and carbonate rocks of the same thickness and even farther away by terrigenous complexes of the continental slope (Klyuzhina, 1985). The central and southern parts of this zone reveal minor manifestations of mafic and intermediate volcanism associated with break-up, extension and destruction of the Earth's crust, signifying commencement of a new stage of development in the Uralian mobile system.

M i d d l e   O r d o v i c i a n. There was some increase in subsidence with a general predominance of uplift and erosion processes. The shallow sea of the Baltic-Moscow basin with clayey and calcareous deposits up to tens of metres thick enlarged; this thickness went up to 180 m in the central part. Carbonates predominated in the Peri-Baltica area and Scandinavia. Conditions were favourable in the Rügen-Pomorie zone for rapid uncompensated subsidence with clay sedimentation. However, in the Peri-Carpathian zone the Middle Ordovician represented an epoch of brief uplift and regression.

The Peri-Urals marginal zone continued to develop in the eastern part of the craton. Subsidence widened here and the sea penetrated into the Pechora basin where shallow-water and littoral formations of predominantly sandstones accumulated. Their thickness sometimes reached 250 to 400 m (Klyuzhina, 1985). As in the Early Ordovician, shallow-water coarse clastic formations were replaced from west to east throughout the Peri-Uralian zone by relatively deeper water formations containing limestones and dolomites with abundant and diverse marine fauna. The thickness of formations in this same direction rose up to 800 m. In the Southern Urals the Middle Ordovician was represented by quartz sandstones resting transgressively and often with a sharp angular unconformity on much older rocks.

L a t e   O r d o v i c i a n. As a result of regression, the Baltic-Moscow basin shrank slightly in size. The role of clay material rose in the composition of its deposits while carbonates were mainly represented by reef facies. In the Scandinavian part of the basin clay deposits predominated, their thickness here not exceeding a few tens of metres. The stage of filling began in the development of the Rügen-Pomorie trough and the thickness of Upper Ordovician clay deposits reaches up to 2.6 km. Subsidence renewed in the Peri-Carpathian zone and predominantly sandy deposits formed in the marginal basin in littoral shallow-water conditions. In the Moesian block the Upper Ordovician is represented by argillaceous rocks about 150 m in thickness.

In the Peri-Uralian zone the conditions did not vary compared to the Middle Ordovician. This margin of the craton continued to subside and clastic material carried away from the central regions was deposited here. Subcontinental alluvial-deltaic facies are replaced more eastward by rhythmic sand-clay and carbonate-terrigenous marine facies.

Thus uplift predominated in the East European Craton in the Ordovician. Zones of subsidence (marine basins) covered no more than 35% of its area and the regime of their development was quiescent with a very low rate of subsidence. An exception was the Rügen-Pomorie zone whose intense subsidence began to be compensated by deposition only in the Late Ordovician; however, it represented the marginal zone of a mobile belt.

2.1.3 Siberian Craton

E a r l y   O r d o v i c i a n. The area of the craton enlarged in the regions of Yenisei range and north-eastern West Siberia where a platform regime replaced the orogenic one. A general redistribution of erosion zones and sedimentation basins occurred towards this epoch and differentiation of tectonic movements intensified (Nikolaeva et al., 1968; *Ordovician Stratigraphy of the Siberian Platform,* 1975; *Phanerozoic of Siberia.* Vol. 1, *Vendian. Palaeozoic,* 1984). In Eastern Siberia regression has been noted in the central part and in the region of Anabar massif and extensive transgression in the north-east. The composition of sediments also changed with the increasing role of terrigenous material. Sandstones and shales predominated in the southern part of the Tunguska basin. Development of the Irkutsk basin concluded with its filling with thick terrigenous formations (up to 1400 m). Evidently this was a consequence of intensified uplift and erosion processes on the southern periphery of the craton.

In the northern part of the Tunguska basin and in southern Taimyr accumulation of terrigenous-carbonate deposits occurred with a predominance of dolomites. Signs of shallowness as well as gypsum admixture in the rocks have been noticed here. The thickness of deposits varied in the range of a few hundred metres. Exclusively carbonate formations of large thickness were deposited in the Southern Verkhoyansk region (Sette-Daban Range). Here, over 1300 m of limestones and dolomites accumulated in a trough far away from the erosion zone under the conditions of a normal marine basin.

M i d d l e   O r d o v i c i a n. The size of the cratonic area increased slightly in the south with completion of development of the Selenga-Upper Vitim zone. The tectonic regime in this epoch was very quiescent. Marine basins maintained their former contours and their areas shrank only slightly (Fig. 2-2) but the role of carbonates decreased in the composition of deposits

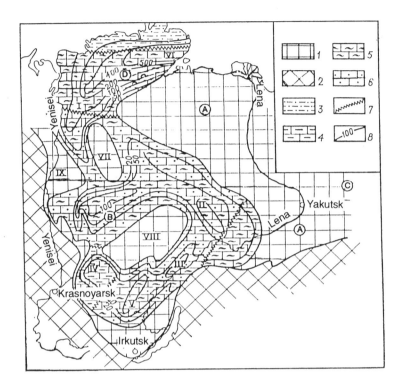

Fig. 2-2. Palaeotectonic scheme of the Siberian Craton for the Krivolutsk stage of the Middle Ordovician (after N.S. Malich In: *Geological Structure of the USSR and Pattern of Location of Mineral Deposits*, Vol. 4. *Siberian Craton*, 1987).

1, 2—uplift zones (1—weak on craton, 2—weak in mobile region); 3 to 6—formations (3—clay, 4—clay limestone, 5—variegated clay limestone, 6—variegated sand-limestone-clay); 7—lateral transition zone of formations; 8—isopachs, m.

Structural regions (circled letters): A—Aldan-Anabar; B—Lena-Tunguska; C—Sette-Daban; D—Taimyr; I to VI—basins (I—Noril'sk; II —Morkokin; III—Peri-Baikal; IV—Peri-Sayan; V—Ilim, VI—Taimyr); VII, VIII—uplifts (VII—Tembenchia, VIII—Katanga); IX—Turukhan salient.

(in the Tunguska and Southern Taimyr basins) and the proportion of terrigenous deposits, often of variegated colours, correspondingly increased (in: *Geology of Yakutsk ASSR*, 1981). In the eastern part of the Tunguska basin bands and lenses of gypsum point to the high salinity of water. In the northern part of Taimyr terrigenous-carbonate complexes were sharply replaced by deepwater clay formations of small thickness. Here, evidently, subsidence was not compensated by deposition. On the whole, the rate of subsidence in most of the basins decreased and the thickness of sediments rarely exceeds 100 m. Only in Southern Taimyr does it rise from south to north up to 1 km but decreases sharply to tens of metres more northward in the zone of uncompensated subsidence.

The predominance of carbonate accumulation and a high tempo of subsidence in the basins continued in the eastern Siberian continent: in Southern Verkhoyansk region (Sette-Daban Range) and in the region of the Omulevka Mountains in the Chersky Range. The thickness here goes up to 1700 to 2000 m.

Late Ordovician. Tectonic development stabilised further. The area of marine basins decreased in the central part of the craton. In the southern part of the Tunguska basin terrigenous marine and continental complexes predominated. Further, gypsum lenses and bands are encountered in several regions pointing to high salinity of water. Northward, sandy and argillaceous rocks were gradually replaced by pure limestones which extended into the Southern Taimyr basin. In the northern zone of the latter carbonates, as in the Middle Ordovician, were sharply replaced by argillaceous rocks. The thickness of formations decreased even more and rarely exceeded a few tens of metres although reaching 500 to 700 m in the axial part of the basin. Land zones were represented in this epoch as flattened peneplains which experienced very weak uplift.

A specific character of development prevailed in the Yana-Kolyma interfluve (Sette-Daban and Omulevka Mountains). Here, as before, limestones predominated in the composition of formations but the rate of subsidence was high, the maximum thickness of the Upper Ordovician exceeding 1.5 km.

On the whole, tectonic activity gradually receded in the Siberian Craton throughout the Ordovician. The area of sea decreased, the thickness of deposits fell and the overall area of predominance of a platform regime increased as a result of extinction of orogeny south of the craton. The quiescent palaeotectonic environment is also supported by the composition of sediments, i.e., absence of large masses of coarse clastics and persistence of facies.

The conditions were more dynamic in the troughs of Southern Taimyr and Chersky Range.

2.1.4 Sino-Korea Craton

Early Ordovician. The overall palaeotectonic environment was inherited from the Cambrian and, as before, much of the territory of this craton experienced subsidence (Kobayashi, 1968). The predominance of carbonate deposition also continued, terrigenous rocks accounting for less than 10%. The thickness of formations usually varied from 500 to 700 m. High rate of subsidence and considerable thickness (up to 1.6 km) have been established in the Pyongnam trough (aulacogen) in the northern Korean peninsula.

Middle Ordovician. The ratio between the regions of uplift and subsidence and also the configuration of marine basins remained as before. In northern China the land area was somewhat more extensive and, as before, carbonates accumulated almost exclusively in the sea. The tempo of subsidence was maintained and the thickness was usually 200 to 400 m. The tempo of subsidence in the Pyongnam trough, however, slowed down about fifty per cent.

L a t e O r d o v i c i a n. Uplift and regression encompassed the craton. A residual marine basin with an accumulation of about 50 m of limestones and sandstones existed only in the Pyongnam trough early in this epoch.

On the whole, throughout the Ordovician uplift increased in the Sino-Korea platform and its development proceeded towards near total regression from the Early Ordovician epoch of maximum transgression.

2.1.5 South China (Yangtze) Craton

In the E a r l y and M i d d l e O r d o v i c i a n, as in the Late Cambrian, the craton was covered by shallow sea with alternation of limestones containing shell fauna and graptolite shales. In the peripheral zones of the craton a transition towards slope flyschoid or even flysch (in the south-east) formations with a predominance of sandy or clay material and with bands of carbonate or siliceous rocks is evident. The Yangtze Sea was separated from the Cathaysian mobile system by the Jiannan uplift which was partly submarine and partly insular.

In the L a t e O r d o v i c i a n the southern part of the craton (Yunnan, Guizhou and Guangxi) experienced uplift and was transformed into land. This coincided with the deformation phase along the south-western and south-south-eastern framework of the craton.

2.1.6 South American Craton

E a r l y O r d o v i c i a n. Almost the whole of the craton experienced uplift and represented an extensive dry land area. But subsidence of the Peri-Andean pericratonic zone continued, its development associated with the growing activity of the mobile system in the adjoining Andes (Aceňolaza, 1982; Dalla Salda, 1982). This zone extended over a long distance from north to south and terrigenous sediments formed in it, their thickness increasing to 500 m or more in a westerly direction. Small volcanic eruptions of intermediate and felsic composition occurred in the Peruvian segment of the zone. Carbonates played a significant role in the deposits of the southern Argentinian part and this represented a unique region of comparatively extensive carbonate accumulation.

M i d d l e O r d o v i c i a n. Subsidence enlarged somewhat in the Peri-Andean zone, especially in its northern and central parts. Here mostly thin terrigenous and argillaceous deposits up to 300 to 400 m in thickness accumulated. In south-western Brazil (in the Serrania de Jacadigo hills) continental sandstones and conglomerates were formed to a thickness of 250 m.

In the neighbourhood of this region the relief was rugged and abundance of clastic material led to continental conditions of sedimentation over an extensive area. The continuing major role of carbonates in the composition of deposits in the southern part of the Peri-Andean zone is a striking feature. In two regions of the latter, in southern Peru and in north-western Argentina, subaqueous andesite and rhyolite-dacite volcanism was manifest. Predominance of weak uplift and erosion processes must be suggested for the entire central part of the platform.

L a t e O r d o v i c i a n. Subsidence of the interior basins—Amazon, Maranâo and Parana—commenced. Continental and, less frequently, marine sandy formations up to 150 m thick were formed in the Amazon basin; only marine sandy-clay formations up to 200 m in thickness were deposited in Maranâo while coarse clastic continental sandstones and conglomerates exceeding 400 m in thickness accumulated in the Parana basin. In the latter case, hills in the neighbouring portions of the Andean mobile system served as a source of clastic material (Dalmayrac et al., 1980). Conditions for marginal marine basins with terrigenous (carbonate-terrigenous in the south) sedimentation of small thickness continued in the northern and southern parts of the Peri-Andean zone. The Upper Ordovician sequences here bear a regressive character since subsidence was replaced by uplift in the concluding part of the epoch.

The Ordovician represented for the South American Craton a period in which uplift predominated and large interior territories were affected by subsidence only at the end of this period. The Peri-Andean zone experienced stable subsidence of small amplitude.

2.1.7 African Craton

E a r l y O r d o v i c i a n. With the completion of Pan-African orogeny, this structure represented one single large craton occupying almost the entire area of the present-day continent as well as the Arabian peninsula and the Middle East. Uplift and erosion prevailed over almost all this territory but subsidence continued along the northern margin of the craton. The boundaries of the marine basin in the Arabian peninsula enlarged but land areas appeared in Iran and Afghanistan (Pyzh'yanov et al., 1980; Bender, 1975; Wolfart and Wittekindt, 1980). Carbonates are no longer present among the Lower Ordovician rocks here while sand facies predominate among terrigenous rocks and continental facies in the littoral zone. The Lower Ordovician rocks are 200 to 400 m thick.

In the northern part of the craton, in the Peri-Atlas zone, subsidence widened and transgression occurred. An extensive sedimentary basin formed here in which predominantly sandy formations accumulated; they occurred in the continental facies over large areas (Aliev et al., 1979; Buggisch et al., 1979). The thickness of the Lower Ordovician is usually 200 to 400 m but in the west (in Ougarta aulacogen) reaches 1 km. A small marine basin also existed in the extreme south of the craton in the Cape province where the lower part of the sandstone suite of Table Mountains was formed in the Early Ordovician (Hobday and Tankard, 1978).

M i d d l e   O r d o v i c i a n. The major portion of the craton experienced uplift as before and the regions of uplift in northern Iran and Afghanistan enlarged in the north-east. Deposits in the marine basins were represented almost exclusively by terrigenous formations. A broad zone of mixed continental and marine sand facies extended along the southern boundary of the Arabian basin. The thickness of the Middle Ordovician was 200 to 400 m but the rate of subsidence in north-eastern Syria was slightly higher and the thickness of terrigenous beds here rose sharply to 2 to 3 km.

In the Peri-Atlas marginal zone transgression increased but subsidence was quiescent and wholly compensated by the accumulation of sandy-clay sediments coming from the eroded central part of the craton. A thickness of 150 to 200 m is most typical here. In the extreme south, in the Cape province, subsidence, small in area and amplitude, was accompanied by the deposition of marine and continental sandy formations.

L a t e   O r d o v i c i a n. Significant changes occurred in the tectonic development of the craton. In its north-eastern part uplift enlarged while the area of the marine basins shrank by one-half. Inversion occurred in the development of the former deep basin, evidently an aulacogen in north-eastern Syria; this region now turned into land. Thin bands of gypsum and rock salt in the Ashgill formations in Jordan point to high water salinity in this part of the basin (Bender, 1975).

Zones of uplift and subsidence were redistributed in the northern and north-western parts of the craton. The major portion of the Sahara basin experienced regression and a rugged relief arose at several places (e.g., Ahaggar region). Marine conditions were only in a narrow Peri-Atlas zone and in the Tindouf and Ougarta basins where coarse clastic strata accumulated (Fig. 2–3). But the sea extended into the western part of the craton. The thickness of coarse clastic formations here did not exceed 150 to 200 m and alternation of marine and continental facies was a characteristic feature. Similar deposits formed in the south, in the Cape province. The entire remaining part of the craton remained dry land throughout the Ordovician. A distinctive

---

Fig. 2-3. Table showing correlation adopted in the text between the characteristic continental formations of the African Palaeozoic from north-west to south (after Fabre, 1988)..

1—sedimentary cover of craton-type (a—sandstones, b—siltstones, c—lutites, d—carbonates); 2—formations correlating with peneplain corresponding to commencement of cratonic sedimentation; 3—molasse; 4—pre- and syntectonic formations of Pan-African orogeny, often in the form of flysch; 5—glacial formations; 6—unconformities; 7—mafic or felsic volcanism; 8—granitoids.

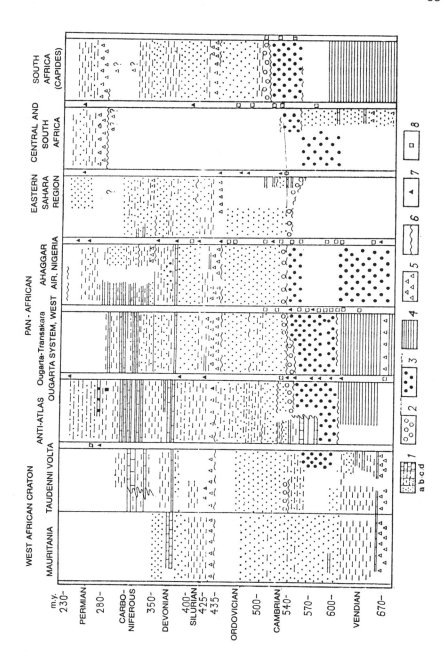

64

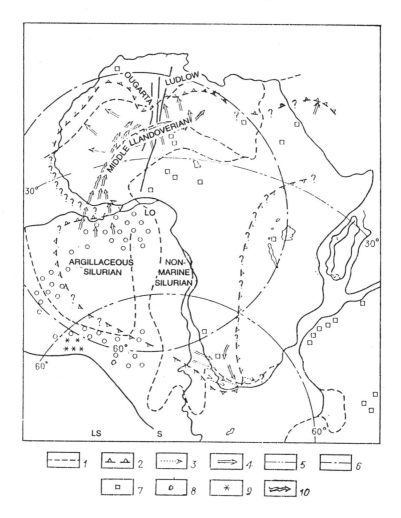

Fig. 2-4. Africa from 500 to 380 (or 370) m.y., roughly from the Ordovician to the Middle Devonian (after Fabre, 1988):
LO and LS—South Pole in Late Ordovician and Late Silurian; S—South Pole based on Air granite (400 m.y.):

1—approximate Silurian coastline in northern Africa and facies boundaries in southern Africa and South America; 2—maximum spread of ice sheet at end of the Ordovician to commencement of the Silurian; 3—main directions of dispersion of sand in the Early Devonian under fluvial (Sahara) or littoral (southern Africa) conditions; 4—main direction of ice flow; 5—palaeolatitudes around 400 m.y. ago; 6—palaeolatitudes by Late Ordovician, around 450 m.y. ago; 7—zones of K/Ar equilibrium or/and granitoid intrusions; 8—Silurian cold-loving (cryophilic) fauna (Malvin-Kafre) in South America; 9—(thermophilic) fauna (Appalachian) penetrating from the north into the west coast of South America; 10—hypothetical cold currents during the Silurian.

feature of the African Craton was the extensive development of the Upper Ordovician tillites, pointing to an ice sheet having encompassed a significant part of the continent (Fig. 2-4). Tillites are known in Morocco, in many regions of the Sahara from Sierra Leone to Tunisia and Egypt and in the Arabian peninsula (Deynoux et al., 1972; Jaeger et al., 1975; Klitzsch, 1978; Legrand, 1974; and Tucker and Reid, 1973). Their age ranges from the end of the Caradocian to the Early Llandoverian. Tillites are also known in the Cape province (Hobday and Tankard, 1978). The thickness of these tillites usually runs into tens of metres but may reach 250 m here and there.

Thus throughout the Ordovician almost the whole of the immense territory of the African Craton experienced uplift. A transgressive-regressive cycle commencing with subsidence in the Early Ordovician and ending with uplift in the Late Ordovician epoch has been established in the northern margin of this craton.

2.1.8 Indian Craton

Ordovician formations are not known in much of the Indian territory and this craton, like the African, evidently experienced uplift. The northern margin of the craton continued to subside (Jain et al., 1980). In the shallow marine basin accumulation of carbonate-terrigenous formations proceeded continuously in the east and terrigenous formations in the west. Their thickness went up by 100 to 250 m after each Ordovician epoch but, nonetheless, the total thickness for the Ordovician does not exceed 600 to 700 m. In the Middle Ordovician a thick complex of alluvial formations, including conglomerates, was formed in the western part of this basin (Garzanti et al., 1986). Its formation was the result of orogeny and erosion of the montane country lying southward.

2.1.9 Australian Craton

E a r l y O r d o v i c i a n. Rapid subsidence of the Fitzroy basin accompanied by accumulation of carbonate and arenaceous rocks of up to 500 m or more in thickness commenced in the west (Webby, 1978). Easterly and north-easterly carbonates were gradually replaced by terrigenous formations. This basin was asymmetric (with a steep north-eastern flank) and joined the Amadeus aulacogen, which continued to subside, in the south-east (Fig. 2-5). The shallow-water origin of arenaceous rocks filling the basin to a thickness of up to 1 km points to the compensated nature of subsidence. Coarse clastic alluvial strata accumulated in the southern marginal part of the basin representing erosion products of a region with rugged relief situated even more southward.

In the Georgina basin too subsidence with regression and rather thin carbonate sedimentation still continued at the end of the epoch. Gypsum bands, pointing to the high salinity of water, are seen here. In the Bonaparte

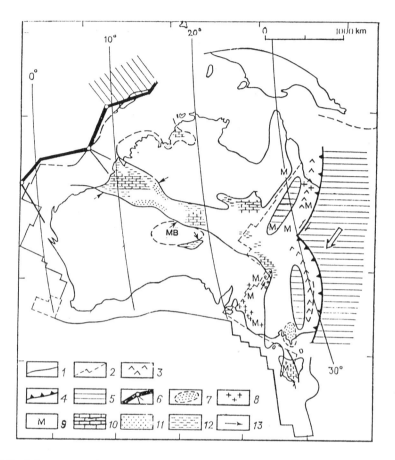

Fig. 2-5. Palaeogeographic reconstruction for end of the Early Ordovician of Australia (from Veevers, 1986)
Broad arrow—assumed direction of plate convergence; MB—Musgrave block.

1—shoreline; 2—Tasman line (assumed eastern boundary of craton); 3—volcanic chain; 4—base of accretionary prism; 5 —marine lithosphere; 6—break-up direction of supercontinent and failed arms of triple junction of rifts; 7—submarine plateaus; 8—granitoids; 9—dated metamorphites; 10—carbonates; 11—arenites; 12—lutites; 13—transport direction of sediments.

basin downwarping occurred over a very small area and gypsum is present in the formations (Plumb, 1979).

M i d d l e O r d o v i c i a n. Uplift covered the former Georgina and Bonaparte basins. Development of the Amadeus aulacogen concluded and much of it also experienced uplift. The marine basin extended from it in a north-westerly direction in the form of a narrow winding strip towards Fitzroy basin where the tempo of subsidence decreased but carbonates, as before, played a major role in the composition of deposits and evaporites (salts) were present. Judging from development in some sections of continental facies, the continuity of this strait was sometimes disturbed. The main portion of Middle Ordovician formations is represented by arenaceous rocks.

L a t e   O r d o v i c i a n. Uplift prevailed in the entire territory of the craton and Late Ordovician formations are not known here.

Thus some increase in subsidence and transgression on the Australian Craton at the beginning of the Ordovician was rapidly replaced by uplift and total regression towards the end of the period.

2.1.10 East Antarctic Craton

This craton continued to rise in the Ordovician but now was surrounded by the newly formed Rosside orogen from the Pacific side.

2.2 REGIONAL REVIEW. MOBILE BELTS

2.2.1 North Atlantic Belt

2.2.1.1 *Appalachian system.* E a r l y   O r d o v i c i a n. Formation continued under the deepwater conditions of submarine-volcanic (mafic, tholeiitic), siliceous and sandy-clay formations. On Newfoundland Island their total thickness was about 4 km and in the Northern Appalachians up to 8 km. In this part of the system, zones of western continental slope with sandy-clay formations (sometimes with turbidites) 2 to 4 km thick and small regions with deepwater uncompensated lutaceous or carbonate-lutite sedimentation are delineated (Osberg, 1978; Ruitenberg et al., 1977; Williams, 1979). In the Southern Appalachians (Piedmont zone) volcanic and terrigenous formations of the Glenarm series belong to the Ordovician. These series are similar to Lower Ordovician complexes of the Northern Appalachians in conditions of formation and also represent a probable relict of the sediments of the Iapetus palaeo-ocean (Higgins, 1972).

In the southern Ouachita-Mexican segment of the system, deepwater conditions of sedimentation are well substantiated by lithological character-istics of Ouachita Lower Ordovician thin shales and carbonate-clay rocks of the Marathon region (Morris, 1974; Young, 1970). Sometimes beds of fine turbidites are deposited among them. In the region of Ciudad Victoria in north-eastern Mexico, judging from the composition of deposits, development conditions in the Early Ordovician were similar (De Cserna et al., 1977) and structural similarity of this region with the Marathon zone can be assumed.

M i d d l e   O r d o v i c i a n. The appearance of significant masses of andesite material in the composition of the products of submarine volcanism suggests changes in the overall palaeotectonic environment and the birth of island arcs. The trough of the axial portion of Newfoundland Island was

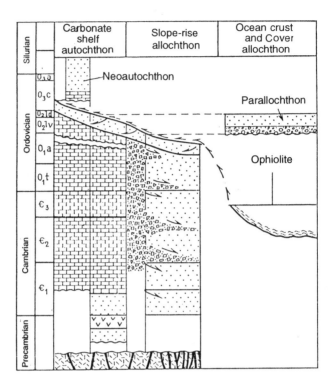

Fig. 2-6. Reconstructed stratigraphic relationships in western Newfoundland (after Williams, 1979, modified by Casey and Elthon).

filled with formations of complex composition that reached about 6 km in thickness. Its main portion is formed of coarse clastic rocks and intermediate volcanics but basalts as well as siliceous cherts are also present (Fig. 2-6). In the deepwater zone of the Northern Appalachians, mafic volcanics predominated over the intermediate ones; coarse clastic rocks were also extensively developed. Their thickness here goes up to 7 km, pointing to an intense filling of the basin. In the shelf basin adjoining the continental margin, carbonate and terrigenous deposits were formed up to a thickness of 3 km. Within the Avalonian block, under comparatively shallow-water conditions, sandy and clay formations were deposited and intermediate and felsic volcanism with a distinct island arc character was manifest (Rast et al., 1983).

Transition from a narrow pericratonic shelf zone with carbonate-terrigenous complexes exceeding 1 km in thickness to a continental slope and rise zone has been reconstructed in the Southern Appalachians where a flysch about 2 km thick has accumulated (Read, 1980; Thomas, 1977). The general environment in the Piedmont zone was evidently similar to that existing in the deepwater zone of the Northern Appalachians.

In the Ouachita-Mexican segment conditions continued to be favourable for the accumulation of pelagic clays, carbonates and sometimes fine turbidites. Their thickness goes up to 1 km.

L a t e   O r d o v i c i a n. Significant changes occurred over much of the area of the system. By the end of the Middle Ordovician, the former deepwater zones, already filled with a complex of sedimentary and volcanic formations, experienced intense uplift in this epoch and became large islands. This process was accompanied by the formation of fold and fault structures and also the intrusion of rather small granitoid massifs. But even narrow marine troughs filled with thick (up to 3 km) coarse clastic formations have been preserved here. Small-scale eruptions of basalts and rhyolites continued on Newfoundland Island and in Nova Scotia. Attention must be drawn to the presence of Upper Ordovician glacial-marine formations in Nova Scotia and in northern Newfoundland Island (Hambrey, 1985).

In the Ouachita-Mexican segment the composition of sediments varied while deepwater conditions continued to prevail. Apart from argillaceous rocks, carbonaceous (predominantly micritic limestones) rocks were manifest here in small quantities and siliceous rocks in the western part of this structure. The thickness of the deposits did not exceed 300 to 400 m.

The territory of eastern Greenland evidently represented the northwestern continuation of the Appalachian system. In the Late Ordovician this territory was affected by processes of epiplatformal orogeny. The Eastern Greenland nappe-fold system formed here.

Thus the Ordovician stage of development of the Appalachian system is characterised by a gradual change from deepwater conditions with submarine volcanism of mafic composition to the filling of deep troughs, formation of island arcs and later to a general uplift with manifestation of folding and metamorphic processes. The latter are associated with closing of the Iapetus palaeo-ocean and collision of continental blocks.

2.2.1.2 *British-Scandinavian system.* E a r l y   O r d o v i c i a n. Commencement of closing of the Iapetus palaeo-ocean and extensive development of island-arc volcanism are assumed in the second half of this epoch. Manifestation of this volcanism has been established along the southern periphery of the ocean from Ireland and Wales in Britain to Scandinavia (Anderton, 1982; Bassett, 1984; Bevins et al., 1984; Leggett, 1980; Oftedahl, 1980; Roberts et al., 1984; Sturt et al., 1980). The earlier volcanic complexes were represented by a basalt-andesite-rhyodacite series whose thickness reached 2 to 4 km. Formation of siliceous and argillaceous rocks continued in the deepwater zones. Shallow-water carbonate deposits of 200 to 300 m thickness predominated in the Lower Ordovician deposits in the north within the Hebridian massif; in the margins affected by subsidence of the English midland massif (Midland microcontinent), shallow-water sandy-clay beds predominated while greywackes dominated the zone of transition to a deep basin.

The Scandinavian margin of the East European Craton began to experience intense subsidence and thick terrigenous beds formed here while felsic volcanism was manifest in its central part. In the north, in the region of eastern Finnmarksvidda (Finnmarken), this zone by and large developed uninterruptedly from the Cambrian and the insignificant pre-Ordovician hiatus and folding were accompanied by the formation of a small volcanoplutonic belt. Tectonic activity died out in the Spitsbergen region. The eastern part of the archipelago experienced an uplift and after a brief interval carbonates formed a deposit up to 1.2 km in thickness in a shallow basin in the west. This pre-Ordovician hiatus was only of local significance.

Middle Ordovician. As a result of uplift the Hebridian massif turned into dry land. On its southern slope inclined towards the axial portion of the deepwater basin, a thick flysch complex (over 5 km) was formed. Further, the axial zone of the system (Ireland and Wales) preserved manifestation of submarine volcanism of contrasting composition (basalt-rhyolite) and pelagic clay sedimentation. The thickness of the Middle Ordovician here was 2 to 3 km. In the northern part of the English Midland massif andesite volcanism predominated and shallow-water sandy and argillaceous rocks accumulated. Extensive eruptions of andesites (island arc complex of calc-alkaline composition) occurred in the Scandinavian part of the belt. Simultaneous with volcanism, sedimentary terrigenous rocks formed here, most of them represented by volcanoclastic rocks. At the same time, turbidites of over 2 km in thickness and redeposited conglomerates accumulated in western Norway.

Uplift encompassed the territory of eastern Finnmarksvidda and folding and metamorphism were manifest here. The entire Spitsbergen territory was drawn into the uplift.

Late Ordovician. Within the system, differentiation of tectonic movements intensified even more and the magnitude of uplift rose. Regions of northern Ireland and northern Scotland turned into islands while Upper Ordovician tillites are known in Scotland and western Ireland (Hambrey, 1985; Harland, 1972; Williams, 1980). On the submarine slopes of these islands, thick (3 to 4 km) terrigenous complexes were deposited. A flysch (about 3 km) was deposited in Wales while rhyolite volcanism was additionally manifest in Ireland. However, a deepwater environment still prevailed in the narrow axial zone of the basin. The Upper Ordovician in southern Ireland and central England was represented by lutaceous facies ranging in thickness up to 1.5 km in the former and about 30 m in the latter region.

The Scandinavian part of the system was almost wholly affected by uplift and processes of tectonic deformation and metamorphism. Here, marginal-continental felsic volcanism was manifest and granitoids intruded on a small scale. A residual trough with coarse clastic sedimentation still survived only in north-western Norway.

Transition to an orogenic regime had concluded in Spitsbergen territory by the Late Ordovician. Evidently this region constituted a single entity with the Eastern Greenland system where, too, an orogenic regime has been established.

After comparatively stable general zoning throughout the Vendian to the Cambrian, a distinct trend is noticed in the Ordovician towards differential tectonic movements and manifestation of magmatism in the development of the British-Scandinavian system. Formation of island arc volcanic complexes concluded with formation of land massifs and the magnitude of deepwater sedimentation decreased.

2.2.1.3 *Innuitian system.* E a r l y   O r d o v i c i a n. The axial zone of the system is characterised, as before, by deepwater conditions and accumulation of siliceous and clay deposits to a small thickness; it extended in the east, to the north-western margin of Greenland (Surlyk and Hurst, 1984; Trettin and Balkwill, 1979). The enlargement of this basin has been assumed and further that its subsidence was not compensated by sedimentation. More northward, there was a relatively elevated block of ancient Precambrian basement in which shallow-water sandy deposits accumulated and eruptions of intermediate and felsic volcanics of the island-arc type occurred. From the south, the system was bound by the slope of the North American Craton whose narrow margin was drawn into rapid subsidence and was compensated by the accumulation of carbonates. The thickness of the latter exceeded 2 km on Ellesmere Island. Regions of northern and northeastern Alaska (Richardson trough and Brooks Range trough) most likely represented the western structural continuation of the belt.

M i d d l e   O r d o v i c i a n. The palaeotectonic environment initially remained as before and a highly extended trough existed. The subsidence of this trough was not compensated by deposits although their thickness on Ellesmere Island was about 700 m and in northern Greenland more than 400 m. As before, there was an island arc zone more northward in which eruptions of intermediate and felsic volcanics (mainly submarine) occurred. Their thickness (together with terrigenous and tuffaceous rocks) exceeded 1 km. In the middle of the epoch fold deformations and granite formation commenced in the northern Canadian Arctic archipelago.

L a t e   O r d o v i c i a n. The conditions in the eastern part of the system underwent change. Uplift and orogenic processes encompassing eastern Greenland at this time intensified erosion and inflow of large masses of clastic material. The eastern part of the basin began to be filled with terrigenous and carbonate flysch. But in the rest of the basin argillaceous, siliceous and carbonate-terrigenous (pelagic) complexes were formed, as before, under deepwater conditions. Within northern Ellesmere and the Axel Heiberg islands, fold deformations, eruptions of andesites and rhyolites of the island-arc type continued.

Thus the comparatively quiescent tectonic regime of development of the Innuitian system survived from the Cambrian throughout the Ordovician (i.e., about 60 m.y.) while subsidence in the deepwater trough remained almost uncompensated. However, an interesting feature is the revival of activity in the north where an island arc formed and in the east where filling of the trough with turbidites commenced in the Late Ordovician.

2.2.2 Mediterranean Belt

2.2.2.1 *West European zone.* E a r l y   O r d o v i c i a n. Ancient massifs within the region evidently maintained their position as island land. In the deepwater basins surrounding them, eruptions of mafic submarine lavas occurred. This was evidently a consequence of further development of extension of continental blocks and rift formation. Volcanic (including bimodal), siliceous and argillaceous complexes were formed within the Armorican massif, in the Saxothuringian zone, eastern and southern framework of the Bohemian massif and west and south of the Moesian block (Belov, 1981; Franke, 1978). The thickness of the Lower Ordovician usually does not exceed 1 km except in the Armorican massif where it reaches 3 km. In the regions of south-western Poland (Sudeten) and eastern Serbia, deepwater lutaceous sedimentation was not accompanied by manifestation of volcanism (Bonchev, 1985; Sergeeva et al., 1979; Cwojdzinski, 1979; Stevanovic and Veselinovic, 1978). In western France, in the Vendee and Limousin regions, a complex of greywackes, shales, falsic and mafic volcanics of the island-arc type was formed while a flysch complex developed more southward (Babin et al., 1976; Boyer et al., 1975; Feist, 1978; Floch et al., 1977).

In the region of the Iberian peninsula subsidence intensified in the Early Ordovician and the magnitude of islands diminished in the Central Iberian zone (Guy Tamain, 1978; Hammann, 1976; Truyols and Julivert, 1976). In the West Asturia zone deepwater conditions recurred and lutites were deposited. In the Cantabrian zone accumulation of sand complexes continued under shallow-water basin conditions.

Conditions in Dobrogea, Crimea, Precaucasus and Caucasus Major are not yet clearly understood.

M i d d l e   O r d o v i c i a n. Predominance of uplift continued in the microcontinental massifs. East and south of the Bohemian massif and within the Moesian block, subsidence of deep troughs continued. Submarine eruptions of basalt and andesite lavas and accumulation of clay deposits occurred in them as before. The thickness of the Middle Ordovician here goes up to 2 km. In south-western Poland accumulation of a deepwater lutite complex continued but the conditions changed in the Saxothuringian zone where a flysch formed. Islands in the regions of the Central and Bohemian massifs provided the material for this flysch. In Brabant and southern England regions a deepwater zone with pelagic sedimentation of small thickness probably existed. This assumption is confirmed by material from the Ordovician formations uncovered by drilling in the London region and from the Ordovician of the Brabant massif. In the western and south-western massif Central subaerial eruptions of andesites and rhyolites occurred under conditions proximate to those of an island arc and small massifs of granitoids intruded. West of this massif, a deepwater zone with accumulation of argillaceous rocks manifested again.

In the Iberian peninsula conditions in the West Asturia zone underwent change and the deep basin began to be filled with flysch, the source material for which was evidently the elevated Cantabrian zone with its islands. The thickness of the flysch exceeded 1 km. The Central Iberian zone experienced subsidence and sea covered the entire area. The thickness of shallow-water sandy argillaceous deposits in this zone constituted hundreds of metres (Julivert et al., 1980). More southward, the Middle Ordovician was represented by graptolite shales whose formation probably occurred under deepwater conditions.

L a t e   O r d o v i c i a n. The tendency towards uplift in the ancient massifs of the belt intensified and the islands enlarged in area, especially in the massif Central (France). In the northern Armorican massif subaerial andesite volcanism and granitoid intrusive magmatism were manifest and tillites formed. Glacial (or glacial-marine) formations have been observed in the northern Bohemian massif (in Thuringia) (Hambrey, 1985). Basaltic submarine-volcanic and siliceous complexes developed in the western massif Central in Vendee and the north-eastern Bohemian massif (Boyer et al., 1975). However, south of the latter, the composition of volcanics, unlike in the Middle Ordovician, became more complex and contrasting: besides basalts, andesites and rhyolites were abundant here (Belov, 1981). One more region with submarine basaltic volcanism lay south-east of the Pannonian basin. The thickness of the Upper Ordovician formations in all these regions usually did not exceed 1 km but most often its evaluation is tentative.

Along the boundaries of the Mediterranean belt with the East European Craton in Tornquist Sea, a deepwater zone with lutite deposition of small thickness existed throughout the Early Palaeozoic. Its relics are traced from southern England and Brabant to Swenty Krzisz Mountains and farther away to southern Bulgaria (Sergeeva et al., 1979; Cwojdzinski, 1979).

In the Iberian peninsula extension conditions arose in the Late Ordovician in the Central Iberian zone where siliceous, shale and submarine basaltic complexes of a total thickness of about 500 m were formed under deepwater conditions. The West Asturia trough continued to be filled with flysch and the thickness of the Upper Ordovician in it went up to 3 km (Julivert et al., 1980). Much of the Cantabrian zone experienced uplift and represented an island. Tillites are known here (Hambrey, 1985). South of the Central Iberian zone, conditions prevailed for deepwater deposition of small thickness since the Upper Ordovician is usually represented here by a lutite suite of small thickness.

Thus no significant reworking occurred in the development of the western part of the Mediterranean belt throughout the Ordovician. In deep troughs separating the ancient massifs, representing large islands, thin sediments were formed and submarine basaltic volcanism manifest. In the margins of the massifs flysch, sometimes shallow-water carbonate facies, was deposited while subaerial eruptions of intermediate composition occurred in the massif Central and granitoids intruded.

2.2.2.2   *Qilianshan system*. E a r l y O r d o v i c i a n. Siliceous-argillaceous deposits formed in the north-eastern zone of this system following a small pre-Ordovician hiatus and submarine eruptions (andesite-basalt) occurred. In the south-western zone continental slope conditions continued from the Late Cambrian with the accumulation of predominantly turbidites to a thickness of about 1.5 km (Xiang Dingpu, 1982). A deepwater trough was preserved in the central zone and was uncompensated by sedimentation.

M i d d l e O r d o v i c i a n. Submarine volcanic activity as well as filling of deep basins concluded. In this epoch the system represented a shallow basin in which carbonate and terrigenous rocks were deposited to a thickness of about 500 m. In the north-eastern part of this structure a segment of the continental slope of the North China Craton was extended and turbidites were deposited to a thickness of over 2.5 km.

L a t e O r d o v i c i a n. Volcanism revived and andesites and dacites predominated in the composition of its products; it was proximate to the island-arc type in chemistry and forms of eruptions. Here, too, terrigenous and tuffaceous strata formed to a total thickness of 2 to 2.5 km.

Thus in the Ordovician filling of deepwater troughs of the Qilianshan system with terrigenous, volcanic and, to a lesser extent, carbonate formations concluded. Volcanism changed from the mafic submarine spreading type to an intermediate-felsic island-arc type.

2.2.3 Ural-Okhotsk Belt

2.2.3.1 *Uralian system*. E a r l y O r d o v i c i a n. Reconstruction of the Sakmara-Lemva zone of the continental slope of the East European

Craton and its rise is most reliable. Here were formed terrigenous-volcanic formations of large thickness (Grubeya series). Volcanic material is represented by bands of alkaline and andesite-basalts and their tuffs. This zone has been most completely studied in the Northern and Polar Urals. In the Central and Southern Urals the zone is exposed only at places and its delineation is rendered difficult in view of the metamorphism of rocks. These are usually phyllite-like shales with bands of green volcanic shales and quartzites and bands of siliceous and carbonaceous-siliceous shales.

M i d d l e   O r d o v i c i a n. On the eastern slope, volcanic activity acquired a very important level and vitally influenced sedimentation. Under submarine conditions, tholeiite-basalts were formed, plagiogranites intruded (Tobol' zone) and andesites are seen in the Salatim zone. Most deposits were formed under relatively deepwater conditions. Among these, silts predominate, remains of benthic fauna and textures characteristic of shallow-water conditions are absent and various siliceous formations, i.e., phtanites, jaspers and cherts, are largely developed. Terrigenous rocks, like siliceous rocks, were closely associated with volcanism. The composition of clasts and textures point to their accumulation on the slopes of submarine volcanoes. The formations are usually hundreds of metres in thickness.

In the Sakmara-Lemva zone of the continental slope, deepwater conditions were more common. Siltstones, argillaceous rocks and clay limestones predominated in the composition of deposits. Siliceous rocks and reticulate limestones were extensively deposited. The thickness of the Middle Ordovician in this zone is 500 to 600 m.

L a t e   O r d o v i c i a n. The environmental conditions in the Uralian system hardly differed in this epoch from those in the preceding stage. In the deepwater zone submarine volcanic eruptions of predominantly mafic composition continued. Small intrusions of gabbro-diorite have been established and granodiorites recorded in the Troitsk zone. Beds of siliceous rocks, genetically associated with volcanism, as well as terrigenous and tuffaceous were formed. Some sedimentary complexes of the Southern Urals and Transurals characterised palaeontologically reveal a distinct shallow-water origin, pointing to complex conditions in the palaeo-ocean.

The continental slope and rise zones of the East European Craton likewise did not experience change in the course of their development. This zone evidently maintained its structural status, comparatively deepwater conditions and the general character of clay, siliceous and, less frequently, carbonate sedimentation with traces of minor volcanic activity.

Thus the Uralian system in the Ordovician was an arena of extensive sea-floor spreading, prevalence of deepwater sedimentation, formation of ophiolites and manifestation of submarine volcanism with a preponderance of tholeiite-basalts. From the west, this basin all along its extent adjoined the East European Craton and lateral facies transition from marginal platform basins to continental slope, its rise and deepwater conditions have been traced. Throughout the Ordovician the general environment hardly underwent any change. Only the magnitude of volcanism enlarged and the continental slope zone deepened.

2.2.3.2 *Kazakhstan-Tien Shan region.* E a r l y O r d o v i c i a n. In the Ishim-Karatau and Chatkal-Naryn zones accumulation of siliceous-terrigenous formations to a small thickness (5 to 300 m) continued under deepwater conditions. A sharp increase in thickness (to 1.5 km) and manifestation of mafic volcanics and jaspers are characteristic of only the northern part of the former zone. An environment of thin siliceous-argillaceous sedimentation also prevailed in the northern Stepniak-Betpak Dala and Junggar-Balkhash zones. Ophiolites were formed in the latter. In the Yerementau-Chu-Ili zone the thickness of the Lower Ordovician siliceous and siliceous-terrigenous formations was significantly greater (up to 2 to 3 km) and submarine eruptions of tholeiite-basalts occurred. These basalts constituted the upper portion of the ophiolite sequence. In the Chingiz-Tarbagatai zone basic submarine volcanism gave way to andesite and andesite-basalt island-arc type. The Kokchetav massif and the southern part of Ulytau-Northern Tien Shan maintained the elevated position of major islands. Turbidite beds 1 km or more thick formed at places on their slopes (e.g., on the eastern Stepniak synclinorium). The Aktau-Junggar massif developed as a submarine uplift and comparatively shallow-water carbonate and terrigenous formations about 1 km in thickness were formed on its surface.

M i d d l e O r d o v i c i a n. In the Ishim-Karatau and Chatkal-Naryn zones the phase of filling of deep troughs with thick terrigenous beds of flyschoid rocks commenced. In the Stepniak-Betpak Dala zone also there was a sharp increase in the rates of sedimentation, mainly represented by flyschoids. Moreover, volcanism was extensively manifest here. In composition, volcanic characteristics and chemistry, it was proximate to an island-arc type. The total thickness of the Middle Ordovician here exceeds 3 km. In this zone the role of volcanics in the sequence decreased from north to south and thickness reduced. In the southernmost part of the zone lava eruptions occurred under subaerial conditions.

Thick terrigenous sedimentation with coarse clastic and turbidite facies also commenced in the Yerementau-Chu-Ili and Chingiz-Tarbagatai zones. Their thickness here is exceptionally variable (500 to 2000 m), suggesting a highly dynamic palaeotectonic environment. Formation of deepwater clay deposits continued on an extensive scale in the Middle Ordovician only in the Junggar-Balkhash zone and it was here that formation of oceanic crust continued.

The Kokchetav massif experienced partial subsidence and transgression and coarse clastic beds accumulated in its basins. Within the Ulytau-Northern Tien Shan and Aktau-Junggar massifs, shelf carbonates and terrigenous facies were formed and regional metamorphism was manifest.

L a t e   O r d o v i c i a n. The main trend in regional development, viz., filling of troughs with thick sedimentary strata, including turbidites, was maintained. This occurred in the Ishim-Karatau, Chatkal-Naryn, Stepniak-Betpak Dala, Yerementau-Chu-Ili and Chingiz-Tarbagatai zones with deposits of large thickness (2 to 3 km) but their deposition occurred under shallow-water conditions, i.e., rapid subsidence of troughs was wholly compensated by clastic material entering from island arcs and ancient massifs. In the Late Ordovician the latter experienced uplift and most of them became islands. Only the Aktau-Junggar massif maintained a submerged position and terrigenous and, less frequently, carbonate deposits continued to be formed on it. As before, wide manifestation of volcanism with differentiated composition (from basalts to rhyolites) of the island-arc type prevailed in the Stepniak-Betpak Dala zone and also commenced in all the other zones except the Yerementau-Chu-Ili. At the same time, in the western part of the belt (from Kokchetav massif to Northern Tien Shan) extensive intrusions of granite-granodiorites occurred, regional metamorphism intensified and folding and thrusting were manifest at the end of the epoch.

Deepwater conditions were maintained in the Late Ordovician only in the Junggar-Balkhash zone and in some zones of Tien Shan (Akhmedzhanov et al., 1979; Makarychev et al., 1981). The thickness of formations here did not exceed a few hundred metres and they are represented by argillaceous and siliceous rocks (sometimes carbonates) and by basalts in Tien Shan. In the north-eastern Junggar-Balkhash system eruptions of basalts and andesites of the island-arc type commenced.

Thus throughout the Ordovician filling of deep troughs with thick beds of sedimentary and volcanic rocks occurred in troughs of the Kazakhstan-Tien Shan region and arc systems with manifestation of calc-alkaline volcanism were formed, this volcanism being most extensive in the Late Ordovician. At the end of the period, folding and granitisation encompassed the western part. A tendency towards intensified uplift and transformation into islands is evident in the development of ancient massifs. The overall environment, judging from the composition and thickness of volcanosedimentary deposi-tions and other manifestations of tectonic activity, was characterised in the Ordovician by high rate and contrast of tectonic movements with a tendency towards increase.

2.2.3.3 *Altay-Sayan region.* E a r l y   O r d o v i c i a n. Filling of former deep troughs with turbidites continued. Clastic material came from interior uplifts (islands) and from the north-east where rugged land of the orogenic zone was elevated (Belousov et al., 1978; Sennikov, 1977; Kheraskov, 1975). Island arc series of volcanics were formed on a small scale (Salair Range).

M i d d l e   O r d o v i c i a n. Flysch accumulated, as before, in the regions of Western Sayan and Kuznetsk Alatau. At other places, the Middle Ordovician was represented predominantly by sandy, often coarse clastic formations (thickness up to 5 km). For them, as well as for flysch, the main source of material was the orogen situated in the north-east. Accumulation of these formations occurred evidently in the zone of its continental slope and rise.

L a t e   O r d o v i c i a n. Filling of the former deepwater Western Sayan trough with flysch concluded, its thickness here attaining 4 km. In other regions sand and conglomerate facies predominated, their formation having occurred under conditions of continental slope and rise. In the Altay segment of the region, at the end of the epoch, folding and, on a small scale, granitisation were manifest.

On the whole, the Ordovician period for the Altay-Sayan region was a period of filling of deepwater basins with clastic deposits. This process preceded the subsequent uplift of this structure and its transition to an orogenic stage.

2.2.3.4 *Sayan-Baikal region.* At the commencement of the Ordovician, the Sayan-Baikal orogenic region greatly expanded its boundaries in a southerly direction. In the E a r l y   O r d o v i c i a n it served as a source of clastic material for the adjoining basins. Coarse clastic beds up to 2 km in thickness were formed in these basins.

In the M i d d l e   O r d o v i c i a n tectonic activity died out in the Baikalian part of the region and it formed a constituent of the Siberian Craton. Elsewhere in the area, uplift intensified and continental molasse as well as marine coarse clastic deposits exceeding 3 km in thickness accumulated in the marginal and intermontane basins.

In the L a t e   O r d o v i c i a n the areal extent of the region increased with the inclusion of central Mongolia, also involved in uplift. The highest and most rugged relief existed evidently in its western part where marine and continental molasse were extensively developed and granitoid intrusive magmatism was manifest.

2.2.3.5 *Gobi-Hinggan system.* The Early Palaeozoic history of the system has yet to be studied well since information on development of deposits of this age is fragmentary. Structurally this system represents the eastern continuation of the Altay-Sayan region. Some outcrops of Ordovician formations in the Shilka and Argun interfluve and thick beds of submarine-volcanic formations of the Greater Hinggan belonging to the Ordovician suggest subsidence here in the Early Ordovician under conditions of a geosynclinal regime (in: *Upper Ordovician Formations of Mongolia*, 1981; Suetenko, 1984). In the M i d d l e O r d o v i c i a n a zone of manifestation of andesite-basalt volcanism of the island-arc type existed south-west of the system. Sandy-clay deposits were also formed in it. A similar environment is assumed in the regions of Greater and Lesser Hinggan. The Middle Ordovician within central and eastern Mongolia, has yet to be thoroughly studied.

In the L a t e O r d o v i c i a n volcanic eruptions ceased (except in the Lesser Hinggan area) and only diverse terrigenous deposits were formed (Dergunov et al., 1980; and Upper Ordovician formations of Mongolia, 1981).

2.2.4 West Pacific Belt

2.2.4.1 *Cathaysian system.* Uplift of the Uishan and Yunkai island arcs continued in the western part of the system and processes of migmatisation and granitisation developed there. More eastward, predominantly graptolite shales and turbidites of 500 to 1000 m thick were formed in the Early and Middle Ordovician (Ren Jishun et al., 1984). In the Middle Ordovician turbidites up to 3 km in thickness were deposited in the south-west (North Vietnam) and uplift of island arcs with weak manifestation of andesite volcanism commenced. In the Late Ordovician uplift intensified and a large island arose in which granitisation, migmatisation and metamorphism developed. The sedimentary basin in which sandy-clay shallow-water deposits were formed bore a remnant, back-arc character.

2.2.4.2 *East Australian region.* E a r l y O r d o v i c i a n. Compared to the Late Cambrian, there were very few changes in the development of the belt. In the Thomson system submarine volcanic eruptions continued in the north and andesites acquired a perceptible role among the volcanic products (Murray and Kirkegaard, 1978). In the region of Molong-Canberra volcanic arc andesite eruptions predominated and flysch possibly formed west of it while a deepwater zone existed to the east (Cas and Jones, 1979). Accumulation of turbidites continued in the Lachlan system, on the continental slope adjoining the Adelaide orogen (Gilligan and Scheibner, 1978; Vandenberg, 1978). The existence of a deepwater zone with pelagic deposits evidently belonging to the Palaeo-Pacific is assumed in the eastern part of the Lachlan system.

In the Early Ordovician folding and intense uplift were manifest in the Adelaide orogen while molasse formed in the east up to 1 to 2 km in thickness (Plumb, 1979). The western part of Tasmania Island belonged to this structure.

Middle Ordovician. The regional structure underwent further complication with active development of new, more eastern regions. Maintenance of former conditions with accumulation of terrigenous and submarine basalt and andesite complexes has been assumed for the Thomson system. In the south, in the Lachlan system, the western zone was characterised by deposition of turbidites under continental slope conditions of the Adelaide orogen. More eastward, in the deepwater Melbourne basin, shales of small thickness were formed. A zone with spilite-siliceous and terrigenous complexes of greater thickness (up to 5 km) lay even more eastward. Formation of these complexes is associated with rifting to the rear of the Molong-Canberra volcanic arc. Within the latter, subaerial eruptions of andesites and rhyolites occurred, their thickness exceeding 4 km, while coarse clastic and carbonate rocks were deposited on the slopes of the arc.

Active transformation of the crust of oceanic type in many regions of north-eastern New South Wales and southern Queensland probably commenced in the Middle Ordovician. The thickness of the Middle Ordovician spilite-siliceous and terrigenous complexes here reaches 5 km. The Middle Ordovician formations of the South Island of New Zealand point to a contrasting dynamic environment. Andesite and terrigenous strata are known in the northern and southern parts of the island and a flysch complex exceeding 2 km in thickness in the central part.

In the Adelaide orogen intense upheaval continued in the Middle Ordovician, granitoids intruded, rhyolite volcanism was manifest in the north and molasse accumulation proceeded in the east in the Ballarat trough. The western part of Tasmania Island experienced an uplift while the north-eastern part of the island represented a continuation of the deepwater Melbourne basin.

Late Ordovician. New significant changes occurred in the development of the region. The Thomson system experienced uplift, folding, metamorphism and granite intrusion in the north and constituted a common entity with the Adelaide orogen as its north-eastern continuation. In the Lachlan system accumulation of coarse greywackes and, less frequently, reef limestones predominated in the west while andesite volcanism was manifest in the north. Farther east, development of thick (over 3 km) submarine-volcanic and terrigenous formations continued in the Melbourne and Wagga deepwater basins. This zone was set off from the open ocean by the Molong-Canberra island arc along which shallow-water formations (coarse clastic and carbonate) and volcanics were deposited. Among the volcanics developed mainly in the northern part of the arc, andesites and their tuffs predominated but more felsic and more mafic rocks are also known. These together form a characteristic island arc association. The end of the Ordovician in this zone was characterised by intense uplift and extensive manifestation of folding (Benambran orogen).

The New England system, east of the Lachlan system, continued to develop. Thin bands of lutites, probably of deepwater origin, formed in this system (Plumb, 1979). Northward, in Queensland, the first to form was a complex of deposits including spilite-siliceous formations and greywackes exceeding 3 km in thickness.

The Adelaide orogen represented a vast zone with a relatively high relief. Compared to the Middle Ordovician, its areal extent decreased since the orogenic regime in the north-west was replaced by a platform one.

Thus throughout the Ordovician complication and differentiation of the structure of the East Australian region occurred with further migration of active processes from west to east. A major island arc (back-arc) basin was formed and back-arc spreading was manifest.

2.2.4.3 *West Antarctic region.* From the beginning of the Ordovician, uplift predominated over much of the area and metamorphism, folding and felsic volcanism were manifest (in: Explanatory Note to the Tectonic Map of Antarctica on Scale 1:10,000,000, 1980; Hoffman and Paech, 1980; Splettstoesser and Webers, 1980) and small granitoid bodies intruded. The overall type of development corresponded to that of an orogenic regime. Subsidence and accumulation of terrigenous deposits continued only in some basins in the region of the Shackleton and Pensacola mountains between the Beardmore and Bird glaciers where these formations bear the nature of a coarse continental molasse. The predominantly coarse clastic composition and fluvial origin have their analogues in Bowers trough where their thickness exceeds 4 km and complete its filling. Only in the Ellsworth Mountains are the deposits of this age very finely clastic and of the shallow-marine type with traces of emersion. Furthermore, manifestation of Ross orogeny has not been noticed here, pointing to a more distal position relative to the Transantarctic mountain range orogen.

2.2.5 East Pacific Belt

2.2.5.1 *Cordilleran system.* E a r l y   O r d o v i c i a n. In the southern part of the Cordillera extensive eruptions of tholeiitic pillow lavas commenced together with an accumulation of siliceous deposits and sandstones (Churkin and McKee, 1974; Stewart and Poole, 1974; Wrucke et al., 1978). Evaluation of the thickness of these formations is contradictory but they are not less than 600 m. A zone with deepwater lutite deposition extended west of the narrow zone with manifestation of volcanism and to the north carbonate-lutite deposition. More westward, mafic submarine-volcanic complexes exceeding 1.5 km in thickness are known in the region of the Alexander archipelago and Sierra Nevada.

In the central Alaska region the Lower Ordovician is represented by micritic limestones and shales of small thickness and at places by products of mafic submarine eruptions (Churkin et al., 1980). These too were formed evidently under deepwater conditions. Siliceous and shale deposits likewise accumulated in northern Alaska (Brooks Range) and in the Richardson trough.

M i d d l e   O r d o v i c i a n. The general zoning of the belt remained as before and accumulation of mafic submarine-volcanic, siliceous and clay complexes continued in deepwater basins. Their thickness in Alaska was 500 to 1000 m. In the region of the Alexander archipelago where the nature of volcanism was close to the early island-arc type, the thickness exceeded 2 km and in California 1 to 2 km. In deepwater regions with no manifestation of volcanism, the thickness of clay and siliceous formations did not exceed 100 to 200 m. A small region in north-eastern Alaska experienced uplift and folding and evidently represented an island. Middle and Upper Ordovician and Silurian are not known here but an angular unconformity has been noticed at the base of the Devonian.

L a t e   O r d o v i c i a n. The regions of the Alexander archipelago, Klamath, Purcell and Pelly mountains probably formed island arcs in this epoch. Eruptions of intermediate composition predominated there and intrusions of small massifs of granodiorites occurred. In the volcanic products of the arc of the Alexander archipelago, rhyolites played an important role and turbidites were deposited on its western slope. Thin clay and siliceous deposits accumulated in the deepwater zones and micritic limestones in Alaska. Upper Ordovician shales of the Richardson trough (aulacogen) in north-eastern Alaska were also formed under deepwater conditions. In the southern (Californian) segment of the system submarine eruption of basalts continued but the magnitude was notably reduced. This segment extended evidently into the north-western part of Mexico. There, in southern Sonora, deepwater cherts with radiolarians and shales with graptolites of the Late Ordovician have been detected (Peiffer-Rangin and Perez, 1980).

On the whole, the Ordovician period for the Cordilleran system represented a period of gradual complication of its structure during protracted formation of island arcs of the Alexander archipelago, Klamath, Purcell and Pelly mountains. To the rear of the latter, deepwater zones were preserved with an accumulation of thin deposits attenuating during manifestation of submarine mafic volcanism.

2.2.5.2 *Andean system.* E a r l y   O r d o v i c i a n. In the Colombia-Ecuador segment the type of volcanism changed and andesite and andesite-basalt island arc eruptions began to predominate in the western zone. In the eastern shelf zone accumulation of terrigenous formations continued (Thery, 1985). In the Peru-Bolivian segment destruction processes developed further in the Arequipa massif and the margin of the South American Craton (Dalmayrac et al., 1980). A characteristic complex of graptolite shales and siltstones up to 3 km thick was formed here under the deepwater conditions of a marginal sea (Aceñolaza, 1982; Mégard, 1978; Zeil, 1979). On the north-eastern flank of this basin, a small manifestation of andesite-rhyolite island arc volcanism is known (Aceñolaza, 1982; Carlier et al., 1982).

In the Chile-Argentinian segment commencement of development of the western zone, where spilites, siliceous and clay shales and sometimes turbidites are known, is associated with this epoch (Coira et al., 1982; Vicente, 1975). It has been assumed that this ensialic trough lay between the South American Craton and a hypothetical continent at the site of the south-eastern part of the modern Pacific Ocean and ophiolites formed and outpouring of tholeiites occurred here from the commencement of the Ordovician. More eastward, in the Sierra Pampeanas region and parts of the Arequipa massif, thick (up to 2 km) beds of sandy-clay deposits were formed under shallow-water conditions. Further, in the southern Sierra Pampeanas massif accumulation of carbonates was widely developed, this being the only region of comparatively extensive carbonate accumulation in the Andean system.

In the Patagonian zone volcanism evidently ceased and filling of troughs with terrigenous sediments transported from the Deseado continental block occurred.

M i d d l e   O r d o v i c i a n. In the western zone of the Colombia-Ecuador segment conditions remained as before while the eastern zone enlarged since intense subsidence encompassed the territory of the Venezuelan Andes, involved in the mobile belt. Within the Peru-Bolivian segment formed before, the deepwater trough between the Arequipa massif and the craton was filled with clay deposits. In the Eastern Cordillera flysch beds began to accumulate in Peru and Bolivia, pointing to the disposition of a continental slope. The thickness of the Middle Ordovician shales and flysch exceeds 3 to 3.5 km. Maintenance of deepwater conditions and thin clay sedimentation have been assumed in north-western Peru. Small-scale manifestation of andesite-rhyolite island arc volcanism continued.

The very same conditions as in the Early Ordovician, i.e., formation of submarine-volcanic (basalt and andesite) and siliceous and terrigenous complexes up to 3 km in thickness in the western zone, were characteristic of the Chile-Argentinian segment. In the east terrigenous and carbonate-terrigenous beds of the same thickness were formed. In the Precordillera of Argentina the thickness of Llanvirnian limestones and dolomites constitutes about 3 km (Cuerda, 1974). In this zone manifestation of andesite island arc volcanism is known and intrusion of granitoids, folding and metamorphic processes have been recognised in the Middle Ordovician of the Sierra Pampeanas massif on the basis of radiometric data (Aceñolaza, 1982; Coira et al., 1982). In the Patagonian zone, separating the Patagonia-Deseado platform block from the South American Craton, subsidence and accumulation of terrigenous beds exceeding 1 km in thickness have been preserved.

Late Ordovician. In the northern, Colombia-Ecuador, segment of the system, levelling of palaeotectonic conditions has been suggested: gradual intensification of uplift and cessation of volcanism. The sedimentary formations of this epoch were represented by only shelf terrigenous complexes of 400 to 800 m thickness. Changes also occurred in the Peru-Bolivian segment where the southern regions were involved in uplift forming a high relief and folding (Aceñolaza, 1982; Dalmayrac et al., 1980; Mégard, 1978; Zeil, 1979). Evidently Mehia granites intruded simultaneous with this phase of uplift in the northern part of the Arequipa massif. The Upper Ordovician is known only in the Western Cordillera of Peru. A deepwater zone filled with graptolite shales and a narrow shelf in which sandstones were formed have been preserved.

Significant changes had occurred by the commencement of the Late Ordovician in the Chile-Argentinian segment. Volcanic eruptions ceased and terrigenous sedimentation of moderate thickness predominated. The deepwater zone had by this time become quite shallow but a narrow zone of continental slope within which flysch was deposited still extended in the west. The zone of development of metamorphic complexes in the littoral part of Chile evidently represented an island throughout the Ordovician. Uplift of the northern Sierra Pampeanas massif occurred and it joined with the Arequipa massif in the form of a large island. Ashgillian deposits are not known in the Chile-Argentinian segment and, considering the generally unconformable attitude of Silurian formations on the Ordovician, a general uplift of the territory and manifestation of folding at the end of the Ordovician should be assumed.

In the Patagonian zone too uplift developed but only in the southern part, as has been established by hiatuses in sedimentation and angular unconformities.

Thus the main events in the tectonic development of the Andean system in the Ordovician were destruction of continental massifs and formation of new deepwater zones in the Peru-Bolivian and Chile-Argentinian segments. Furthermore, extension and destruction of continental crust were accompanied in the latter segment by outpouring of tholeiite-basalts. In the Middle Ordovician island arc complexes developed and granitoids intruded while volcanism ceased throughout the system in the Late Ordovician, several regions experienced uplift and the general environment became more contrasting.

2.3 PALAEOTECTONIC ANALYSIS

In the Ordovician development of the main structural elements of the Earth's crust (lithosphere) noticed in the Vendian-Cambrian continued, i.e., Gondwana supercontinent, North American, East European, Siberian, North China-Korean continents, and mobile belts, i.e., palaeo-oceans of the North Atlantic (Iapetus), Ural-Okhotsk (Palaeo-Asian), Mediterranean (Palaeo-Tethys) and western and eastern Pacific (Palaeo-Pacific). After the decline of tectonomagmatic activity at the end of the Cambrian to commencement of the Ordovician, a phase of maximum enlargement of palaeo-oceans set in at the end of the Early to commencement of the Middle Ordovician (Fig. 2-7). Transgression maximal for the entire Palaeozoic, encompassing significant areas within the continents, also occurred during this stage. But even in the middle of the Middle Ordovician (Fig. 2-8), a new burst of tectonomagmatic activity commenced, reaching culmination in the Late Ordovician. Closure of the Iapetus ocean commenced from this time but its final closure came at the end of the Silurian (see Section 3.3.3).

2.3.1 Continental Cratons

2.3.1.1 *Gondwana.* In the Ordovician much of the Gondwana super-continent still continued to experience some upheaval and remained dry land. But nevertheless the sedimentation area steadily enlarged. This is noticeable, first of all, in northern Africa and Arabia, as well as in the zone of pericratonic subsidence in the western and southern South American and southern African cratons. In the former region transgression extended from the Mediterranean belt and in the latter from the East Pacific. In the Late Ordovician downwarping and sedimentation commenced in the Amazon syneclise, in the Maranâo and Parana syneclises, and also in the adjoining region of north-western Africa, from southern Morocco to Ghana (Senegal, Guinea and Volta syneclises). In Australia, compared to the Cambrian, con-figuration of the zones of downwarping changed; these zones, together with the meridional zone, acquired a west-north-west orientation by joining the Canning syneclise (Fitzroy trough) through the Amadeus aulacogen with the Tasman (Lachlan) geosyncline. In the Late Ordovician in Australia, unlike in South America and Africa, regression was manifest and the whole of Aus-tralia stood out as land. Ordovician deposits of Gondwana are represented by shallow-water marine sandy and sandy-clay sediments, limestones hav-ing developed only in the Fitzroy basin. Continental, evidently alluvial-deltaic formations accumulated at places along the periphery of shelf seas.

In the Late Ordovician and Early Silurian, Arabia, northern and western Africa and adjoining parts of South America (northern Brazil) were encom-passed by a thick ice sheet (see Fig. 2-4). Glaciation is associated with the migration of this segment of Gondwana into high, polar latitudes (Caputo

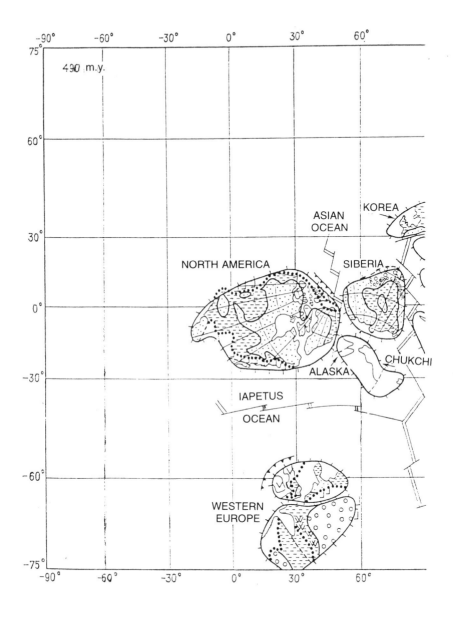

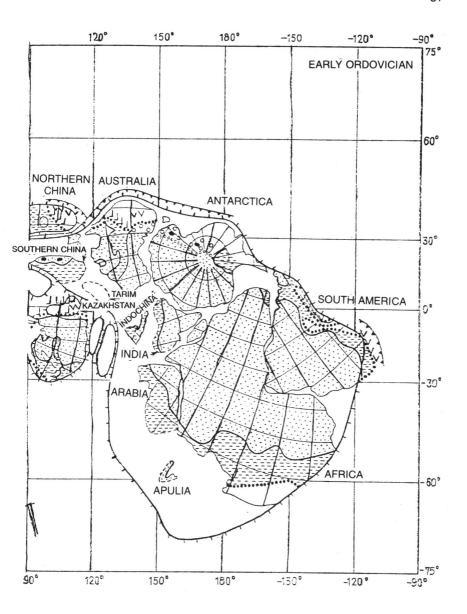

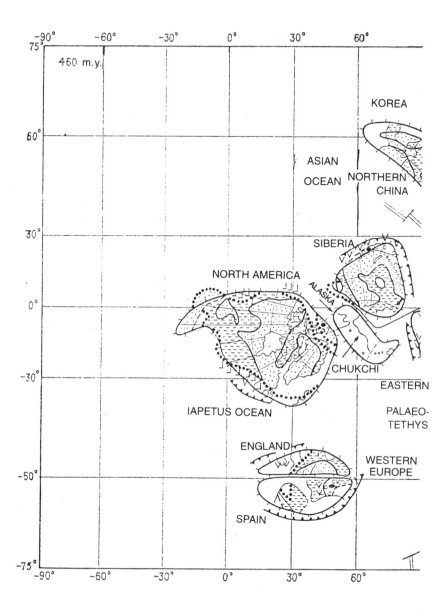

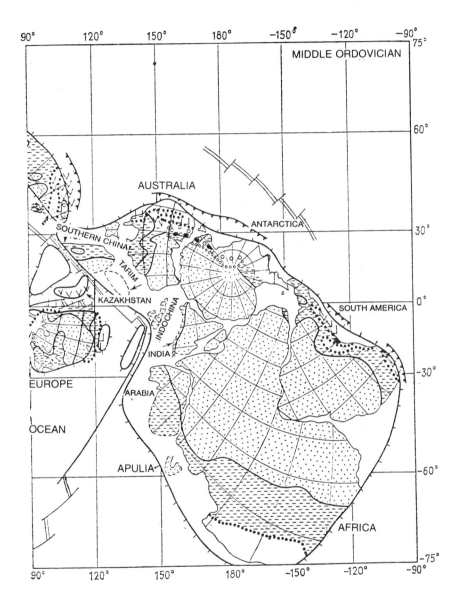

and Crowell, 1985; Hambrey, 1985). The material of glacial origin spread into Western as well as Central Europe and also the Atlantic margin of North America (Avalonia), confirming their affinity to Gondwana.

2.3.1.2 *Northern row of cratons.* These cratons, as in the Cambrian, continued to lie in the lower latitudes. As a result, carbonate deposits accumulated predominantly, as before, in the sea covering them. The area of sea in this group of cratons exceeded the corresponding area of the Gondwana group. The North American Craton experienced a particularly extensive transgression which was maximum in the Late Ordovician. In Eastern Europe the maximum spread of sea occurred in the north-west, i.e., the western part of the Baltic Shield and the Baltic-Moscow syneclise, and in the regions of south-western and eastern margins of the continent. In Siberia the sea-covered area was somewhat less than in the Cambrian; in the Ordovician, the Siberian Craton with its Verkhoyansk margin was separated from the Omolon microcontinent and the Hyperborean Craton, which until then constituted a single entity (?). The Sino-Korea Craton, almost wholly inundated in the Early and Middle Ordovician, began to experience uplift and regression in the Late Ordovician; it remained as land until the middle of the Carboniferous. Thus, as in the Cambrian, continental cratons of the northern series

---

Fig. 2-7. Palaeotectonic reconstruction of the Early Ordovician (Mercator projection with centre at 0°N lat. and 90°E long.). See Fig. 1-7 for legend.

The disposition of some continental blocks proposed in this scheme by L.P. Zonenshain and coauthors raises objections since it does not accord with geological data. Due to the proximate disposition of Siberia and North America, the eastern part of the Siberian Craton, well known to be continental in nature, is about 1000 km short. In the geological history of this Siberian region, there are no traces of the continental Greenland block passing abreast of it during the period Late Cambrian (see Fig. 1-7) to Early Ordovician. In the Uralian system the Early Ordovician was a period of extension and formation of ophiolites but, in this scheme, the Kazakhstan block very closely adjoins the East European block and a comparison with the Late Cambrian scheme suggests the possibility of their collision, which, however, does not conform to geological data (see Section 2.3.2.2). In the southern part of the East European block there is no place for the Scythian platform territory. The profiled features of palaeocontinents for eastern Australia, north-western South America and western Antarctica are not accurate.

Fig. 2-8. Palaeotectonic reconstruction of the Middle Ordovician (Mercator projection with centre at 0°N lat. and 90°E long.). See Fig. 1-7 for legend.

In our opinion, L.P. Zonenshain and coauthors have, in this scheme of disposition of continental blocks, unduly approached the Arctic region and Siberia and left no space for Verkhoyansk territory with continental Earth's crust in the Middle Ordovician.

The Innuitian system of North America and the Patagonian zone in South America have not been depicted. Interpretation of several small blocks—Amuria, Kolyma, northern Scotland, etc.—is extremely problematic.

separated by mobile belts, i.e., palaeo-oceans, revealed individual developmental characteristics; progressive subsidence and transgression of North America are counterbalanced by uplift and regression in China and Korea.

Transition from carbonate shelves to shale (graptolite shales), siliceous shale and sandy (greywacke) shale formations of continental slopes, rises and floor of marginal seas is distinctly delineated on the periphery of the continental cratons of the northern series. In North America such a transition is traced in the west along the Cordillera from Alaska to California, in the north in the Canadian Arctic archipelago and northern Greenland, in the east in the Appalachians, Ouachita and Marathon (transition is highly complicated here by major overthrusts), in Eastern Europe along the Teisseyre-Tornquist line in Polish Pomorie and in Eastern Siberia along the Chersky Range.

2.3.2 Mobile Belts

2.3.2.1 *North Atlantic belt.* Maximum opening of the Iapetus palaeo-ocean to a width of 1000 to 3000 km occurred, according to various accounts, at the beginning of the Ordovician to the Tremadocian. This is also supported by radiometric age determinations of most ophiolites. But already in the Arenigian conditions had begun to change: in the ocean and (or) on its south-eastern periphery, volcanic arcs were formed—ensimatic or ensialic respectively. The first type, with isolation of marginal seas, is characteristic of Newfoundland and central and southern Norway while the second type is characteristic of the British Isles and south-western Norway; it calls for the existence of a subduction zone inclined under the continental margin. In the Tremadocian a subduction zone of opposite, north-western dip arose in Scotland and northern Ireland, while in northern Norway this led to formation of a volcanoplutonic belt (Seeland magmatic province) with evolution of magma over time from tholeiitic (gabbro) to calc-alkaline and more alkaline at the end of the Cambrian to commencement of the Ordovician (Finnmarken tectonomagmatic epoch, roughly simultaneous with the Grampian). The deformation of this epoch with obduction of ophiolites was manifest in central Norway also. In this same epoch, in the Arenigian-Llanvirnian, obduction of ophiolite nappes commenced in the western part of Newfoundland in connection with a seismofocal zone of the same orientation. In the Llandeilian and Early Caradocian this process extended southward within the Appalachians. Thus between Grampian and Taconian tectonomagmatic epochs, continuous transition prevailed in the North Atlantic belt. Dewey and Shackleton (1984) pointed to prolonged tectonic processes in the British Caledonides (Metamorphic zone), Newfoundland and the Appalachians where, in the Ordovician (O_{1-2}), a major ophiolite nappe was formed and soon underwent erosion. Only small klippen, the largest occurring in western Newfoundland (Bay of Islands and Humber Arm), represent its outliers. But total closure of the Iapetus was quite far away even at the end of the Ordovician as it set in at the end of the Silurian to commencement of the Devonian.

2.3.2.2 *Innuitian system.* In this system, in the first half of the Ordovician, as in the Cambrian, an outer shelf and slope of the Hazen trough basin are delineated. But the oceanic crust of this trough commenced to subduct under the Northern Ellesmere microcontinent (or possibly the margin of the Hyperborean continent) with formation of a marginal volcanic belt. In the middle of the Middle Ordovician, the following events took place: intense compressive deformation causing sharp unconformity between the upper Middle Ordovician and the various underlying formations, including ophiolites and Precambrian rocks; filling of Hazen trough with thick terrigenous flysch; some metamorphism of pre-upper Middle Ordovician formations and intrusion of granitoids. Deformation repeated before and during the Late Ordovician.

2.3.2.3 *Ural-Okhotsk belt.* The structure of the belt experienced further complication in the Ordovician and its development, as in the Cambrian, proceeded differently in different segments and even in different parts of the same segment.

The main event in the northern Ural-Siberian segment was the formation of the Urals basin with an oceanic crust. This basin was wholly formed in the Late Ordovician; its formation was preceded by an intense continental rifting in the Early-Middle Ordovician (Fig. 2–9). In much of the Urals the spread of the Urals Palaeozoic geosyncline fairly conforms to that of pre-Ordovician structures, the 'Preuralides' of N.P. Kheraskov and A.S. Perfil'ev, but in the Polar and Peri-Polar Urals, the Uralides intersected the Preuralides at a high angle. These Preuralides are traced from the basement of the Timan-Pechora Platform through Precambrian salients in the Urals within the basement of the West Siberian Platform. In turn, the Urals structure forms a tectonic junction at a right angle with the Pay-Khoy structure whose geosynclinal development concluded only in the Jurassic.

The conditions under which the Urals Hercynian geosynclinal system originated in the Ordovician are not clear on two important points. Firstly, what was the nature of the Preuralides preceding it? Secondly, what was the role of the salients of Early Precambrian continental crust in the eastern Urals uplift zones, including Mugodzhary, in the development of the Urals?

There are presently two concepts competing to answer the first question. According to one, in the Late Precambrian the Urals underwent a geosynclinal stage of development that was completed by orogeny and molasse formation in the Vendian (Asha molasse). This view prevailed until quite recently and is still supported by several researchers. According to another view, substantiated by S.N. Ivanov (1979) and supported to some extent, the Urals experienced only continental rifting in the Late Precambrian. It must be acknowledged that the facts known to date are inadequate for making a conclusive choice between these two concepts but nevertheless do help in evaluating their relative probability. Data on the Polar and Peripolar Urals and on the basement of the Pechora Platform (Belyakova, 1985) suggests the geosynclinal development of this territory in the Riphean and orogenic development in the Vendian and Cambrian. Data on the western slope of the rest of the Urals confirms the concept of the rift development of this Uralian zone in the Early and commencement of the Middle Riphean. But for the much later Riphean, a comparison of Bashkirian and Uraltau anti-clinoria suggests more a transition from shelf conditions in the former to continental slope conditions in the latter: wedging of carbonate bands with stromatolites and turbidite manifestation is discernible. The composition of the Maksyutov complex in southern Uraltau suggests it to be a formation of basin floor with oceanic or suboceanic crust. Another flank of this basin should fall on the western slope of the Kazakhstan microcontinent and its western Siberian continuation (Uvat-Khanty-Mansy massif.) Angular uncon-formities between the Riphean and Ordovician point to folding at the end of the Riphean or Early Vendian. Finally, the Late Vendian Asha series actually bears the form of a molasse (lower), as substantiated by Bekker (1982). It is most probable therefore, from Bekker's point of view, that the greater part of the Urals experienced geosynclinal-orogenic development in the second half of the Riphean and Vendian although this was not wholly typically man-fest. Reliable data on the Riphean for the eastern slope of the Urals is as yet inadequate for a final solution to this problem.

In so far as the second question is concerned, in the light of seismic data pointing to continuation of an Early Precambrian basement of the East European Craton, not only up to Uraltau but also more eastward, under the Magnitogorsk 'synclinorium', and also geological data on the allochthonous state of formations filling this synclinorium (megasynform), and finally on analogy with the Appalachians, it is more probable that Early Precambrian salients in the Eastern Urals zone represent granite-gneiss domes of remo-bilised ancient continental crust of the same East European Craton overlain by ophiolite-island-arc-marginal sea rocks overthrust from the east. In other words, the opinion of M.A. Kamaletdinov and T.T. and Yu.V. Kazantsev (in: Kazantseva and Kamaletdinov, 1986) could be supported. Thus the primary basin with oceanic crust should have been situated east of the Eastern Urals 'anticlinorium' but evidently west of the Kazakhstan and Khanty-Mansy microcontinents. Since the Riphean sequence of the Transuralian uplift is similar to the sequence of the latter, the Eastern Urals synclinorium could be regarded most probably as performing the role of a relict of a primary ocean basin. From this, ophiolites could have been translated to the west as well as east and, in particular, preserved in the Denisov synclinorium (synform).

South of the Kazakhstan microcontinent development of a similar basin, a marginal sea of the Palaeo-Asian ocean continued in Tien Shan in the

94

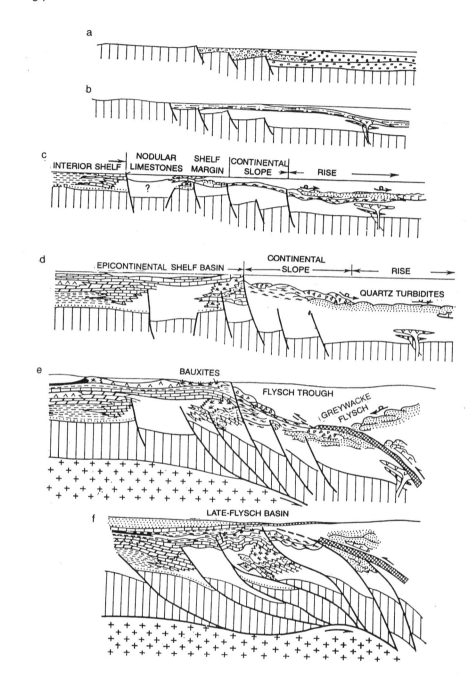

a

b

c

INTERIOR SHELF | NODULAR LIMESTONES | SHELF MARGIN | CONTINENTAL SLOPE | RISE

?

d

EPICONTINENTAL SHELF BASIN | CONTINENTAL SLOPE | RISE

QUARTZ TURBIDITES

e

BAUXITES

FLYSCH TROUGH

GREYWACKE FLYSCH

f

LATE-FLYSCH BASIN

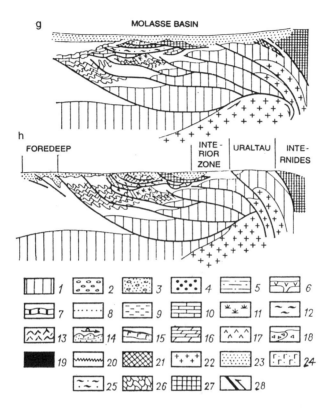

Fig. 2-9. Main developmental stages of the Urals continental margin in the Palaeozoic (after Zhivkovich and Chekhovich, 1985).

a—Late Cambrian-Early Ordovician; b—Middle-Late Ordovician; c—Ludlovian-Lochkovian; d—Daleian-Eifelian; e—Late Devonian; f—Middle-Late Carboniferous; g—Early Permian; h—Late Permian

1—Riphean-Vendian basement; 2 to 5—Lower Palaeozoic terrigenous formations of the nascent rift margin; 2—coarse clastic; 3 and 4—quartz-sandy (3—sandy, 4—coarse clastic formations); 5—sandy shale; 6—basaltoids and their tuffs; 7—alkaline rhyolites; 8—basal sandstones of quartz and subarkosic composition; 9—carbonate-terrigenous formations of littoral shelf zones; 10—formations of laminated limestones; 11—reef formations; 12—black shales and siliceous formations; 13—deepwater sand formations; 14—turbidite formations; 15—massive and pillow diabases; 16—formations of magnesian carbonates; 17—evaporites (anhydrites, dolomites and gypsum); 18—siliceous breccias; 19—Domanik formation; 20—bauxite-bearing unconformities; 21—melange; 22—crystalline basement; 23—flyschoid and lower molasse formations; 24 and 25—slivers of the Biryagmo-Kirgishan nappe (24—diabases and sandstones, 25—shales, cherts and sandstones); 26—dislocated formations of laminated limestones; 27—oceanic and island arc complexes; 28—fault dislocations and direction of migration along them.

Ordovician. As before, the greater part of this ocean lay east of the Kazakhstan microcontinent. It experienced considerable expansion in the Ordovician due to new formations of oceanic crust in the space between the microcontinent and the Chingiz-Tarbagatai island arc arising as early as the Middle Cambrian. Development of this marginal sea was associated with the existence of a subduction zone inclined from the east under the Chingiz-Tarbagatai arc. Its activity also caused the formation of nappes with an eastern vergence within the arc. The Kazakhstan microcontinent experienced partial destruction with the formation of latitudinal and meridional rift troughs of the Red Sea type, i.e., Kalmyk-kul' and Stepniak. The axial zone of the Palaeo-Asian ocean in this segment evidently extended between the Chingiz-Tarbagatai and Salair island arcs. Even more eastward, between Salair arc and the western margin of the Siberian continent that enlarged in the second half of the Cambrian, filling of the residual basin with clastic material continued. This material came in abundance from the Salair mountain system and partly accumulated on the eastern margin of the main ocean basin, in the Gornyi (High) and Mongolian Altay.

In the Late Ordovician, the Kazakhstan and Altay sections of the Palaeo-Asian ocean experienced intense compressive deformation. In central Kazakhstan and Northern Tien Shan, extensive granitoid plutonism occurred, evidently associated with the subduction zone now inclined to the west under the microcontinent. A similar zone, but with an opposite dip, should have existed along the Altay periphery, which evidently represented a non-volcanic arc associated with the Salair volcanic arc.

Late Ordovician folding and uplift did not affect some parts of Kazakhstan and Altay-Sayan regions where deepwater basins were preserved and acquired the status of interior seas. The largest of them, Junggar-Balkhash, is associated with the open ocean in the south-east; Anuy-Chu and Western Sayan represent smaller basins.

The eastern, Gobi-Hinggan segment of the Palaeo-Asian ocean developed in a relatively quiescent regime during the Ordovician (Fig. 2-10). From the north, in the Transbaikalian section, it was bound by an active Andean-type margin advancing southward after collision of the Central Mongolian microcontinent with the Siberian continent. Within this margin, formation of the Barguzin granite batholith, the largest in the world, continued (Fig. 2-11). Eastward, both the margins of the basin continued to remain passive.

2.3.2.4 *Mediterranean belt.* In the western, European part of the belt a deepwater basin of one of the main zones of the Central European Hercynides, i.e., Saxothuringian, was formed with continuation into the central Armorica and Ossa-Morena zone of the Iberian Meseta. Formation of the basin commenced with rifting and accumulation of thick beds of 'Armorican' sandstones (apart from the Ossa-Morena zone) were replaced by black

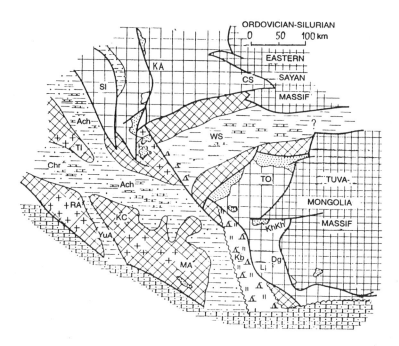

Fig. 2-10. Palaeotectonic scheme of Central Asian Caledonides for the Ordovician-Silurian (after A.B. Dergunov. In: Mossakovsky and Dergunov, 1983). See Fig. 1-10 for legend.

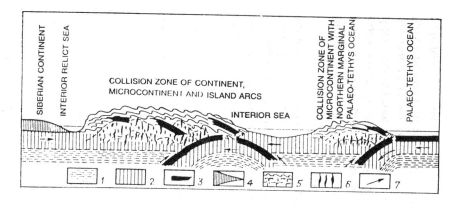

Fig. 2-11. Schematic reconstruction of the Sayan-Baikal-Mongolia region for the period of formation of Late Caledonian (Ordovician-Silurian) calc-alkaline granitoids (after Gordienko, 1987).

1—asthenosphere; 2—lithosphere; 3—blocked ocean plates absorbed in subduction zones; 4—continent; 5—front of granitisation, gabbro and granite formation in nappe-fold structures; 6—streams of fluids, heat and magmatic melts and solutions; 7—direction of stress and motion in plates.

shales upward along the sequence as the opening of the basin increased and by radiolarites in Bavaria. The rift origin of the basin has been confirmed by manifestation of bimodal volcanism (Thuringia etc.). This rifting was most probably associated with subduction of the Iapetus ocean lithosphere under the northern margin of the Armorican microplate (subduction from the south did not commence before the end of the Ordovician; Behr et al., 1984). The presence of ophiolites in the Münchberg massif and Gory-Sowie may suggest that the extension in the southern part of the basin could have proceeded until total extension of the continental crust and new formation of oceanic crust. This was the Middle European palaeo-ocean or Rheicum. Its existence has been confirmed by differences between fauna of the Baltic region and the rest of Europe.

An uplift extended north of the Saxothuringian basin and was associated in the west with northern Armorica. In the east it was known as the Midgerman (Mitteldeutsche) uplift. It is mainly made up of metamorphosed Upper Riphean formations. An accumulation zone of shallow-water terrigenous deposits of the Cambrian-Silurian extends even more northward and experienced Caledonian folding at commencement of the Devonian. These are evidently the deposits of the outer shelf of the margin of the East European continent but they girdle the Baikalian Midland massif in the west. In the east their more deepwater analogues extend along the Teisseyre-Tornquist line filling the Rügen-Pomorie trough of continental slope and rise.

The axial (Moldanubian) zone of Hercynian orogen of Europe is represented in its eastern part, i.e., Bohemian massif, by a block of continental crust formed in the Cadomian (Baikalian) epoch and reworked in the Hercynian. The Barrandian trough or Prague synclinorium filled with Vendian-Middle Devonian strata overlies it. The shallow-water nature of the Ordovician formations in this trough, their small thickness and absence of metamorphism confirm the stable position of the massif in the Ordovician and its microcontinental nature. More complicated is the western continuation of the zone which includes the massif Central and southern part of Armorica as the most prominent parts. This segment of the Moldanubian zone underwent such an intense and repeated reworking in the Hercynian epoch, commencing with the Middle Devonian, that it is not clear which part of the pre-Devonian sequence pertains to the Ordovician here. It is possible that this segment of Moldanubicum also experienced uplift in the Ordovician, albeit relative. This is quite evident for the Central Iberian zone, a homologue of the Moldanubian in the Iberian Meseta.

Some researchers (for example, Matte, 1986) regard the metamorphosed ophiolites known in the Armorican and Central massifs and eastward in Belledonne in the French Alps as Ordovician. This has been confirmed by radiometric age determinations in the latter massif at 496 ± 6 m.y. (Ménot et al., 1988).

The western Palaeo-Tethys stretches south of the Moldanubian zone. Terrigenous flysch of the Pyrenees, south of the massif Central, Provence, Corsica and northern Sardinia may represent deposits of the southern slope of the Moldanubian microcontinent while the dark-coloured clay deposits of the Cordillera Betica, Rif, Tell and southern Sardinia represent deposits of the bottom of this relatively deepwater basin. The absence of ophiolites and the faunal similarity of Maghreb and southern Europe witness against the large width and depth of this Palaeo-Tethys, which was most probably underlain by a transitional type crust, i.e., continental crust that had experienced destruction before the Ordovician.

Eastward, as a result of the 'wedging' of the Moldanubian microcontinent, the Saxothuringian basin merged with the Palaeo-Tethys. This deepwater basin encompassed the contemporary Eastern Alps, Carpathians, Dinarides-Hellenides and Balkanides having been underlain by a crust of transitional and oceanic types. Ophiolites overlain by deepwater carbonates, silicites and lutites of the Late Ordovician have been detected in the Severogemeride zone in the Western Carpathians. It is possible that the ophiolites of the Vardar zone too are of the same age. The Carpathian-Balkan basin extended eastward in the direction of Caucasus Major where ophiolites are most probably of Ordovician age as in the Urals. The Transcaucasus salient of Gondwana served as the southern boundary of the Caucasian section of the Palaeo-Tethys. Here the Devonian lies directly on the Proterozoic-Lower Cambrian. In the Dinarides-Hellenides sandy-shale Ordovician formations with lenses or olistoliths of neritic limestones may belong to the southern flank of the Palaeo-Tethys.

The southern part of the Alpine belt of Eastern Europe and Western Asia, including Anatolia, Iran and Transcaucasus, continued to remain as the Epibaikalian periphery of Gondwana. Ordovician formations take part in the cover sequence of this platform and are represented by shallow-water and thin terrigenous deposits. In the east the platform encompasses central and southern Afghanistan, the Pamirs, Karakoram, the Himalayas and Sinoburma. As in the Cambrian, only more to the east, in Qilianshan and Qinling, and possibly in the far northern Pamirs, the Ordovician was again marine geosynclinal in character. As in the adjoining segment of the Central Asian belt, this was a period of development of volcanic arcs (Zhang Zhijin, 1984) and accompanying flysch troughs.

2.3.2.5 *West Pacific belt.* In the north-west, in Penzhina ridge, development of pre-Late Ordovician ophiolites has been established. These ophiolites evidently represent the Palaeo-Pacific crust. The reef limestones of Upper Ordovician overlying them may correspond to the cover of seamounts but later an island arc arose here in the Devonian. More westward the boundary of the ocean basin, as in the Cambrian, is represented by shelf, predominantly carbonate, partly clastic facies of the Omolon and Okhotsk massifs and also the Aya and Shevli zones of the eastern and south-eastern framework of the Aldan-Stanovoy Shield. But the Omolon (Kolyma-Omolon) massif had already begun to be separated in the Ordovician from the Siberian continent by a deepwater basin with thin graptolite facies and manifestation of alkaline-basalt volcanism, confirming the rift origin of this basin. Its width, judging from the presence of bands of clastic rocks, could not have been significant but the recently discovered pre-Ordovician ophiolite-clastites in the Omulevka uplift point to new formations here of oceanic crust even in the Cambrian.

The easternmost Amur-Okhotsk segment of the Palaeo-Asian ocean should have extended south of the Aldan-Stanovoy Shield, as before, but its existence in the Ordovician has not been practically established although more than likely.

South of the Amur-Okhotsk basin, in Sikhote-Alin', signs of the presence of oceanic crust are again manifest. These are ultramafics of the Anuy region intruded by granodiorites aged 488 m.y. Voznesensk granitoids of the Khanka massif intruding the Cambrian formations and probably Birobidzhan granitoids of the Hinggan-Bureya massif overlain by Silurian are of proximate age (482–423 m.y.). Subaerial dacite-rhyolite formations with ignimbrites are associated with the latter. All this demonstrates the Andean type active character of the Bureya-Khanka continental margin of the Palaeo-Pacific.

Recently, fairly extensive development of ophiolites with basalts directly below cherts containing Late Devonian fauna of conodonts has been demonstrated in the main anticlinorium of Sikhote-Alin' (Khanchuk et al., 1988). It is possible that these represent the youngest ophiolites of this system since the ophiolites in its continuation into northern Honshu are of pre-Silurian age.

Farther southward, this margin continues into south-eastern China and north-eastern Vietnam after a break associated with the confluence of the Palaeo-Tethys and Palaeo-Pacific. Here, as in the Cambrian, transition from shelf carbonates to flyschoid greywacke-shale formations of continental slope are distinctly manifest. A decrease in volcanic activity of the island arc situated more south-east, at the boundary with the open ocean, occurred in the Ordovician. The area of the marginal sea shrank in the Late Ordovician due to accretion along the continental slope.

Extending between southern China (Yangtze) and the Indosinian continental massif, the Viet-Lao geosynclinal system served as yet another connecting link between the Palaeo-Tethys and Palaeo-Pacific. Uplift commenced in the Late Ordovician in its north-eastern part, and accumulation of thick volcanic-terrigenous strata continued in the south-west (Chyongshon range).

A typical environment of active continental margin of the modern western Pacific Ocean type existed in the Ordovician in eastern Australia (Scheibner, 1987).

The continent-ocean boundary here advanced considerably eastward compared to the Early Cambrian. The easternmost salient of the ancient continental crust on the mainland, i.e., Williama (Brocken Hill) massif and western Tasmania, already lay in the shelf zone surrounded by the Ballarat trough of continental slope and rise with thick siliceous-greywacke-shale sedimentation. The wide Wagga marginal sea separated from the open ocean by the Molong-Canberra volcanic arc extended farther east. A seismofocal zone inclined westward evidently formed at their boundary. At the end of the Ordovician to commencement of the Silurian, filling of the marginal sea underwent compressive deformation with accretion of Ballarat trough formations, origin of non-volcanic Heathcot and Wellington arcs and outcrop of ophiolites along them onto the surface. Deformation of this Benambran epoch (in local terminology) was accompanied by formation of anatectic bodies of granite-gneisses and initial metamorphism.

On the South Island of New Zealand, which should have adjoined Australia and Tasmania as before, continental slope formations, marginal sea and volcanic arc, similar to the Australian, are known.

In Antarctica the zone of Transantarctic mountains experienced the orogenic stage (Rossorogeny) and most probably represented an active Andean type margin.

2.3.2.6 *East Pacific belt.* In the type region Chile-Argentinian Andes (29–33° S lat.), referred to in Section 1.3.2.5, conditions in the Early Ordovician did not differ from the Cambrian (Ramos, 1988) but in the Middle and Late Ordovician changed significantly. The eastern half of the Precordillera experienced uplift with accumulation of shallow-water clastic sediments, including fairly thick conglomerates at places. The clastic material is of local origin. The western half of the Precordillera represented the slope of a deepwater basin with deposition of clastic formations, among which sheets of tholeiite-basalts of oceanic type, differing slightly in petrochemistry from basalts of mid-ocean ridges, are observable. The basin evidently represented a marginal sea; somewhere west of it lay the microcontinent of the modern Cordillera Coastal, i.e., Chilenia. East of the Precordillera, an outburst of granitoid plutonism in Late Ordovician of age (450–440 m.y. ago) has been observed in the Sierra Pampeanas massif. This is mainly represented by tonalites, granodiorites and later normal granites as well as migmatites. Metamorphites of amphibolite facies are associated with them. If this massif and the Precordillera did adjoin each other in this epoch, then magmatism could be explained by the existence of a subduction zone inclined eastward.

The Ordovician deepwater basin on oceanic or suboceanic crust, partly set off from the open ocean by microcontinents (Chilenia and Arequipa), extended along the Andes north up to the Caribbean coast (Ramos, 1988). It attained maximum width in the Middle Ordovician. In the Eastern Cordillera of Peru and Bolivia and in the Subandean foredeep zones, the Ordovician rests transgressively on the Cambrian and Precambrian and is composed of conglomerates and quartzites in the lower part. From the Arenigian to Early Caradocian, black shale sedimentation predominated here while regression associated with uplift at the end of the Ordovician commenced later. Quartzites forming flyschoid interstratification with sandy shales are again extensively developed. In the Northern Andes and the Andes of Bolivia, the Silurian rests on the Ordovician with angular unconformity.

The Ordovician fauna of the Central and Southern Andes reveals similarity with the fauna of Australia and New Zealand while North American and European elements are seen in the Northern Andes (Dalmayrac et al., 1980). This confirms the connection of the Iapetus with the the Pacific ocean, in any case with the Pacific margin of Gondwana.

In the North American Cordillera transition from east to west, from thin shelf carbonate or terrigenous-carbonate deposits to slope clastic or shale (with graptolites) deposits, at places with manifestation of basaltic volcanism, and later to deepwater carbonate-siliceous-shale formations and tholeiite-basalts of marginal sea (Ordovician ophiolites are also known in the Klamath Mountains; Dobretsov, 1988), and finally to volcanic arc rocks, has been firmly established. However, in the light of contemporary date on the structure of the central and inner zones of the Cordillera and the concept of exotic blocks or terranes, this sequence should be regarded as only approximating the original disposition of elements of the active margin of the North American continent. Only shelf formations, and too not wholly, preserved their primary disposition. Continental slope deposits were shifted along major transcurrent displacements, e.g., Tintin and Denali in Alaska (Churkin et al., 1982). Island-arc volcanics with intrusions into them of Late Ordovician granites travelled, according to palaeomagnetic data, 1800 km (!) from south to north; these have been exposed in the Alexander archipelago along the coast of south-eastern Alaska. This extremely significant remark will concern the much younger Palaeozoic also.

One more enigmatic event in the Ordovician of the Cordillera needs to be mentioned, i.e., the detection of Ordovician granitoids (monzonites, diorites and granite-gneisses) in the composition of the Omineca polymetamorphic complex on the very margin of the continent (Okulitsch, 1985). Their host complex also evidently experienced horizontal displacement relative to the main part of the continent.

If the Vendian and Cambrian represented periods of formation of main geosynclinal belts of the Palaeozoic era, the Ordovician commenced with the first compressive impulse within them. Simultaneously, however, the palaeo-ocean space continued to enlarge, reaching maximum width in the Arenigian-Llandeilian. In the second half of the Ordovician, compression clearly predominated over spreading. Manifestation of Taconian folding and accompanying phenomena in the Appalachians and Newfoundland, Benambran in eastern Australia, deformations of the same age in the Innuitian system, south-eastern China and Northern Andes and granite formation in the Sierra Pampeanas and North American Cordillera reveal the global scale of this epoch of tectonomagmatic activity. At the same time, tectonic activity continued to weaken during the Ordovician within the Gondwana supercontinent[1] and remained at a low level in the group of Laurasian continents while magmatic activity within the continents almost wholly died out. The ice sheet of Gondwana was associated with displacement of much of Gondwana, Africa, South America and Western Europe, into high latitudes of the Southern Hemisphere by the end of the Ordovician. The disposition of the rest of the continental blocks generally changed little compared to the commencement of the Ordovician and even the Late Cambrian (see Figs. 1-7 and 2-7).

---

[1] It must be admitted that the Late Ordovician glaciation was caused not only by displacement of part of Gondwana into the Polar region, but also by some uplift of the corresponding region.

3

Silurian-Early Devonian. Completion of Development of the Caledonian and Early Stage of Development of Hercynian Mobile Belts

3.1 REGIONAL REVIEW. CONTINENTAL CRATONS AND THEIR MARGINS

3.1.1 North American Craton

E a r l y   S i l u r i a n. After maximum Palaeozoic transgression in the Late Ordovician to Early Silurian, uplift again began to predominate and the total sea area on the craton decreased significantly (in: *Stratigraphic Atlas of North and Central America,* 1975). Uplift was of the nature of a gentle rise and resulted in formation of extensive land expanses, especially in the region of the Midcontinent, representing plains and smoothened peneplains. This feature as well as warm climate made for the prevalence of carbonate accumulation in all marine basins of the craton (often in reef facies). The thickness of the Lower Silurian usually does not exceed 200 m. In the Michigan basin subsidence was not compensated by deposition. There were almost no terrigenous rocks on the craton with the exception of the Peri-Appalachian zone, in whose troughs sandy, clay and conglomerate beds accumulated in marine and continental facies eroded from uplands in the orogenic zones of the Southern Appalachians. The thickness of the beds here reaches 1 km. Greater thickness has been established in the Peri-Innuitian and Peri-Cordilleran zones but there are almost no terrigenous rocks.

L a t e   S i l u r i a n. Uplift continued to expand and marine basins were transformed into small interior semi-enclosed seas. Areas of all marginal basins of the craton shrank. As earlier, carbonates were the predominant type of formations, their thickness usually not exceeding a few tens of metres. The Michigan basin constitutes an exception: here, after an epoch of uncompensated subsidence, intense deposition of evaporites of the Cayuga series containing carbonates, sulphates and salt took place. The thickness of the Upper Silurian in this basin was about 1 km. Considerable thickness (up to 750 m) and the significant role of terrigenous rocks (partly continental) are noticeable in the Peri-Appalachian zone.

Hudson basin is the second region with greater water salinity. Dolomites, sands (sometimes under continental conditions) and also gypsum and anhydrite (Canoghamn River formation) accumulated in it. The third evaporite basin was located north of the craton in Sommerset island region. Gypsum bands are known there in the Leopold formation.

Intense deposition of clastic material occurred in the central segment of the Peri-Cordilleran zone. Here it replaced formation of reef carbonate deposits and thus this complex of sedimentary rocks in the form of a sedimentary wedge built up the area of the craton westward.

E a r l y   D e v o n i a n. Subsidence enlarged in the north-western part of the craton and an extensive West Canadian basin formed whose dimension increased during this epoch. Accumulation of carbonates predominated in it and a salt-bearing complex up to 400 m thick formed in the southern part of the bay. However, in spite of this transgression, the total sea area on the craton, compared to the Late Silurian, decreased since the Peri-Innuitian basin almost disappeared and, as a result of uplift markedly shrank the Peri-Appalachian basin and others. The role of terrigenous rocks grew in the composition of the deposits of the Michigan and Hudson relict basins. Only sandy-clay formations accumulated in the basins adjoining the Appalachian mobile system.

The broad and elongated Peri-Cordilleran marginal basin continued to experience uninterrupted subsidence compensated by the deposition of shelf carbonate and terrigenous facies. Their thickness gradually increased westward to 400 to 500 m.

In the southern part of the craton, in the zone adjoining the Ouachita-Mexican segment of the Appalachian system (Marathon region), small manifestations of Early Devonian basalt volcanism are known (Weyl, 1980).

Thus the Silurian-Early Devonian stage of development of the North American Craton represented a period of successive intensification of uplift and regression. The arid climate and low altitude of this continent as well as the smoothened relief made for a predominance of carbonates in the accumulated deposits and extensive development of evaporites. A feature of marginal cratonic zones is the prolonged stable development of subsidence compensated by the accumulation of reef facies. This unique feature of the North American Craton was also manifest, though to a lesser extent, in its Early Palaeozoic history.

3.1.2 East European Craton

E a r l y   S i l u r i a n. At the beginning of the epoch negative movements caused transgression in the western and south-western parts of the craton and resulted in southward expansion of the Baltic-Moscow basin and formation of the Volyn-Podol basin. In these sediments shallow-water carbonate silt enriched with remnants of diverse fauna predominated. The thickness goes up to 150 to 200 m, rising to 700 m in southern Scandinavia. This increase reflects the last phase of subsidence and final filling of the Rügen-Pomorie trough.

In the Pechora basin too some transgression has been observed and marine conditions extended up to the Kanin peninsula. The presence of gypsum and lagoonal carbonate-terrigenous facies in the Llandoverian and Wenlockian formations points to high water salinity.

The eastern margin of the platform continued to experience subsidence. In the Peri-Uralian zone terrigenous sedimentation was almost exclusively replaced by carbonates and in littoral areas by terrigenous carbonates with cherts. Reef facies formed in the eastern Peri-Uralian zone.

L a t e   S i l u r i a n. The general magnitude of uplift rose. The area of the Baltic-Moscow basin shrank severely, the basin turned shallow and lime-dolomite silts were deposited in it (Fig. 3-1). The Volyn-Podol basin maintained its former boundaries and carbonate and terrigenous strata formed in it and even sulphate rocks in the eastern part. Argillaceous rocks up to 200 to 300 m in thickness predominated in the Upper Silurian formations within Moldavia and the Moesian block.

A characteristic complex of coarse clastic formations of the molasse type was formed in southern Scandinavia in a zone adjoining the Scandinavian orogen. The thickness of the complex here exceeds 1 km.

In the north-eastern part of the craton almost total regression occurred in the Pechora basin in the Late Silurian as a result of uplift and predominantly carbonates accumulated in the remaining small bay up to 200 m in thickness. The Peri-Uralian zone somewhat enlarged. As before, predominantly lime-dolomite deposits formed in it except in the littoral part in which terrigenous, at places alluvial-deltaic facies predominated. A particularly extensive system of such littoral complexes has been traced in the Southern Urals. The rest of the craton experienced uplift and represented levelled land.

E a r l y   D e v o n i a n. An even more intense uplift on the craton led to almost total regression of marine basins in the west and east. The Baltic-Moscow basin turned into an intracontinental, periodically drying up salinised water reservoir in which sandy-clay alluvial and carbonate-terrigenous lagoonal deposits were formed to a thickness of tens of metres. Filling of basins in southern Scandinavia (in the region of Oslo) with coarse clastic material concluded under continental conditions. The thickness of these formations goes up to 900 m.

Sandstones, clays and carbonates with variable, sometimes significant thickness accumulated in the Volyn-Podol basin with a notably reduced area in the south-western part of the craton. The Moesian block, with Lower

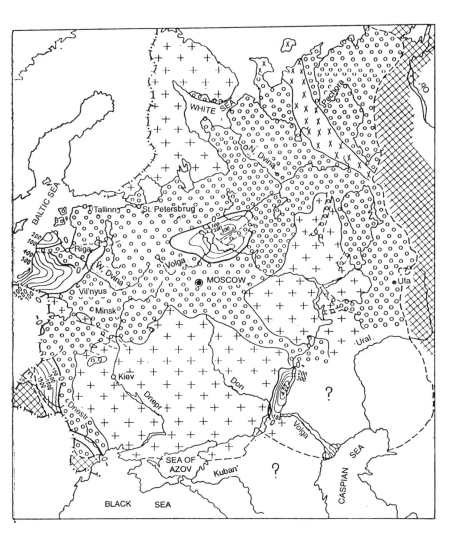

Fig. 3-1. Scheme showing distribution and thickness variations of regressive and emersive stage formations of the Early Palaeozoic stage of development of the East European Craton, Late Silurian-Early Devonian (after I.I. Bykova and N.S. Igolkina. In: *Geological Structure of the USSR and Pattern of Location of Mineral Deposits.* Vol. 1, *Russian Craton,* 1985). See Fig. 1-2 for legend.

Devonian formations 700 to 900 m in thickness, experienced very marked subsidence (Spassov et al., 1978).

The Pechora basin underwent total regression and the width of the Peri-Uralian zone shrank. Within the latter limestone, dolomite and terrigenous shallow-water deposits formed, among which brecciated formations were distributed, which points to unstable subsidence and erosion zones (see Fig. 2–9).

On the whole, during the Silurian-Early Devonian the East European Craton experienced a tendency towards uplift and regression. Almost its entire territory was a land with poorly dissected topography which led to silt or carbonate deposits in most of the sedimentary basins. Stable subsidence and accumulation of carbonates in the Peri-Uralian zone are notable, their development being similar to the 'carbonate' marginal zones of the North American Craton.

3.1.3 Siberian Craton

E a r l y   S i l u r i a n. Subsidence intensified and extensive transgression is noticed in the central part of the craton. Maximum areas of subsidence of the Tunguska and Taimyr basins corresponded in time to the Late Llandoverian. All this extensive territory represented an open sea basin with carbonates rich in benthic fauna predominating in its deposits. Graptolite shales too accumulated in the marginal areas. A periodically drying-up semi-enclosed salinised water reservoir existed in the southern Tunguska basin. Red-coloured sands, gypsum-bearing dolomites and gypsum were deposited in this reservoir. Uplift occurred even in the Wenlockian stage and the reservoir ceased to exist. Continuous slow subsidence of the Tunguska basin proper led to the accumulation of carbonate strata up to 300 to 400 m in thickness.

Special conditions inherited from the Ordovician prevailed in the Taimyr basin. In the axial part of the trough, characterised by prolonged stable tectonic regime of uncompensated subsidence, formation of graptolite facies continued. In the marginal zones of this trough predominantly carbonate (in the south) and carbonate-terrigenous (in the north) deposits formed to a thickness of a few hundred metres.

The boundaries of the Sette-Daban and Yana-Kolyma basins in the eastern part of the craton and carbonate composition of formations remained unchanged. The role of terrigenous sediments rose only in the Sette-Daban and evidently filling of this basin concluded. The thickness of formations in the Yana-Kolyma basin is significant, for example, up to 1.6 km in southern Verkhoyansk region.

L a t e   S i l u r i a n. Intensification of uplift resulted in regression, the area of Tunguska basin shrank, it became markedly shallow and had limited contact with the open sea (Fig. 3–2). Accumulation of variously coloured dolomites, at places bearing gypsum, predominated in it, pointing to high water salinity. The former palaeotectonic environment prevailed in the Taimyr basin more to the north, marine conditions were normal and argillaceous (graptolite-bearing) strata formed in the central part of the basin and carbonate complexes in its marginal zones.

Uplift and regression fully encompassed the Sette-Daban basin and the intensity of subsidence in the area of the Yana-Kolyma basin decreased. Limestone-dolomite deposits formed in the Yana-Kolyma while evaporites (gypsum and anhydrite) and small basalt sheets as well were formed in the north-east. Basalt sheets point to commencement of active tectonic movements (rift formation?) in the Yana-Kolyma interfluve.

Early Devonian. The north-eastern part of the craton (Yana-Kolyma interfluve and Omolon River basin) underwent tectonic reactivation as a result of rifting and became a constituent of the Verkhoyansk-Kolyma mobile region. Uplift intensified elsewhere in the territory. The size of the

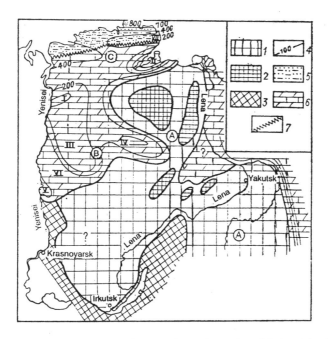

Fig. 3-2. Palaeotectonic sketch of the Siberian Craton in the Late Silurian (after N.S. Malych. In: *Geological Structure of the USSR and Pattern of Location of Mineral Deposits.* Vol. 4, *Siberian Craton,* 1987):

1 to 3—zones of uplift (1—weak on craton, 2—moderate on craton, 3—weak in mobile region); 4—isopachs, m; 5—dolomite-carbonaceous paragenesis; 6—sulphate-carbonate formation; 7—lateral zone of transition of formations.

Structural regions (circled letters): A—Aldan-Anabar; B—Lena-Tunguska; C—Taimyr.

I to V—basins (I—Norilsk, II—Khatanga, III—Vivin, IV—Morkokin, V—Vorogovka); VI and VII—uplifts (VI—Bakhta, VII—Byrranga).

Tunguska basin shrank even more, it was dismembered by low islands and maintained the character of a semi-enclosed shallow reservoir, at places with high water salinity and deposition of thick beds of gypsum and sometimes rock salt, apart from other sediments. The greater part of the deposits here is represented by terrigenous (continental and marine facies) and carbonate rocks a few tens of metres in thickness. Similar basins were formed in the lower course of the Khatanga and in the Severnaya Zemlya archipelago.

Subsidence of the central zone ceased in the Taimyr basin and graptolite shales were replaced by comparatively shallow-water carbonate deposits up to 500 m thick. Limestone deposits prevailed even in the marginal zone adjoining the Verkhoyansk-Kolyma mobile area, the intensity of subsidence being considerably greater and the thickness of deposits here reaching 600 to 700 m.

Thus in the Silurian-Early Devonian the Siberian Craton escaped Early Silurian transgression and experienced intense uplift. Land areas did not exhibit a rugged topography and carbonates predominated in marine basins. The warm climate and semi-enclosed basins led to high water salinity and at places the deposition of evaporites. Reactivation of tectonic movements in the Early Devonian in the north-eastern part of the craton and replacement there of a platform regime by conditions of a mobile belt are noteworthy features.

3.1.4 Sino-Korea Craton

E a r l y S i l u r i a n. The Lower Silurian is not known in this cratonic area. An uplift and erosion process predominated here and possibly a rugged topography was formed (in: *Atlas of the Palaeogeography of China,* 1985; Lin Baogu, 1979). The same is true of the Tarim and Qaidam blocks. Only the northern part of Tarim and a small section in the south-west of the craton were covered by a shallow sea in which sand and clay deposits of small thickness were formed in the Tarim and carbonate-terrigenous deposits in the south-western margin of Ordos.

L a t e S i l u r i a n. There was no significant change compared to the Early Silurian except for the near-total drying up of the Tarim block.

E a r l y D e v o n i a n. Consequent to the disappearance of the Qilianshan geosynclinal basin, Tarim and Qaidam joined with the Sino-Korea Craton to form a single large continental mass. The eastern Tien Shan orogen adjoined north-eastern Tarim.

The Silurian-Early Devonian time for the Sino-Korea Craton was a period of prevalence of uplift and formation of a large single continent from Tarim to Korea.

3.1.5 South China (Yangtze) Craton

E a r l y S i l u r i a n. At the commencement of this period, much of the craton was covered by a semi-isolated marine basin with an anoxic environment and accumulation of black graptolite shales. Jiannania Island separated it from the Cathaysian geosyncline; a different island occupied the southern part of the craton while the Kam-Yunnan meridional uplift served as a barrier between the epicontinental sea and the south-eastern branch of the Palaeo-Tethys (in: *Atlas of the Palaeogeography of China,* 1985). Later in the Early Silurian conditions in the Yangtze basin normalised and carbonates were deposited in the shallow sea with bands of clastic rocks in its north-western part.

L a t e   S i l u r i a n. In the middle of the Silurian uplift and fold deformation occurred in the region east and south-east of Jiannania. Following this, uplift extended to much of the Yangtze Craton; accumulation of argillaceous and argillaceous-carbonate deposits occurred in the residual basins.

E a r l y   D e v o n i a n. In this epoch some redistribution of land and sea occurred in Southern China. The marine basin now lay in the southern part of the craton. Its deposits were represented mainly by carbonates, replaced by continental clastic red-beds along the northern and western periphery. Similar formations filled the trough in the lower course of the Yangtze. Clastic supply was provided from the rest of the craton as well as from the newly formed Cathaysian orogen. As a consequence of the Late Silurian folding, Devonian beds lay even on the craton transgressively and unconformably over much older formations.

3.1.6 South American Craton

E a r l y   S i l u r i a n. Subsidence of the Amazon, Maranâo and Parana basins made for development of the first extensive transgression on the craton in the Palaeozoic. The Amazon basin intersected the craton from west to east and sandy-clay sediments deposited in it, their thickness reaching 1 km. In the Maranâo basin sand facies predominated and fluvioglacial facies are known on its western flank at the base of the Lower Silurian (see Fig. 2–4). Coarse clastic sand deposits also predominated in the Parana basin and the Peri-Andean zone of Chaco; further, the inflow of clastic material here superseded subsidence and continental facies played a significant role. Uplift within the land was comparatively powerful and an elevated undulating relief was formed.

L a t e   S i l u r i a n. Following the Early Silurian transgression, uplift and regression again encompassed the craton. Marine conditions probably continued to prevail only in the Maranâo basin but subcontinental terrigenous facies predominated even here. The Chaco and Parana regions represented the second region of subsidence (pericratonic). However, subsidence in this region was wholly compensated by accumulation of clastic material and hence continental conditions prevailed. The thickness of formations here is usually 200 to 400 m.

E a r l y D e v o n i a n. The beginning of the Devonian is characterised by a new wave of subsidence and transgression covering the Amazon, Maranâo and Parana basins and the southern part of the Peri-Andean zone. Sandstones predominated everywhere in the deposits and continental facies were extensively developed. Subsidence affected only the eastern part of the Amazon basin in this epoch and the affected portion had yet no contact with the marine basin of the Andean system. The thickness of the Lower Devonian is usually in tens of metres, sometimes 200 to 300 m, reaching 1 km in the Peri-Andean zone.

In this period, as a whole, extensive transgression encompassed the South American Craton twice, i.e., its tectonic regime was quite mobile and contrasting. Abundance of clastic material points to powerful uplift in land regions with development probably of a moderately dissected topography.

3.1.7 African Craton

E a r l y S i l u r i a n. The northern part of the craton experienced major subsidence and transgression penetrated far southward (see Fig. 2-3) in the Sahara Platform region up to the coast of the Gulf of Guinea (Aliev et al., 1979; Buggisch et al., 1979). Sandy-clay formations of 100 to 200 m thickness were deposited all over the extensive Sahara shallow-water basin. Continental facies too are known in these formations. Interior islands and southern elevated parts of the craton represented the sources of clastic material. A zone of intense subsidence was particularly prominent in the Anti-Atlas region where Early Silurian deposits are represented only by argillaceous rocks up to 1 km in thickness. There were small manifestations of rhyolite volcanism in eastern Algeria.

In the Arabian peninsula a high rate of subsidence has been identified in the central part of the basin, with the thickness of deposits reaching 500 m. Subaerial basalts are known in northern Iran while carbonate-terrigenous deposits up to 500 to 700 m thick were formed in southern Afghanistan (Pyzhyanov et al., 1980; Wolfart and Wittekindt, 1980). Subsidence continued uninterruptedly in the southern part of the craton in the Cape province and the thickness, predominantly of sandstones, goes up to 700 m there.

L a t e S i l u r i a n. Subsidence continued in the vast area of the Sahara basin and the boundaries of the basin remained as before. The relief was smoothened and exclusively clay deposits a few tens of metres thick covered a large surface. Islands subsided and subcontinental coarse clastic facies were replaced outside the littoral zone by increasingly lutaceous marine facies. The relatively compensated intense subsidence of the Upper Silurian in the island sections and deep in the littoral basin was 800 and 900 m respectively.

Uplift intensified in the Arabian peninsula, the area of sea shrank, sandstones predominated in deposits and continental facies played a significant role. As before, a carbonate-terrigenous complex of formations was deposited in eastern Iran and southern Afghanistan and subaerial eruptions of basalts up to 300 m in thickness evidently continued in northern Iran.

The earlier conditions of continuous subsidence and accumulation of sandstones were maintained in the Cape province.

E a r l y D e v o n i a n. Uplift and regression encompassed significant parts of the Sahara basin and a major island appeared in the region of Reguibat massif. The supply of coarse clastic material increased, indicating intensified uplift in the central regions of the craton. Continental facies were extensively developed in the marginal zones of the Sahara basin. The growing dynamics of tectonic movements was reflected in the isolation of independent basins and troughs with a much greater rate of subsidence. Thus the thickness of the Lower Devonian in the Ougarta basin exceeded 800 m and in the Tindouf basin reached 1.5 km.

In the Arabian peninsula too uplift intensified; the marine basin decreased in area and represented a partially sealed reservoir with high water salinity. Apart from sandy-clay deposits, carbonates and gypsum not exceeding 150 m in total thickness were deposited here. A typical narrow trough extended from south-eastern Turkey to south-eastern Caucasus Minor. Carbonate-terrigenous deposits (sometimes subcontinental) formed in this trough and their thickness in Caucasus Minor reaches 1.5 km.

In the Cape province in the southern part of the craton, subsidence intensified somewhat and a minor transgression occurred. A sandy-clay suite up to 800 m in thickness here belongs to the Lower Devonian.

Thus the extensive subsidence encompassing the North African Craton in the Early Silurian was replaced by uplift and regression in the Late Silurian. This tendency developed further in the Early Devonian, i.e., in this epoch, as in the preceding one, a transgressive-regressive cycle was distinctly manifest. The extensive central part of the craton represented the largest land massif on the Earth and the erosion products of this massif compensated subsidence of the Sahara, Arabian and Cape basins. The near-total predominance of terrigenous rocks among marine formations is noteworthy.

3.1.8 Indian Craton

E a r l y S i l u r i a n. Almost the whole of the craton continued to experience upheaval throughout the Silurian-Early Devonian. Formations of this age are not known except for the northern region covering the Himalaya and southern Tibet. As before, uninterrupted subsidence occurred here while terrigenous and carbonate rocks were deposited in the shallow marine basin.

Their thickness varied somewhat in different regions. Lower Silurian formations are usually 200 to 300 m thick but reach 700 m in Nepal (Jain et al., 1980); the Upper Silurian is considered to be 100 to 200 m thick (up to 300 m in Nepal) while the Lower Devonian does not exceed 100 to 150 m. Thus this marginal cratonic zone underwent continuous stable subsidence for tens of millions of years.

3.1.9 Australian Craton

In the Early Silurian the area of the craton increased somewhat due to completion of development of the Adelaide orogen, which became a constituent of the craton. The territory of the craton continued to heave and only in the west did subsidence lead to transgression and formation of the small Carnarvon basin. Terrigenous rocks were deposited here under marine and continental conditions. Their thickness is not much (50 to 100 m) but rises to 1.8 km in a small graben, evidently a rift.

In the Late Silurian the Carnarvon basin turned shallow and salinised. Apart from dolomites, sandstones and clay deposits, evaporites (gypsum) too were deposited in it. In this same epoch subsidence soon renewed in the Amadeus aulacogen (Fig. 3-3). Aeolian and alluvial sands of the Merini formation were deposited to over 400 m in thickness in a large interior basin. The bulk of these sands came from the north where an elevated topography was formed and Lower Palaeozoic formations experienced weak folding.

In the Early Devonian the territory of this continental basin was drawn into an uplift and also experienced small dislocations. Lower Devonian formations are not known for this craton and its entire territory was evidently dry land.

3.1.10 East Antarctic Craton

Silurian formations have not yet been identified in the Antarctic region and predominance of uplift has been assumed for the craton. By the beginning of the Devonian, the orogenic process (Rossorogeny) had concluded in the western Antarctic mobile system and these territories became constituents of the craton. In the Early Devonian extensive subsidence and transgression manifested here. Lower Devonian marine sandstones lie on all the much older rocks with angular unconformity and their thickness reaches 400 to 600 m.

3.2 REGIONAL REVIEW. MOBILE BELTS

3.2.1 North Atlantic Belt

3.2.1.1 *Appalachian system*. Early Silurian. Palaeotectonic conditions differed in various segments of the system. In the Eastern

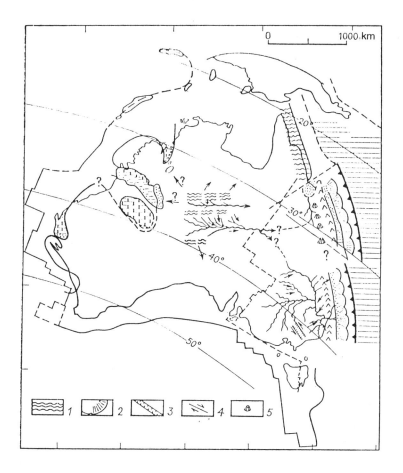

Fig. 3-3. Palaeogeographic reconstruction of Australia in the Silurian-Devonian (420–380 m.y.) and for eastern Australia in the Early Devonian (400 m.y.) (from Veevers, 1986).

1—deformations in the preceding stage; 2—zones of flysch accumulation; 3—grabens; 4—displacements; 5—volcanoes. Rest of legend the same as in Fig. 2–5.

Greenland system development of uplift, folding and thrusting and metamorphic processes continued under the conditions of an orogenic regime (Henriksen, 1978). In the Northern Appalachians and on Newfoundland Island zones of rapid uplift with manifestation of folding, metamorphism and granitisation (the last on a small scale) adjoined deep troughs in which submarine eruptions of mafic and intermediate lavas and also accumulation of tuffaceous and terrigenous formations up to 3 km or more in total thickness occurred (Osberg, 1978; Williams, 1979). Subaerial eruptions also occurred in the Avalonian zone. On the whole, volcanism in the Northern Appalachians corresponded to the calc-alkaline, island-arc type.

Much of the territory of the Southen Appalachians was covered by uplift, metamorphism, folding and granitisation in the Early Silurian. These events are associated with the rapprochement of the Florida Platform block with the North American Craton and piling up of Early Palaeozoic formations of the Southern Appalachians (Thomas, 1977). Along the north-western boundary of this segment extended a trough in which thick beds of coarse clastic deposits formed in marine and continental facies.

In the Ouachita-Mexican segment deepwater conditions and thin sedimentation predominantly containing silt-clay material continued to prevail. In the western part of this segment (within Mexico), micritic limestones also accumulated (De Cserna et al., 1977). The thickness of the Lower Silurian runs into tens of or a few hundred metres.

L a t e S i l u r i a n. In Eastern Greenland uplift and nappe movements occurring under conditions of an orogenic regime attained maximum in the Ludlovian epoch. Uplift and orogenic processes prevailed from the Late Silurian in Newfoundland Island. Subaerial and subaqueous volcanism of intermediate and felsic composition manifested here and the thickness of volcanics exceeds 2 km. In the Northern Appalachians narrow islands were divided by troughs in which an extremely complex set of formations was deposited. These comprised greywackes, turbidites, reef limestones or volcanic formations of the island-arc type from mafic to felsic in submarine and subaerial facies. The thickness of these formations varies widely (up to 2.5 km). Processes of folding and metamorphism continued in island arcs. In its general tectonic activity, the Late Silurian of the Northern Appalachians represents a transitional stage to an orogenic regime.

The preceding conditions prevailed in the Southern Appalachians during this epoch too: uplift predominated, topography formed and metamorphism was manifest. Coarse clastic formations and reef limestones accumulated in a small residual marine trough adjoining the Michigan cratonic basin.

No changes whatsoever occurred yet in the development of the Ouachita-Mexican segment of the Appalachian system. A deepwater zone with uncompensated subsidence containing argillaceous and, less frequently, argillaceous-carbonate sediments of small thickness is preserved here. The Central American massif adjoining the above structure from the west was involved in tectonothermal reactivation processes (Weyl, 1980).

E a r l y D e v o n i a n. Uplift, formation of complex fold-thrust structure and high thermal activity manifested almost all along the mobile system under consideration from north-eastern Greenland to Central America. Very thick (2 to 4 km) carbonate-terrigenous and submarine-volcanic strata still formed in narrow troughs of the Northern Appalachians. However, in general, here too land areas with a very rugged topography and high activity of folding and metamorphic processes predominated. In island arcs sub-aerial eruptions of intermediate and felsic composition occurred and granitoids intruded. On Newfoundland Island and in New England molasse accumulated to a thickness of up to 2 to 3 km. Molasse is also known in the Southern Appalachians but a marine trough with terrigenous sedimentation is preserved here (Mesolella, 1978; Thomas, 1977).

The Ouachita-Mexican segment was a unique constituent of the system in which no major changes occurred and argillaceous and siliceous shales up to 100 to 120 m in thickness were deposited under deepwater conditions. A branch of this structure located in the Early Palaeozoic and Silurian in north-western Mexico evidently concluded the cycle of subsidence and became a constituent of the Central American massif.

Thus in the tectonic evolution of the Appalachian system, formation of fold structure, metamorphism and granitisation at places were the important features throughout the Silurian and Early Devonian. In the Early Devonian, manifestation of island arc volcanism proceeding from the Middle Ordovician ceased. Marine sedimentation and fairly thick residual submarine volcanism were confined only to narrow zones of subsidence of the Northern Appalachians. As·before, an exceptionally protracted deepwater environment with thin sedimentation characterised the Ouachita-Mexican part of the system.

3.2.1.2 *British-Scandinavian system.*

E a r l y S i l u r i a n. The palaeotectonic environment acquired an even more contrasting nature than in the Ordovician. In northern Ireland and Scotland intense uplift continued and folding and metamorphic processes were manifest in this island area (Bluck, 1985). In the axial zone of the system, on the contrary, deepwater conditions remained, as before, and graptolite silts accumulated. Very thick beds of turbidites (up to 8 km in Ireland) were formed on the northern slope of the basin. On the southern slope, adjoining the English Midland massif, a very thick terrigenous complex (over 3 km) including turbidites was deposited (Bassett, 1984).

The Scandinavian segment of the system experienced brief subsidence and transgression in the Llandoverian stage. The marine basin extended into the south-western part of this segment where coarse sandy and conglomerate strata were formed. A small manifestation of andesite volcanism is known here. In the Wenlockian stage subsidence was succeeded by uplift, its intensity growing over time. In the north-eastern part (eastern Finnmarksvidda) a regime of uplift, folding and thrusting and possibly granitoid intrusive magmatism has been preserved. Similar conditions existed in Spitsbergen also.

L a t e S i l u r i a n. Uplift became even more widespread and land massifs now predominated within the system. The upheaval of these massifs was accompanied by manifestation of fold-thrust and metamorphic processes as well as the intrusion of granites (Bluck, 1985; Oftedahl, 1980;

Sturt et al., 1984). All this suggests development here of an orogenic regime which encompassed the Eastern England (Midland) massif and adjoining north-western part of the East European Craton, in addition to the British and Scandinavian segments. Spitsbergen, where granite plutons were formed at the end of the epoch, also undoubtedly became a constituent of the orogenic belt. In the British part of the system, however, conditions of intense subsidence were still preserved in very narrow zones while thick strata of marine and continental coarse clastic molasse (up to 3 km in northern England) accumulated in the Southern Scotland as well as Wales residual troughs (Leggett, 1980). By the end of this epoch, these troughs had become filled, experienced uplift and ceased to exist.

E a r l y   D e v o n i a n. The development of fold-thrust and metamorphic processes and also granitisation continued in this epoch and uplift intensified sharply, leading to the appearance of a rugged topography. Thick complexes of molasse formed in intermontane basins. Thus the arched uplift in Scotland and northern Ireland was complicated by the Caledonia basin which represented a deep (thickness of Lower Devonian, about 6 km) rift trough. In Scotland, the Midland Valley graben was formed with sedimentary strata reaching 3 km in thickness. In the basins of the Scandinavian segment too continental red conglomerates and sandstones up to 2 km thick accumulated, their thickness in Spitsbergen reaching 5 km. It is in this epoch that the main deformation of Lower Palaeozoic deposits of Scandinavia occurred but commenced slightly earlier, i.e., in the Late Silurian, in the interior zones.

Thus in the tectonic development of the British-Scandinavian system a sharp change occurred and transition to an orogenic regime was achieved by the Late Silurian, which involved small portions of the East European Craton. In the Late Silurian-Early Devonian this was a vast region (from Spitsbergen to southern England) of manifestation of highly contrasting tectonic movements, folding, thrusting, regional metamorphism and granitisation.

3.2.1.3 *Innuitian system.* E a r l y   S i l u r i a n. Filling of the deepwater basin with turbidites continued. This process, begun in the Late Ordovician in the east, gradually migrated westward as the trough filled (Surlyk and Hurst, 1983; Trettin and Balkwill, 1979). Turbidites were deposited in this epoch even in the central sector of the belt but their thickness is comparatively small at 0.5 to 1 km. The material came from an island arc located northward and from the eastern Greenland Caledonides but not from the North American Craton with its shelf basin in the marginal zone; here carbonates were deposited almost exclusively. Within the island arc, however, coarse terrigenous and tuffaceous strata were formed and eruptions of felsic lavas occurred. In the western part of the belt deepwater conditions continued to prevail with pelagic clay deposits.

L a t e S i l u r i a n. Filling of the eastern sector of the former deep-water basin concluded and the composition of formations changed there: turbidites were replaced by arhythmic sandy-clay deposits containing bands of conglomerates. The Eastern Greenland Caledonides that had undergone the orogenic stage were the source of material. As before, the central sector was filled with turbidites to variable thickness (500 to 2600 m), a deepwater environment was preserved in the west and pelagic silts including lime muds accumulated.

In the northern island arc the area of islands enlarged and volcanic eruptions ceased. Formation of shallow-water sand deposits and limestones in reef facies up to 1 km in thickness predominated here.

E a r l y D e v o n i a n. A large uplift covered the eastern part of the belt including northern Greenland and northern Ellesmere Island. Complex fold and thrust dislocations occurred in this region and the belt represented elevated land. Intense sedimentation shifted to the western sector of the belt where sandy-clay deposits (including turbidites) were formed at the site of a former deepwater basin under unstable tectonic conditions.

Early Devonian spilite-siliceous associations, including terrigenous rocks up to 3 km in thickness, are known in the northern part of Axel Heiberg Island. This suggests the probable manifestation here of destruction of the continental crust. However, the magnitude of this process was not significant.

Thus the Silurian-Early Devonian stage of development of the Innuitian belt was crucial, much of the deepwater basin was filled with turbidites, development of the island arc concluded and orogenic processes, i.e., uplift and mighty tectonic deformation, encompassed the eastern part of the belt (including the island arc).

3.2.2 Mediterranean Belt

3.2.2.1 *West European region.* E a r l y S i l u r i a n. The overall tectonic environment stabilised. Intense uplift and erosion of the French massif Central, evidently constituting a single block with the Bohemian and Armorican massifs, continued. Intrusion of granitoid massifs occurred here and the general tectonic regime was close to orogenic. The area of the marginal shelf basins of this large massif increased due to the inclusion of former zones of continental slopes and deepwater zones filled with sedimentary prisms. In the Early Silurian shallow-water carbonate-terrigenous deposits 200 to 300 m thick formed in these basins. Eruptions of lavas and felsic tuffs, partly subaerial, occurred in the northern Normandy, Vendee and Limousin plateaus. In deepwater troughs whose relicts have been preserved to the north and east of this elevated massif, the Lower Silurian is represented by usually condensed sequences of pelagic clay and carbonate-clay facies (Belov, 1981; Bonchev, 1985; Cwojdzinski, 1979). In such troughs submarine eruptions of andesites and basalts continued to a thickness of 2 to 3 km east of the Pannonian basin and west of the Moesian block.

A new deepwater trough (rift ?) arose evidently as a result of extension to the north of the massif Central and extended to southern England (Autran, 1980; Behr et al., 1984 and 1980). Formation of this trough in Brabant and Ardennes was accompanied by submarine eruptions of tholeiite-basalts and andesite-basalt lavas.

In the Cantabrian zone of the Iberian peninsula and in the eastern West Asturia zone, argillaceous deposits accumulated under shelf conditions. In all the other zones of this region clay silts of small thickness were formed under deepwater conditions (Nedjari, 1981). Moreover, submarine mafic eruptions and chert accumulation continued in the Central Iberian zone.

In the Caucasian system the northern shelf zone is characterised by subsidence and terrigenous sedimentation while pelagic argillaceous rocks were probably deposited more to the south.

Late Silurian. The area of the central part of the island joining the Armorican, Central and Bohemian massifs was enlarged even more by the inclusion of regions of southern France. A high intensity of uplift was maintained, granites intruded and tectonic deformation developed, i.e., the tectonic regime was close to orogenic. Subaerial rhyolite volcanism manifested in the south-eastern part of the Bohemian massif. As before, this massif was surrounded from the north, east and south by deep troughs with argillaceous and sometimes siliceous sedimentation (Saxothuringian, Barrandian and Southern Alps) of small thickness. Submarine basalt eruptions occurred in the Barrandian trough and the southern Carpathian regions. The deep Brabant trough (southern Belgium) was probably almost wholly filled with sediments and experienced compression by the Late Silurian. Accumulation of shallow-water carbonate-terrigenous deposits was interrupted here by eruptions of a bimodal volcanic rhyolite-basalt complex.

Deepwater conditions prevailed in the Iberian peninsula and mainly graptolite shales of small thickness were deposited. Islands existed in the Cantabrian zone, the only zone in which shallow-water shelf facies were extensively developed, and in the eastern Central Iberian zone. Elsewhere in the latter zone volcanic eruptions of basalts continued from the Early Silurian and rhyolites also erupted at the end of the Silurian.

In the northern Caucasian system, uplift, manifestation of subaerial andesite volcanism and shelf terrigenous deposition have been assumed. More southward, deepwater basin conditions probably persisted.

Early Devonian. Marginal parts of the ancient massifs were again involved in subsidence and transgression and mainly shallow-water terrigenous deposits accumulated there. In the south-western massif Central and in the southern Armorican massif, carbonate-terrigenous strata were deposited. Volcanism practically ceased almost everywhere but intrusion of granites continued in the massif Central. On the northern slope of the latter and also in the regions of southern England and Ireland, thick terrigenous complexes (up to 3 to 5 km) comprising material coming from the north, i.e., from the British-Scandinavian orogen, were deposited on the continental slope. Deepwater troughs were preserved in the Early Devonian only along the boundaries of the belt with the East European Craton. It is possible to trace the palaeotectonic relations between northern and central European structures from this epoch but these have been separated from southern Europe by tectonic sutures coinciding with the Alpine mobile belt.

Terrigenous and carbonate-terrigenous formations 300 to 400 m in thickness accumulated in the Cantabrian zone of the Iberian peninsula and in the eastern West Asturia zone under shelf conditions. Uplift of a small island continued in the eastern Central Iberian zone and deepwater conditions probably prevailed elsewhere in the peninsula where the Lower Devonian is represented by pelagic shales and carbonates.

Within the Caucasian system, the Lower Devonian was made up of clay shales of 1.2 km and more in thickness. Their accumulation is thought to have occurred in a deep basin which extended south-east and farther, deep into Central Asia.

Thus in the Silurian-Early Devonian stage of development of the West European region, fusion and uplift of ancient massifs and also the disappearance of deep basins occurred in the Silurian. However, in the Early Devonian these tendencies underwent change and subsidence enlarged.

3.2.2.2 *Kuen Lun-Indochina region.* E a r l y   S i l u r i a n. The central axial uplift, representing a major island massif separated the Qilianshan system into two zones: northern and southern. In the northern zone intense filling of the deep trough with coarse terrigenous strata to a thickness exceeding 2 km proceeded. In the south-eastern part of this zone, calc-alkaline andesite-basalt volcanism, clearly of the island-arc type, took place.

The southern zone also comprised a marine residual trough but its filling with coarse clastic deposits brought in from the central cordillera had already ceased. The thickness here is not great (about 400 m) and deposition occurred under shallow-water conditions. In Kuen Lun and Qinling accumulation of clayey and lime-clay silts continued under deepwater conditions but flysch formed in the western part of the Qinling system adjoining the Qaidam block.

Within the southern branches of the Palaeo-Tethys running from China into Laos and then on one side into Vietnam, on the other into Thailand and Malaya and constituting the Yunnan-Malaya system, relatively deepwater conditions with accumulation of terrigenous-carbonate deposits and clay deposits more to the south continued to prevail. In northern Yunnan was an island with an aureole of shallow-water carbonates (*Atlas of the*

Palaeogeography of China, 1985). In the ancient massifs, i.e., Shan-Thai and Indosinian microcontinents, fine clastic pelagic deposits accumulated, reaching hundreds of metres in thickness, with bands of black graptolite shales and shallow-water limestones (Gatinsky, 1986). A deep trough with thin clay deposits lay east of the Shan-Thai massif and judging from the manifestation of andesite volcanism, development of a volcanic island arc commenced even farther eastward. Mighty andesite volcanism is also known north of the Indosinian massif. Here the preceding deepwater conditions were probably replaced by an island arc environment.

L a t e   S i l u r i a n. Uplift gradually extended to almost all of the territory of the Qilianshan system. Along its south-western boundary with the Qaidam ancient block, volcanism of complex basalt-andesite-rhyolite composition was manifest. In residual troughs of the system, continental molasse and sometimes marine conglomerate-sand beds up to 2 km in thickness were formed. Angular unconformity at the base of the Upper Silurian suggests folding. An orogenic regime was established here from this epoch.

At the same time, Kuen Lun held the very same environment as in the Early Silurian but a burst of submarine basaltic volcanism occurred in the eastern Kuen Lun system. Pelagic sedimentation was substituted by flysch in the Qinling system.

Within the Yunnan-Malaya system, the Indosinian massif evidently experienced uplift while clay and carbonate formations reaching a few hundred metres in thickness were deposited in the Shan-Thai massif as before. Much thicker terrigenous deposits were formed, possibly under continental slope conditions, north of the Indosinian massif (up to 2.5 km) and east of Shan-Thai massif (over 1 km) where carbonate reef complexes likewise developed. East of the Shan-Thai massif a deepwater trough with graptolite silts was preserved in the Late Silurian and an island arc with andesite volcanism continued to develop farther eastward.

E a r l y   D e v o n i a n. Lower Devonian formations are not known in the Qilianshan system (*Atlas of the Palaeogeography of China,* 1985). An orogenic regime was evidently preserved here and folding, metamorphism and possibly granitisation were manifest. Conditions in the Qinling system changed with the formation of carbonate deposits in shallow-water facies. But in the Kuen Lun system, deepwater conditions were still preserved.

In the south-eastern branch of the Palaeo-Tethys (Yunnan-Malaya system), the Indosinian microcontinent was again submerged and shallow-water carbonate and terrigenous deposits formed here as well as on the Shan-Thai microcontinent. Deep troughs with pelagic clay deposits existed within the frame of these massifs. Submarine eruptions of basalts and deposition of siliceous rocks occurred in the regions of North Vietnam and eastern Malaysia, evidently on an oceanic crust. Conditions east of Sumatra Island can be reconstructed from this epoch: Lower Devonian carbonate and terrigenous rocks, as well as flysch more to the west, are known here.

Thus the most important event in this stage of evolution of the eastern part of the Palaeo-Tethys was transition to an orogenic regime from the Late Silurian in Qilianshan and shoaling in the Qinling region, while no special change occurred in Kuen Lun and within the south-eastern and southern branches of the Mediterranean belt. Some deepening of basins in the Early Devonian may have occurred in the latter regions.

3.2.3 Ural-Okhotsk Belt

3.2.3.1 *Uralian system.* E a r l y   S i l u r i a n. The magnitude of submarine volcanism reached maximum. Its most active manifestation occurred in the Tagil and Magnitogorsk zones. Basalts, andesite-basalts and also andesite tuffs, basalts and some contrasting basalt-rhyolite series predominated in the volcanic products. Their thickness reaches 1.5 to 2 km. Jasperoids were extensively deposited. In the eastern part of the system the thickness of volcanics decreased and accumulation of siliceous, lime-clay and clay-graptolite shales acquired great significance. Accumulation of the latter occurred under deepwater conditions and their thickness reached only 10 to 100 m. Compared to the Ordovician, the structure of the system was complicated and islands and reefs appeared. The history of development of the easternmost regions of the Uralian system (Transural zone) can be reconstructed from the beginning of the. Silurian. Siliceous and carbonaceous-siliceous muds and silts accumulated here and submarine eruptions of lavas and mafic tuffs occurred, evidently under deepwater conditions.

L a t e   S i l u r i a n. Throughout the extent of the eastern zones of the system, intense submarine volcanic activity continued and fairly thick beds of lavas and tuffs of basalts, andesite basalts, andesites and andesite-dacites were formed. Accumulation of clay, carbonaceous clay, siliceous rocks and jasperoids played a very subordinate role. In some regions (e.g., in western Magnitogorsk zone) only deepwater lutites were deposited and their condensed sequences are about 100 m in thickness while that of the Upper Silurian volcanics in the adjacent zones reaches 1 to 1.5 km. The deepwater trough of Kara zone (Pay-Khoy) was fringed from the Late Silurian by reef structures whose remnants are known in the regions of Sukhoy Nos Cape on Vaigach island and Pyrkov Nos Cape in western Pay-Khoy. Replacement of the persistent Upper Ordovician-Lower Silurian spilite-diabasic series by the differentiated Upper Silurian consistent andesite-dacite-quartz-albitophyre or contrasting diabase-quartz-albitophyre series with a significant role of explosions, points to the formation of volcanic island arcs in the Silurian.

E a r l y   D e v o n i a n. The overall conditions were better differentiated. Reef limestones 300 to 600 m thick (e.g., in Yemanzhelin-Polyakov,

Alapayevsk and other regions) were deposited on extinct submarine volcanoes. Coarse clastic terrigenous rocks including tuff-sand and tuff-conglomerates up to 1 km in thickness accumulated here. Volcanism continued in other island arcs but andesite-dacites and their tuffs and often highly alkaline rocks, i.e., trachyandesites, predominated in volcanic products. Deepwater conditions with uncompensated sedimentation were preserved in the Sakmara-Lemva zone and tholeiitie-basalts erupted in the Sakmara region. The thickness of the Lower Devonian phtanites, carbonaceous-clay-siliceous shales and reticulate limestones is only about 30 m here. But in the Belaya-Yelets zone adjoining from the west, shallow-water reef limestones up to 600 m in thickness were formed.

Thus the intensity of volcanic manifestation in the Uralian system was maximum in the Silurian-Early Devonian and the composition of volcanic products changed successively. Major submarine fissure lava flows predominated in the Early Silurian and the role of volcanoes of the central type and explosive activity intensified in the Late Silurian epoch with progressive formation of volcanic island arcs. Also, the volcanic products became more felsic. Alkaline lavas and tuffs were manifest in the Early Devonian, the importance of dacites increased and many arcs entered a mature stage of their development. Thus the extension conditions which had predominated at the beginning of this stage changed to compression conditions. This is reflected in particular in the composition of terrigenous sedimentary rocks. Coarse clastic facies began to play a major role among them.

3.2.3.2 *Kazakhstan-Tien Shan region.* E a r l y S i l u r i a n. Almost the whole of this region was affected by powerful uplift from the beginning of the Silurian and marine troughs were preserved only in the Peri-Balkhash region. A rugged topography was formed in the central and southern parts of Kazakhstan and Northern Tien Shan while the products of its erosion accumulated in the intermontane basins. The thickness of the coarse clastic (mainly continental) formations in them reached 5 km but continental conditions of deposition alternated in some basins with marine conditions. Volcanism was manifest on a limited scale but developed over time and was confined to the north-western Peri-Balkhash, Chingiz range and in the south to Ketmen range. Lavas and tuffs of andesites and rhyolites predominated in the composition of subaerial eruption products. In the Kokchetav-Northern Tien Shan zone folding occurred in this epoch and granite-granodiorite massifs intruded. All the available data suggests the existence of orogenic regime conditions within the Kazakhstan-Tien Shan range from commencement of the Silurian.

L a t e S i l u r i a n. The area of residual marine troughs decreased and its portions characterised by hilly relief enlarged. In the relict marine trough of north-western Peri-Balkhash region and the Ketmen range, reef limestone massifs and littoral conglomerate-sandstone strata up to 1.5 km in thickness were formed. In the Chingiz-Tarbagatai zone subaerial eruptions of andesites and rhyolites occurred and molasse accumulated. Folding too was manifest and a granite-granodiorite complex intruded here.

E a r l y   D e v o n i a n. The main event was commencement of formation of a continental marginal volcanic belt fringing the Junggar-Balkhash basin from the north-west. Eruptions of intermediate-mafic and intermediate composition are characteristic of the initial stage of development of the belt. Volcanism was manifest most completely and powerfully in its western part. Sedimentation occurred simultaneously in some regions but was sporadic. It was concentrated in a few small basins disposed at the rear of the volcanic belt. Lower Devonian continental formations filling these basins represented a typical red-coloured molasse in whose composition volcanic rocks took part. The total thickness of the Lower Devonian reaches 1.5 to 2 km.

Thus from the commencement of the Silurian, conditions of orogenic regime prevailed in the territory of Kazakhstan-Tien Shan, i.e., uplift, tectonic deformation, and effusive and intrusive crustal magmatism. Over time residual marine troughs decreased in area, the magnitude of volcanism enlarged and the topography became complicated.

3.2.3.3 *Altay-Northern Mongolia region.* E a r l y   S i l u r i a n. Orogenic regime conditions were maintained over an extensive expanse of this region. Further, its area increased in the south-west where subsidence was replaced by uplift and land with a rugged topography predominated. Residual troughs, however, still existed here and coarse continental molasse exceeding 3 km in thickness formed in them. This molasse contained andesites at places as also lagoonal carbonate-terrigenous beds (Dergunov et al., 1980; Dodin, 1979). The intensity of intrusive magmatism decreased perceptibly. In the north-western Mongolia region the activity of manifestation of an orogenic regime was much less than in Altay proper.

L a t e   S i l u r i a n. Almost all the residual marine basins in the western part of the region were filled with sediments. Formation of a rugged topography continued here and molasse exceeding 2 km in thickness was deposited in the intermontane basins. Granodiorites intruded in the Kuznetsk-Kobda zone and in its south-eastern continuation in Mongolia (Gordienko, 1987; Leont'ev et al., 1981; Tsukernik et al., 1986). Fold and fault deformation also developed. The relief smoothened northward towards the boundary with the Siberian Craton and transition to a platform regime was gradual.

E a r l y   D e v o n i a n. In the west, in the Altay-Sayan region, a rugged topography was formed and large masses of coarse clastic continental, at places marine, formations (molasse) accumulated in the intermontane basins. Their thickness reaches 2 to 3 km. Extensive volcanic activity renewed from the commencement of the Devonian. Subaerial eruptions, mainly of intermediate but also felsic and mafic alkaline-basalt composition occurred. The thickness of the lavas and tuffs was 1 to 2 km.

From the Early Devonian deep subsidence and manifestation of submarine volcanism commenced as a result of rifting in the Mongolia-Amur zone and in its eastern continuation in the Amur-Okhotsk zone (Kuz'min and Filippova, 1979). Basalts erupted here over large areas and siliceous and greywacke beds of 4 to 5 km in total thickness were formed. The Kerulen and Bureya massifs preserved their status as islands and coarse clastic terrigenous rocks and, less frequently, reef carbonates were deposited in their margins under shelf conditions.

Thus in the tectonic development of the Altay-Northern Mongolia region, orogenic regime conditions prevailed, being most distinctly manifest in its Altay-Sayan part. In the Northern Mongolia-Amur part the orogenic regime was very subdued in the Silurian and was substituted in the Early Devonian by rifting, crust destruction and very strong subsidence.

3.2.3.4 *Ob'-Zaisan and Junggar-Balkhash systems.* E a r l y   S i l u r i a n. Development of the Ob'-Zaisan system and a new stage of development of the Junggar-Balkhash system commenced during this period. Submarine volcanism of mafic and, less frequently, intermediate composition occurred in both regions and clay and siliceous deepwater formations as well as sandy-clay strata were formed. The thickness of the Lower Silurian was 3 to 4 km in the Ob'-Zaisan and 5 to 6 km in the Junggar-Balkhash system. In the marginal zone bordering the Kazakhstan orogen, under continental slope and rise conditions, a thick (up to 5 km) terrigenous complex, including turbidite members, accumulated. The north-western continuation of the Ob'-Zaisan system has been traced from drilling and geophysical data up to the lower course of the Ob' River.

L a t e   S i l u r i a n. In the Ob'-Zaisan system deepwater conditions were preserved, submarine basalt eruptions continued and deep uncompensated troughs with accumulation of clay and siliceous deposits prevailed. Unlike this situation, in the Junggar-Balkhash system palaeotectonic conditions underwent change. Volcanic eruptions ceased and the deep trough was filled with thick terrigenous strata including turbidites. The clastics sources were the uplands of the Kazakhstan orogen encircling this trough. In the east deepwater conditions still prevailed in the Transaltay zone but extremely thick submarine lava eruptions of mafic and intermediate composition commenced. Their thickness went up to 3 to 4 km (*Atlas of the Palaeogeography of China,* 1985).

E a r l y   D e v o n i a n. The magnitude of submarine volcanism decreased in the Ob'-Zaisan system. It was manifest only in its south-eastern part with andesites predominating in the volcanic products. Further, terrigenous deposits were formed and the thickness of the Lower Devonian ranged from 1 to 2 km. In the northern part of this system, in the zone adjoining the Altay orogen, reef carbonate complexes were extensively developed. In the Junggar-Balkhash system filling of basins with coarse deposits brought in from the Kazakhstan orogen proceeded as before. The thickness of the Lower Devonian here went up to 6 km or more while continental facies were somewhat developed. Subsidence was replaced at many places by uplift and islands formed consequently. Volcanic eruptions ceased in the Transaltay zone.

So the Junggar-Balkhash system underwent evolution during the Silurian and Early Devonian from conditions of extension and deep trough to compression, filling of basins and partial uplift. In the Ob'-Zaisan system, on the other hand, extension conditions (spreading) prevailed throughout this stage and submarine volcanism and deepwater sedimentation were manifest.

3.2.3.5 *Gobi-Hinggan system.* E a r l y   S i l u r i a n. This system, extending directly into the eastern Ob'-Zaisan system, fringed the Altay-Northern Mongolia orogen from the south. It developed in the Early Silurian in the same way as the Ob'-Zaisan system and was characterised by extension, submarine eruptions of mafic and intermediate composition and accumulation of clay, siliceous and also sand beds up to 1.5 to 2 km in thickness. The strongest volcanism was manifest in the region between the Kerulen and Bureya massifs, which preserved their elevated position as major islands. In the marginal northern zone adjoining the Northern Mongolia orogen, flysch (up to 2 km) and reef limestones were formed.

L a t e   S i l u r i a n. Volcanic eruptions continued only in the western part of the system; they ceased in the eastern part where the extension conditions evidently changed into compressive ones (*Atlas of the Palaeogeography of China,* 1985) with formation of island arcs. This is suggested by the shallow-water character of the Upper Silurian formations in south-eastern Mongolia and minor manifestation of andesite-rhyolite volcanism. In the rest of the Gobi-Hinggan system formations of this epoch were represented by shelf reef complexes and terrigenous beds whose origin is often not very clear. From the end of the Silurian, formation of the Southern Mongolia sea commenced in the south and separated the Southern Gobi microcontinent from the Caledonian continent.

E a r l y   D e v o n i a n. Compressive conditions and formation of island arcs extended into the area between the Kerulen and Bureya massifs. Eruptions of intermediate and felsic composition occurred there and sandy-clay and tuffaceous formations accumulated. The mighty andesite-basalt volcanism of Southern Mongolia evidently also manifested under compressive conditions. These processes obviously combined with extension of the Mongolia-Amur system, which lay 500 to 600 km to the north. The two systems joined in the east and palaeotectonic conditions in the Amur-Okhotsk zone corresponded to the extension conditions of the Mongolia-Amur system. Here, submarine mafic (tholeiitic) volcanism was very powerful and the general thickness of the Lower Devonian reached 4 km or more.

Thus from the Early to Late Silurian, especially in the Early Devonian, a change in character of tectonic development took place in the Gobi-Hinggan system. Extension changed into compression and formation of volcanic island arcs. This process was combined with extension occurring more northward in the Mongolia-Amur system and manifested in the form of spreading in the Southern Mongolia system (Ruzhentsev et al., 1991).

3.2.4 West Pacific Belt

3.2.4.1 *Verkhoyansk-Kolyma region.* E a r l y   S i l u r i a n. Formation of thick beds of carbonate and clay rocks inherited from the Ordovician continued uninterruptedly. Deposition of carbonates occurred in the shelf basins under shallow-water conditions and the thickness of calcareous rocks here reaches 1.5 to 2 km. In some zones the Lower Silurian is represented predominantly by shales of considerable thickness (up to 1.5 km). Their accumulation occurred in deep basins with subsidence uncompensated at the beginning of the epoch. The general location of these troughs and the nature of their boundaries with adjoining structures are not yet clearly known.

L a t e   S i l u r i a n. The palaeotectonic environment underwent almost no change and deposition of carbonates and shales continued to considerable thickness (up to 1.3 km), with limestones predominating. Evidently filling of the Early Silurian deepwater troughs occurred initially. Later, they shoaled and carbonate deposition gradually extended into them.

E a r l y   D e v o n i a n. Perceptible reactivation of tectonic movements occurred in this epoch over an extensive area, from Verkhoyansk region to the Omolon River basin. Subsidence of priorly uplifted structures caused major transgression and formation of new deep troughs with terrigenous and carbonate-terrigenous deposition. As a result of this reactivation, the territory of the Verkhoyansk-Kolyma mobile zone markedly enlarged towards the more western region, which had earlier been part of the platform. In most sedimentary basins formation of carbonate, often reef formations almost wholly replaced accumulation of terrigenous rocks. But the thickness of limestones is comparatively low and does not exceed 1 km.

3.2.4.2 *Koryak system.* The oldest faunistically proven deposits belong to the Ordovician in Koryakia. Finds are also known of Silurian and Devonian formations but as they are very rare hardly help in reliably judging tectonic development of the region. Available information suggests that submarine-volcanic complexes of mafic and intermediate composition and sandstones and shales were formed here in the E a r l y S i l u r i a n in island arc and marginal sea conditions. Similar conditions could be assumed in the L a t e S i l u r i a n too but in the E a r l y D e v o n i a n volcanism possibly ceased and the Lower Devonian terrigenous formations in the western part of the system were most probably deposited in the continental slope zone.

3.2.4.3 *Nippon-Sakhalin system.* E a r l y S i l u r i a n. The Primorie, Sakhalin and Japanese islands fall in this system whose tectonic evolution has been studied from the Silurian-Early Devonian stage since much older formations are almost unknown. The western shelf zone with early (Precambrian) continental crust has been reconstructed for the Early Silurian within Japan. Here a shallow-water carbonate-terrigenous complex was formed and small subaerial eruptions of rhyolites occurred. A deepwater zone falls in the east in which pelagic silts and carbonates of small thickness have formed. In Primorie, south and east of the Khanka massif, submarine basaltic eruptions occurred and siliceous and terrigenous formations accumulated to a total thickness of not less than 2 km.

L a t e S i l u r i a n. In western Japan shallow-water conditions prevailed and reef limestones, sandstones and clays were deposited, reaching a few hundred metres in thickness. In this zone eruptions of lavas and tuffs of intermediate and felsic composition continued under subaerial and submarine conditions. The Upper Silurian of the eastern deepwater zone was represented by siliceous and clay shales and sandstones up to 200 m in thickness. As before, spilitic-siliceous deposits formed in the southern Primorie region. Further, the Khanka massif experienced uplift and probably represented part of an island (microcontinent).

E a r l y D e v o n i a n. In western Honshu Island shallow-water conditions prevailed and terrigenous formations accumulated. Felsic volcanism of this age is known in the northern part of the island. In eastern Japan manifestation of submarine volcanism of mafic composition is known from the Early Devonian onwards and siliceous and clay rocks were deposited here. It may be assumed that a similar palaeotectonic environment existed in this epoch in the central zones of Sakhalin and Hokkaido islands and in southern Primorie. In these regions submarine-volcanic eruptions (of mafic composition) occurred and siliceous formations and greywackes accumulated (Brodskaya et al., 1979). The elevated position of the Khanka microcontinent was preserved.

3.2.4.4 *Cathaysian system.* E a r l y S i l u r i a n. Almost the entire area of this system experienced rapid and powerful uplift and was transformed into land with a rugged topography (*Atlas of the Palaeogeography of China,* 1985). The south-eastern regions of the adjoining Yangtze Craton were involved in orogenic processes. Uplift was accompanied by folding of Late Precambrian and Early Palaeozoic formations and the development of nappes. Small residual marine troughs were preserved in the southern and northern parts of the system. Their subsidence was compensated by the deposition of terrigenous formations up to 2 km in thickness.

L a t e   S i l u r i a n. Uplift prevailed as before and the size of residual marine basins shrank. These basins were filled with coarse clastic formations in marine and continental facies. Their thickness reaches 1 km or more. Development of fold and fault deformations continued and intrusion of granitoids commenced.

E a r l y   D e v o n i a n. That the intensity of uplift increased is supported by extensive formation of molasse. The Cathaysian orogen was joined with the Qinling orogen in this epoch. Manifestation of granitoid magmatism, tectonic deformation, crumpling of sedimentary strata into a series of isoclinal folds and metamorphism enlarged.

Thus from commencement of the Silurian, the Cathaysian system experienced conditions of an orogenic regime, which were preserved throughout the stage under consideration. The scale of manifestation of orogenic processes grew over time and, during commencement of the Devonian, the marginal part of the Yangtze Craton underwent reactivation. But the tectonic style of development of the system later changed sharply and orogenic conditions were replaced by platformal at the end of the Early Devonian and the former mobile system became a constituent of the Yangtze Craton.

3.2.4.5 *East Australian region.* E a r l y   S i l u r i a n. Benambran orogeny in the western part of the region at the end of the Ordovician to commencement of the Silurian caused powerful uplift and compression in the region of former Ballarat trough while its sedimentary filling underwent folding. With these processes are associated the formation of granite-gneiss bodies and metamorphism in the Wagga-Omeo zone, tectonic intrusions of ultramafites westward in Heathcot and Wellington zones and formation of anatectic granites at the end of the Early Silurian, after which the intensity of orogenic processes decreased. Consequently, a narrow orogenic zone continued to exist along the boundary of the Australian Craton. This orogenic zone was displaced eastward somewhat compared to the Late Ordovician (Murray and Kirkegaard, 1978; Plumb, 1979). Early Silurian manifestation of subaerial rhyolite volcanism and molasse are known in its southern part.

East of the orogenic zone, within the Lachlan system, subsidence and sedimentation continued uninterruptedly in the Melbourne trough set off from the open sea by the Caperty volcanic arc (Cas and Jones, 1979; Plumb, 1979), building up the Molong microcontinent. In the Melbourne trough, flysch exceeding 3 km in thickness was formed and could be traced to the north-eastern regions of Tasmania in the south. In the volcanic arc subaerial and sometimes submarine eruptions of andesites and basalts alternated with deposition of clastic rocks and reef limestones. In the northern continuation of the Melbourne trough, in Cobar basin, deposits of clay-greywacke predominated. The presence of biogenic limestones and conglomerates points to shallow-water conditions.

East of the Melbourne trough conditions in the Silurian were complex and varied. At the site of the Wagga marginal sea, the metamorphic and concomitantly plutonic Wagga-Omeo belt arose as a result of Benambran diastrophism at the beginning of the Silurian. This belt shifted westward later in the direction of the Melbourne trough along the Kieva suture. In the Middle-Late Silurian this belt represented dry land. A wide strip of complex structure and topography with volcanic arcs of the ensialic type representing products of destruction of the Molong microcontinent (Gunumbla, Molong and Caperty) extended eastward along the aforesaid land and divided the arcs into troughs on transitional (Hill End and Cowra) or oceanic type crust (Tumut, where ophiolites are known). On the whole, this region may be regarded as one of diffuse spreading (Scheibner, 1985). Troughs were filled with turbidites, initially quartzose, and later by increasingly litho- and volcanoclastic turbidites. Granitoids intruded into the uplift. All of this strip should have been encircled by a zone of subduction with a westward inclination.

Formation of a new, evidently ensimatic, Calliope volcanic arc is assumed in the New England system. At the back of this arc, between it and the volcanoplutonic Molong-Canberra belt (Caperty), a marginal sea should have existed. West in this sea flysch accumulated (up to 2 km) while deepwater conditions with pelagic argillaceous formations were still preserved towards the east. In the north-east Australian region, rapid subsidence of the eastern periphery of the Georgetown marginal massif of the craton along the Palmerville fault continued in the Early Silurian. Terrigenous, siliceous and submarine-volcanic formations 2 to 3 km in thickness accumulated here. The sublatitudinal Brock River rift graben, located south of the Georgetown massif, subsided as before. The presence of typical ophiolites of Late Ordovician age in this rift points to total break-up of the continental crust. More than 3 km of deposits, predominantly of the flysch type with subordinate limestones and mafic volcanics, accumulated here during the Early Silurian.

Late Silurian. As a result of the decrease in intensity of tectonic processes in the western Lachlan system, the orogenic regime here transited into a platform one but orogeny probably still continued northward in the Thomson system. Consequently, the dimension of the orogenic zone and the mobile zone as a whole decreased somewhat. In the eastern Lachlan system filling of the Melbourne trough with flysch (to about 3.5 km in thickness) was completed. Uplift of the Caperty island arc continued. Subaerial eruptions of rhyolites occurred within this island arc zone and coarse clastic and tuffaceous formations were deposited on the slopes (Gilligan and Scheibner, 1978; Vandenberg, 1978).

In the New England system in the west, flysch accumulation was replaced by formation of sandy shallow-water sediments of the shelf type (thickness about 2.5 km) and uplift commenced here. As before, a deepwater zone with pelagic deposits existed eastward. In the Queensland system, in the shelf adjoining the Georgetown massif and in the Brock River graben, subsidence continued and an extremely thick sandy-clay complex including turbidites accumulated. East of the shelf, the Hodgkinson trough was filled with mafic and intermediate submarine volcanics, greywackes and argillaceous deposits exceeding 3.5 km in total thickness. A palaeotectonic regime similar to that of Queensland has also been suggested during the Silurian on New Guinea Island, although data on development of this region is yet very meagre.

At the end of the Silurian, the East Australian zone experienced a new phase of compression and intensification of tectonic and plutonic activity, viz., Bowning. It manifested mainly within uplifts and island arcs in the form of fold and fault dislocations, in the formation of granite batholiths from the southern regions of the Lachlan system and Tasmania up to the Ravenswood block of the Queensland system, and in high-pressure metamorphism in the Wagga zone where Lower Palaeozoic sedimentary formations were crumpled even before into complex isoclinal folds.

E a r l y   D e v o n i a n. The orogenic zone separating the platform from the zone of powerful subsidence on the eastern mainland enlarged, mainly because platform territories were drawn again into uplift and involved in granitisation processes. By commencement of the epoch, epiplatformal orogeny had spread to the western part of Tasmania Island and regions adjoining the Queensland system. In the Lachlan system large areas were still occupied by residual marine troughs (Melbourne and Trandle). However, in the composition of their deposits, aside from marine terrigenous rocks, continental molasse played a steadily increasing role, reaching a thickness of 4 km. At several places east of the system, felsic subaerial volcanism took place and andesite, also subaerial, volcanism in the Drummond and Adavale basins. In the region of the Condobolin uplift, an island arc andesite-basalt volcanism occurred and reef limestones were formed. By the end of the Early Devonian, subsidence in the Cobar-Melbourne trough had ceased, the sea regressed from this zone and freshwater lake deposits had already formed in the eastern part of the former Cobar trough.

Thick sedimentation and submarine volcanism continued from that time within the New England system; further, these processes migrated eastward into zones where open ocean basin conditions could be reasonably assumed for the Devonian epoch. In its western shelf zone (forearc basin) accumulation of carbonate-terrigenous (including coarse clastic) deposits proceeded. Subaerial felsic and intermediate volcanism was manifest in the southern part of this zone, turbidites were formed and the thickness went up to 5 to 7 km. In the deeper water eastern zone, in the Yarrol-Tamworth and Wollomin troughs (continental slope, accretionary wedge), pelagic clay and siliceous muds, sometimes greywackes, were deposited and submarine basaltic volcanics were widespread. The thickness of the Lower Devonian was also high here (over 4 km). In the Queensland system intense sedimentation built up a prism of sedimentary and volcanic rocks on the edge of the continent. The Brock River graben continued to be filled with terrigenous, now coarse clastic deposits about 2.5 km in thickness. Volcanism ceased in the Hodgkinson trough and its filling with flysch and clay strata proceeded to a thickness of over 3 km.

Thus in the East Australian region longitudinal zoning was distinctly manifest during the Silurian-Early Devonian stage. It is seen in the permanent development of an orogenic regime along the boundary with the platform zone and, in the east, of a system of marginal seas and island arcs. The intensity of orogenic processes abated in the Late Silurian but was restored and prevailed over the Early Silurian scale in the Early Devonian. A characteristic feature is the continuation of processes of gradual attenuation of tectonic activity in the western part of this region, resulting in enlargement of the Australian Craton. These processes have already been recorded for the Cambrian and Ordovician stages. Contrarily, an intensification and differentiation of tectonic movements and magmatism took place in the eastern regions, representing probably the former ocean bottom.

3.2.4.6 *West Antarctic region*. Uplift of the Transantarctic mountain orogen and accumulation of molasse in the basins continued in the Silurian and Early Devonian. These were followed by a new epoch of intense tectogeny called the Shackleton orogeny. It was accompanied in the Robertson Bay terrain, in the region of Victoria Land, by intrusion of Admiralty I-type granites aged 390 to 360 m.y. and later by island arc volcanism. Granitoids of proximate age are known in Mary Byrd Land and as orthogneisses within the Antarctic peninsula, in Palmer Land and Graham Land. In Mary Byrd Land there are signs of a much earlier Late Ordovician regional metamorphism (450–444 m.y.).

3.2.5 East Pacific Belt

3.2.5.1 *Cordilleran system*. E a r l y S i l u r i a n. In the volcanic island arc regions of the Alexander archipelago and Klamath Mountains, manifestations of andesite-basalt (calc-alkaline) volcanism continued (Churkin and Eberlein, 1977). The thickness of volcanic and terrigenous formations (sometimes with reef carbonates) exceeds 3.5 km. Within the Purcell-Pelly arc region, volcanic eruptions ceased and sand and clay rocks accumulated on its western slope (including turbidites). In the deepwater zone traced from Alaska to California, thin clay and, less frequently, siliceous (in the southern sector of the belt) and carbonate pelagic complexes were formed. In the region of Brooks Range in Alaska carbonate-terrigenous formations of the shelf-type were deposited in the western part and there was probably an uplift in the form of an island in its eastern part.

Late Silurian. The general conditions of development of the belt did not change compared to the Early Silurian. In the zone of deepwater basins, extending from the Richardson trough and central Alaska in the north to California in the south, there were only thin deposits of clay and at places micrific limestone. The island zone enlarged in northern Alaska. Within the volcanic arc of the Alexander archipelago, eruptions of andesites were combined with accumulation of terrigenous rocks and reef carbonates. The total thickness of formations here exceeds 2 km. Eruptions of andesites and, less frequently, of basalts continued in the volcanic arc of the Klamath Mountains.

Early Devonian. Deepwater sedimentation conditions changed into shallow-water ones over large areas of the southern and central parts of the belt. Clay deposits were replaced here by shelf carbonate-terrigenous and terrigenous rocks of 400 to 800 m thickness. Volcanic activity was pre-served only in the island arc of the Klamath Mountains where volcanics of mafic, intermediate and felsic composition were formed. Accumulation of a flysch complex proceeded on the rear slope of the island arc region of the Alexander archipelago but predominantly sand formations at other places. Uplift in the form of islands existed in eastern Alaska and the Vancouver Island region. Development of marginal seas, their deposits represented by pelagic clay rocks (micritic limestones too in Alaska), continued between these islands and the shelf of the North American continent.

Inversion of the deep Richardson trough in north-eastern Alaska occurred in the Early Devonian. Formations filling this trough were dislocated and an island arose in this region.

On the whole, there was no major transformation in the tectonic develop-ment of the Cordilleran belt in the Silurian and Early Devonian. Development of most island arcs came to an end and their volcanic activity attenuated. The former deepwater zones enlarged the area of the periplatform shelf basin. Further, this enlargement occurred not so much because of gradual filling of the marginal zone of deepwater basins adjoining the continent by sedimentary prisms, but rather its uplift as a result of tectonic accretion.

3.2.5.2 *Andean system.* E a r l y   S i l u r i a n. The Northern Andes constituted an arena of manifestation of orogenic activity in this epoch, characterised by uplift, folding, metamorphism and also intrusion of granitoids into the Eastern and Central Cordillera of Colombia (Zeil, 1979). Development of these processes commenced and concluded earlier in the Venezuelan Andes, evidently even in the Ordovician; during commencement of the Silurian a brief phase of subsidence and transgression occurred, followed by a regression after the Wenlockian. The Lower Silurian terrigenous Canaro suite formed here was weakly affected by deformation, unlike the highly folded Cambrian shales.

In the Peru-Bolivian segment of the Andes powerful subsidence renewed at the end of the Ordovician after a brief phase of uplift, accompanied here and there by dislocation and granitisation. Only the Arequipa and Pampeanas massifs were preserved as major islands of high relief. Within Peru, Lower Silurian shales (Cabanales group) conformably lie on Upper Ordovician rocks (Contaya suite), pointing to the inherited development of the basin. In the Cordillera of Bolivia and northern Argentina, however, Sapla tillites formed at the base of the Silurian unconformably overlie the Middle Ordovician and much older formations.

The peculiarities of the lateral spread of Late Llandoverian tillites, the change in their thickness and their relationship with marine Lower Silurian formations point to the advance of Early Silurian glaciers from west to east and south-east from the Arequipa massif. The type of glaciation has not yet been decisively established but were it continental, it was small in area.

In the central zone of the Peru-Bolivian Andes subsidence successive to formation of tillites was most intense (Dalmayrac et al., 1980). Possibly, a new phase of extension (rifting?) manifested between the craton and the Arequipa massif. As a result, a deep trough arose with uncompensated sedimentation. The Lower Silurian is represented in it by graptolite shales. Sand formations up to 1.5 to 2 km in thickness played a major role on the slopes of this trough.

The eastern part of the Chile-Argentinian Andes experienced uplift in the Early Silurian and evidently fold and fault deformation developed here. In the western zone formation of marine sedimentary strata continued. Shelf sand formations of comparatively small thickness alternated in a westerly direction with a flysch complex formed obviously under continental slope conditions.

In the Patagonian zone, after Late Ordovician uplift and folding, subsidence renewed at places in the Early Silurian. This is suggested by angular unconformities and basal conglomerates at the base of the Silurian at those places where it is developed. Thus in the Rio Negro province, the Lower Palaeozoic Ligende suite is unconformably overlain by the Silurian Sierra-Grande suite made up of quartzites and conglomerates.

L a t e S i l u r i a n. In the Northern Andes the residual Venezuela marine basin experienced a regression and now all of its area turned into a region of uplift and manifestation of various dislocations, metamorphism and granite formation, i.e., predominance of a typically orogenic regime.

Uplift of the Arequipa and Pampeanas massifs intensified, the sea strait separating them regressed and they represented a single large continental block. Intrusion of granitoids and fold deformation occurred in the southern Sierra Pampeanas in this epoch. A deepwater trough with thin clay deposition was preserved in the central zone of the Peru-Bolivian Andes and coarse clastic strata up to 2 km or more in thickness were deposited on the slopes of the Arequipa massif and the craton. At the end of the Silurian, the north-eastern periplatform flank of the trough experienced a brief uplift. As a result, the upper portion of the Silurian formations was eroded.

In the Chile-Argentinian Andes the western zone experienced uplift and now almost the whole of its area, except for a small residual basin in the eastern central segment of Chile, became a region of land and manifestation of orogenic processes. Granodiorites intruded in the north on a small scale. In the Patagonian zone an orogenic regime was also established from the Late Silurian. This is confirmed, in particular, by the formation of red sandstones of the molasse type in the southern part of this zone.

E a r l y D e v o n i a n. The nature of development of the Northern Andes in this epoch remained essentially as before, with an orogenic regime prevailing over large areas. However, the eastern part covering Sierra de Perija, the Eastern Cordillera and the eastern Central Cordillera was affected by subsidence at the end of the Early Devonian, which led to the formation of semi-enclosed shallow-water marine basins in which sandy, clay and, less frequently, carbonate deposits accumulated to a thickness of a few hundred metres. They overlie the Ordovician and much older formations with angular unconformity.

In the Peru-Bolivian Andes the area of subsidence diminished and redistribution of sediment supply sources for the central trough occurred. The threshold of its north-eastern flank, uplifted at the end of the Silurian, experienced a partial subsidence that restricted the inflow of clastic material from the Brazilian Shield; hence the trough was filled with sediments coming from the Arequipa massif. Strong uplift of the latter led to commencement of filling of the central uncompensated trough with terrigenous flysch whose thickness exceeded 2 km in the Early Devonian (Jacobi and Wasowski, 1985).

In the Chile-Argentinian Andes zones of subsidence enlarged and a fairly wide transgression was manifest. By the time it attained maximum, almost one-half of the area of the segment was covered by sea. This subsidence was brief, however, and a regression commenced even at the end of the epoch; thus it did not, by and large, disturb the prevalence of an orogenic regime, given the fact that molasse-type strata of continental con-glomerates and sandstones played a significant role in the composition of Lower Devonian formations. Brief subsidence affected the Patagonian zone too where the Lower Devonian formations are represented by littoral-continental sandstones and conglomerates unconformably lying over much older formations.

Thus in the tectonic development of the Andean mobile zone, a distinct tendency towards transition from subsidence to uplift and orogenic regime is evident in the Silurian-Early Devonian. In the Silurian, folding, metamorphism and granitisation were manifest in many regions while the comparatively extensive Early Devonian transgression was, nevertheless, of a local nature. Only in Peru and Bolivia, in the central segment of the Andes, did the phase of extensions in the zone between the Arequipa massif and the Brazilian Shield ensure a high tempo of subsidence and preservation of deepwater basin conditions uncompensated by sedimentation. But at the end of the epoch, in the Early Devonian, subsidence slowed down here too (or even ceased) and filling of the trough with flysch commenced.

3.3 PALAEOTECTONIC ANALYSIS

The primary tendencies in the development of mobile belts, i.e., palaeo-oceans, in the Silurian were the build-up of compression, uplift and moun-tain formation in their peripheral sections (in the Iapetus and also in its central part) on the one hand, usually referred to as the Caledonides and, on the other hand, continuous spreading with new formation of oceanic crust in the axial zones belonging to the region of future Hercynian folding. The former regions included the Northern Appalachians with Newfoundland, eastern Greenland and Spitsbergen, Britain and Scandinavia and also the Caledonides of Brabant, of the south-western framing of the East European Craton, Central Kazakhstan and Northern Tien Shan and the Altay-Sayan-Mongolia region. The second category included the Middle European Her-cynides, western and eastern part of the Palaeo-Tethys, the Palaeo-Asian ocean and West Pacific belt, in particular the East Australian region. Cale-donian diastrophism was extensively manifest even in the Andes, without, however, leading to their total cratonisation.

No significant changes occurred in the development of continental cra-tons. Late Ordovician regression was followed by Early Silurian post-glacial transgression but not everywhere (Fig. 3-4). Throughout the Silurian period uplift of cratonic blocks increased in a very general manner. It coincided with the tendency towards development of considerable sections of mobile belts entering the main epoch of Caledonian orogeny at the end of the Silurian (Fig. 3-5).

138

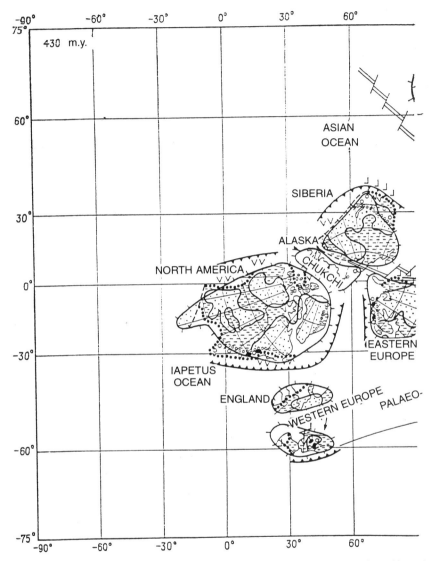

Fig. 3-4. Palaeotectonic reconstruction of the Early Silurian (Mercator projection with centre

In this scheme, based on the reconstruction of L.P. Zonenshain and coauthors (1987), in our
more, unlike in the Middle Ordovician reconstruction, here there is no continental block of nor

3.3.1 Continental Cratons

3.3.1.1 *Gondwana.* Cratons of the Gondwana group were concentrated
mainly in the Southern Hemisphere while north-western Africa and western

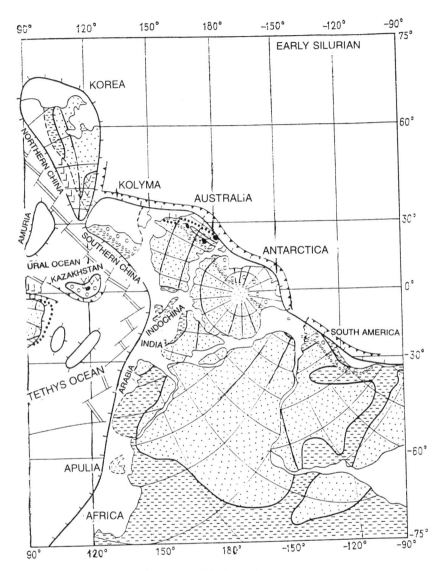

at 0° N lat. and 90° E long.). See Fig. 1-7 for legend.

view the contours of the continents of North America and northern China are inaccurate. Further-
thern Scotland and Ireland.

South America reached Polar latitudes. This explains the presence of tillites and tilloids in Llandoverian-Wenlockian formations of the Amazon basin and the central Andes (Hambrey, 1985). It is possible that glaciation was partly montane, taking into account the Late Ordovician uplift (see Section 2.3.2.5).

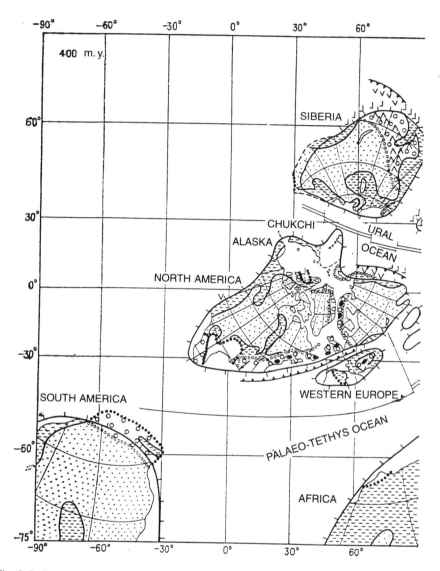

Fig. 3-5. Palaeotectonic reconstruction of the Early Devonian (Mercator projection with centre

In this scheme, unlike in the preceding (see Fig. 3-4 and others), L.P. Zonenshain and co-depicted for the Kazakhstan continental block, as well as its extent, including the entire territory

Another point of significant interest pertains to the Ouachita-Mexican system, which older epochs since deepwater basin conditions prevailed there.

India remained dry land throughout the Silurian (apart from the northern periphery) and Australia in the Early Silurian. In the Late Silurian marine

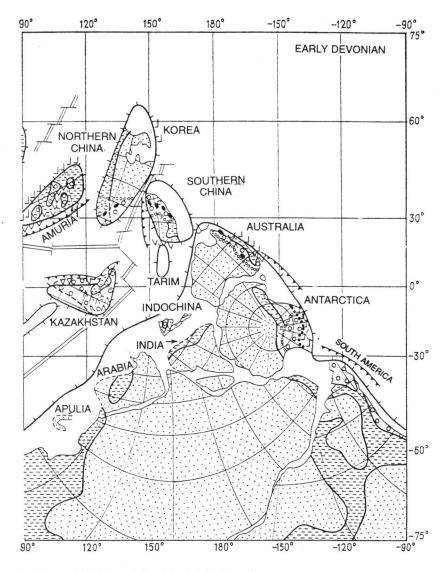

at 0° N lat. and 90° E long.). See Fig. 1-7 for legend.

authors (1987) used a degree grid for the first time, which reveals the inaccuracy of the contours of the Uralian system.

territory cannot be included in the North American continent for the Early Devonian nor for much

deposits were manifest in the Carnarvon trough which was subsequently transformed into a rift and continental formations into the Amadeus aulacogen inherited from the Late Precambrian. Australia, Antarctica, India, Tarim and the Yangtze Craton were located in tropical and subtropical belts.

The main zones of relative subsidence and sedimentation in the Silurian in Gondwana appeared in the northern margin of the supercontinent turned towards the Palaeo-Tethys and Palaeo-Asian oceans. In the Sahara, where epicontinental sea was most extensive, black shales were considerably widespread. The rift troughs (aulacogens) Tindouf and Ougarta appear here in the background of moderate or even weak subsidence.

Deposits of epicontinental Gondwana basins are almost exclusively of sandy-clay composition, which is explained by the position of much of the supercontinent in a zone of cold or moderately humid climate. The entire periphery of Gondwana turned towards the Palaeo-Tethys represented a passive margin while considerable segments of the Pacific margin, i.e., south-eastern China, eastern Australia and North American Cordillera, belonged to the active type.

3.3.1.2 *Northern row of cratons.* Cratons of the Laurasian group changed in disposition very little compared to the Late Ordovician. They lay as before at low and moderate latitudes. By the end of the Silurian, the width of the Uralian ocean had increased and correspondingly the distance between the Siberian and East European continents increased. Collision occurred later between the East European and North American continents with the formation of a single continental mass called Laurussia (or Euramerica).

As before, a quiescent tectonic regime was characteristic of the Laurasian cratons. Eruptions of plateau basalts are observed only in the northeastern Verkhoyansk-Kolyma region of the Siberian continent and in the northern Iranian margin of Gondwana. These basalts erupted in the first region against a background of differentiated block movements. Quiescent development is suggested by the small thickness of the Silurian deposits. In the inner regions of the cratons their thickness runs into a few hundred metres, reaching 1 to 2 km at places only in the outer shelf (Chersky Range, Nevada and others). In correspondence with the climatic conditions, carbonate deposits predominated. In the Late Silurian they were partly or even wholly replaced by continental clastic formations, pointing to the growth of uplift. In many regions evaporites appear as in the North American Craton where the Michigan basin represents the main region of salt accumulation and also in the south-western margin of the East European Craton, in the Tunguska basin and the eastern margin of the Siberian Craton.

Transition from shelf to continental slope and rise was manifest no less distinctly in the Laurasian group of cratons in the Silurian than in the Ordovician: along almost all of the periphery of the North American continent, along the south-western, Baltic-Podolian lineament and eastern periphery of the East

European, and the northern (Taimyr) and eastern (Chersky Range) Siberian Craton, predominantly in the replacement of shallow-water dolomites and limestones by dark-coloured layered limestones and shales and later by graptolite shales, cherts and turbidites and, less frequently, by carbonates with the participation of mafic volcanics.

Throughout the Silurian the Sino-Korea Craton continued to remain elevated above ocean level. Silurian formations are known only in the eastern margin of the Tarim microcontinent.

After studying the continental cratons as a whole, three groups could be distinguished among them based on the nature of vertical movements in the Silurian and Early Devonian (440–390 m.y. ago). These groups are: 1) throughout this interval, stable upheaval of the Sino-Korea and Indian cratons; 2) North American, South China and Australian cratons distinguished by a growing tendency to uplift; and 3) in the East European, Siberian, South American and African cratons, transgression observed in the Early Silurian was later succeeded by a regression. Further, in the South American and East Antarctic cratons a new transgression occurred in the Early Devonian; in the Antarctica it was the first in the period under study.

3.3.2 Mobile Belts

3.3.2.1 *North Atlantic belt.* In the Silurian-Early Devonian the palaeoocean Iapetus entered the concluding phase of its development. Its northern segment, including eastern Greenland and Spitsbergen, was the earliest to close. Fold-thrust deformation affected this segment even at the end of the Ordovician and movements of nappes continued in eastern Greenland in the Silurian. These movements reached their culmination at the end of the Wenlockian and in the Ludlovian. They also affected the eastern extremity of the Innuitian belt in the Peary Land region (Pearya) and ensured an enormous inflow of clastic material in the Early Silurian. This flow formed a flysch series which covered the carbonate shelf and filled a deepwater trough (Surlyk and Hurst, 1983). In Spitsbergen the end of the Silurian to commencement of the Devonian was the main epoch of regional metamorphism and granite formation. Uplift, deformation, granitisation and metamorphism extended into the Late Ordovician to the Scandinavian segment of the belt also. West of it an ocean basin still existed in the Silurian. Within this system, a marginal volcanoplutonic belt arose and a marginal sea basin to the back of it on ancient continental crust that underwent destruction (Brekke and Furnes, 1984). At the end of the Silurian this basin as well as the ocean itself ceased to exist. In the British segment during the Silurian and Early Devonian a two-way subduction of the residual Iapetus occurred towards the north-west under Scottish uplands where a complex of so-called Newer Granites was formed (Soper, 1986) and towards the south-east under the English Midland microcontinent on the edge of which a volcanoplutonic belt was formed with the marginal Wales basin in the back on thinned but not wholly destructed continental crust (some resemblance to Scandinavia except that destruction of the crust of a similar basin there attained the stage of substitution by oceanic crust while deformation of corresponding complexes did not proceed until the formation of nappes).

The Iapetus residual basin in Britain was filled during the Silurian with flysch deposits of continental slope and rise advancing from both sides, especially from the north. The terminal events at the end of the Silurian to commencement of the Devonian were caused in Scandinavia by collision between the North American and East European (Baltica) cratons, followed by collision between North America and the Avalon-Midland microcontinents. Two generations of so-called Newer Granites correspond to these two phases of collision (Soper, 1986).

By the Middle Devonian, Iapetus oceanic crust was wholly absorbed here. Simultaneously, from the end of the Silurian development of sinistral strike-slip faults commenced, in particular the prominent Great Glen strike-slip fault in Scotland. The total displacement along them and their continuation in Newfoundland and in the Appalachians was estimated to be 1800 km by Dewey and Shackleton (1984).

In the Early Devonian the Midland Valley rift of Scotland arose at the back of the northern subduction zone where formation of the Old Red sandstone, Caledonian molasse began to be deposited. The accumulation of this formation was accompanied by manifestation of volcanism of a type intermediate between subduction and rift volcanism. The Caledonian Scotland and northern Ireland basin and Orcade in north-eastern Scotland originated from a similar rift.

The same ultimately occurred in Newfoundland too but here subduction proceeded mainly under the Avalonian microcontinent along the margin of which a volcanic belt formed. In the Late Silurian calc-alkaline volcanism extended into the central part of Newfoundland while accumulation of molasse commenced in the Early Devonian in residual troughs, pointing to the commencement of mountain building and collision of the Avalonian microplate with Laurentia.

In the Appalachians folding, metamorphism and uplift increased from the Early Silurian to the Early Devonian. In the Late Silurian residual troughs with clastic and volcanoclastic sedimentation were still preserved between island arcs; in the Early Devonian deposition had already commenced of typical molasse in the Northern Appalachians as well as in Newfoundland. In the Southern Appalachians, in the outer Ridge and Valley province, at the boundary with the platform, a trough was preserved with the accumulation of clastic formations reflecting the growth of uplift in the inner zones caused by the collision of the Florida salient of Gondwana with Laurentia.

These processes did not at all affect the Ouachita-Mexican system adjoining the Appalachian at a right angle, however. Here, deepwater siliceous-clay sedimentation continued.

I n n u i t i a n   s y s t e m. In the course of the Silurian filling of the deepwater basin in the axial part of the belt with turbidites of eastern origin occurred: it touched the eastern sector in the Early Silurian, extended into the central part in the Late Silurian and into the western part in the Early Devonian. These clastic formations were replaced here by pelagic silts. Volcanic activity ceased in the northern land (Pearya) in the Late Silurian.

3.3.2.2 *Mediterranean belt.* Processes of diastrophism leading to formation of the North Atlantic Caledonides extended in the Early Devonian to the northern margin of the western part of the Mediterranean belt in the strip from south-eastern England through the southern part of the North Sea and in the south through Brabant, northern Ardennes and Rheinische Schiefergebirge in the direction of northern Germany and Poland right up to the northern Carpathian front. These processes were evidently associated with opening of the Central European basin with oceanic crust ('Rheicum ocean') representing the marginal sea of the Palaeo-Tethys, the main zone of which extended south of the Moldanubian microcontinent. Calc-alkaline Brabant volcanics point to subduction of the crust of the Central European basin under its northern margin, where eruptions had commenced even in the Late Ordovician.

The Central European basin encompasses two zones of Hercynides—Rhenohercynian and Saxothuringian—divided by a narrow Midgerman High. In the Silurian the former zone together with this uplift still belonged to the Caledonian margin of the Mediterranean belt while the latter already experienced extension and in its southern part represented a deepwater basin with the accumulation of radiolarites and turbidites and also outpourings of mafic lavas. The northern part of the Saxothuringian basin with its shallow 'Thuringian' facies was disposed as early as in the Ordovician on continental, most probably Epibaikalian crust while the southern deepwater facies could have arisen much earlier—at the end of the Riphean according to some data. The presence of ophiolites, nowhere providing a full section, shows that this part of Saxothuringicum is underlain by oceanic crust but indirect evidence shows that the width of the strip with oceanic crust could not have been significant. The ophiolite belt of the southern Saxothuringian zone is traced from the western extremity of the Armorican massif (Audierne bay) to Gory Sowie in the Polish Sudeten mountains.

The Rhenohercynian basin (subbasin) represents an Early Devonian new formation on the extreme margin of the Caledonian continent (Laurussia or Euramerica), the frontal salient of which was possibly the Midgerman

High. Much of the northern part of the Rhenohercynian basin, like the Saxothuringian, has a continental basement (Baikalian-Caledonian) and is made up of shallow-water formations but deeper water formations, including radiolarites, appear in the extreme south while Devonian mafic volcanics reveal proximity to the MORB type; this prompts the assumption that the crust changed into an oceanic type or, in any case, one similar to it. Ophiolites emerge in the western continuation of the zone, on Cape Lizard in Cornwall. The age of these ophiolites has been determined as 375 m.y., i.e., as Middle Devonian. They could possibly be the detached fragment of the crust of this portion of the Rhenohercynian basin.

Both northern zones of the Central European Hercynides on the Atlantic margin are known to have initially turned south and southeast forming the Iberian-Armorican arc and continued into the structures of the Iberian Meseta, i.e., Rhenohercynian in Southern Portugal and Saxothuringian in its more northern zones. The Iberian-Armorican arc ran round the Aquitaine-Cantabrian massif, i.e., the western continuation of the Moldanubian microcontinent. The Iberian Hercynides should have joined with the Hercynides of the Moroccan Meseta (they were later separated by Neotethys structures—Cordillera Betica and Rif) and the latter with the Mauritanides. Further, the Mauritanides reveal only marginal, shelf facies of the Middle Palaeozoic geosyncline; during opening of the Atlantic deeper water formations appeared along its western, i.e., American, side.

The eastern continuation of the Rhenohercynian zone, buried under a Mesozoic-Cenozoic cover, intersects Germany and Poland and again emerges on the surface in Czechia where it is known as the Moravo-Silesian zone. It extends from south-west to north-east, its interior thrust front shearing the Saxothuringian and Moldanubian zones. In the south-western zone, in turn, it is sheared by the Carpathian front and its continuation appears displaced and reworked in the course of Alpine deformation.

The deep Alpine reorganisation undoubtedly makes difficult the restoration of conditions within the Palaeo-Tethys proper. The outer crystalline massifs of the Western Alps together with Montagne Noire, Provence, the Pyrenees, Corsica and Sardinia pertain to the southern slope of the Central European zone of uplifts while the Southern Alps together with the birthplace of the Austro-Alpine nappes belong to the Gondwana margin. The axial zone of the Palaeo-Tethys passed evidently in the intervening zone; in the south-west it is observed in the Devonian in the region of modern Er-Rif and the Balearic islands (Bourrouilh et al., 1980). Bourrouilh and coauthors place the opening of the Palaeo-Tethys in the western Mediterranean in the Devonian but point out that a basin with extensive distribution of graptolite shales and sporadic occurrence of Orthocera limestones existed here in the Silurian.

In the far south-west, according to Boulin (1991), the axis of the Palaeo-Tethys extended in the Cambrian-Silurian through the Atlas zone of north-western Africa, between the African Craton (a part of Gondwana) and the Moroccan Meseta block tending towards North America (Avalonia ?—V.Kh. and K.S.). Further continuation of this basin is seen in the Southern Appalachians and west of Mauritania in northern Africa.

East or south-east of the Alps the axial zone of the Palaeo-Tethys could be traced through the Balkanides (Bonchev, 1985). Even more eastward this zone should be looked for in the Caucasian region. The northern slope of Caucasus Major in the Silurian to Early Devonian represented a passive margin of the East European continent. The position of the axis of the Palaeo-Tethys in the Caucasus is not clear: it may pass between the Fore and Main Ranges of Caucasus Major, along the southern slope of the latter; according to some (Adamia, 1984), it may even lie in the central part of Caucasus Minor. Ophiolite outcrops in the Resht and Meshhed regions of northern Iran may help trace this zone farther in the east in the direction of the northern Pamirs and Kuen Lun.

Development of the eastern segment of the Mèditerranean belt, including the Pamirs, Kuen Lun, Nanshan and Qinling, continued in the Silurian with features inherited from the Ordovician. Deepwater sedimentation continued here: sandy-clay-carbonate flyschoid in western Kuen Lun, sandy-clay flysch in eastern Kuen Lun and sandy-clay with carbonates and cherts in Qinling. The Qilianshan geosynclinal system represents, as before, the connecting link between Tien Shan and Altay on the one hand and Kuen Lun and Qinling on the other, dividing the Tarim-Qaidam and Sino-Korea continental blocks. The Qilianshan deepwater basin was already dissected by an island arc into two partial basins. In the Late Silurian the arc was actively volcanic while the north-eastern subbasin shoaled and began to be filled with coarse clastic material, molasse. By the beginning of the Devonian the Qilianshan geosyncline was fully closed, evidently due to collision of the continental blocks confining it. In the Early Devonian the western Kuen Lun basin with terrigenous sedimentation as well as the southern Qinling basin with carbonate sedimentation and alternation of limestones and dolomites shoaled.

Farther south-east, the Silurian and Lower Devonian Kuen Lun and Qinling as well as the eastern Yunnan, Laos, North Vietnam and eastern Thailand deepwater deposits help trace a branch of the Palaeo-Tethys to the east, south-east and south. This branch of the Palaeo-Tethys separated the Sino-Korea, South China (Yangtze) and Indochina continents and then joined with the Palaeo-Pacific (*Atlas of the Palaeogeography of China,* 1985). The zone of Tibet, Myanmar and western Thailand still constituted the Gondwana margin. The northern (Qilianshan) branch of the Palaeo-Tethys in the Late Silurian began to experience compression due to the approach and later collision of the Tarim-Qaidam and Sino-Korea continents and by the Early Devonian had transformed into a fold edifice. Shoaling of the basin in the Qinling region echoes these events.

3.3.2.3 *Ural-Okhotsk belt.* The various structural elements of this extensive belt developed quite differently in this stage. For the Uralian system, this was the main period of ensimatic island arc volcanism and broadening of marginal-sea basins. In its Southern Tien Shan continuation, deepwater sedimentation with formation of flysch on the slope of the northern continent predominated. The Kazakhstan-Northern Tien Shan continent had widened significantly in an easterly direction by the commencement of this stage. This led to isolation of the Junggar-Balkhash basin bound in the east by the Chingiz-Tarbagatai volcanic arc, underlain by a seismofocal zone inclined westward. The Ob'-Zaisan deepwater basin extended farther eastward. This basin played the role of the axial zone of the Palaeo-Asian ocean from this period almost to the end of the Palaeozoic. The Junggar-Balkhash and Southern Tien Shan basins opened into this ocean from the northwest and west respectively, while the eastern continuation of the axial zone extended in a latitudinal direction through Southern Mongolia. In all these basins accumulation of deepwater sediments was accompanied by submarine volcanism of mafic and intermediate composition. Northern Mongolia together with the eastern part of the Altay-Sayan region had accreted to the Siberian continent by the beginning of this stage (Fig. 3–6). Troughs preserved from the preceding stage were filled in the Early Silurian with marine and in the Late Silurian to Early Devonian with increasingly coarse continental deposits, pointing to development of mountain building along the periphery of the Siberian Craton, evidently representing a margin of the Andean type with manifestation of granitoid magmatism. A mighty explosion of subaerial alkali-basalt volcanism, probably of rift origin, occurred here in the Early Devonian (Fig. 3–7).

The axial Gobi-Hinggan system of the Palaeo-Asian ocean was accompanied from the north by a volcanic arc at the back of which a riftogenic deepwater trough arose north of the Kerulen-Argun microcontinent in the Early Devonian. This trough extended from north-eastern Mongolia into Transbaikalia. Eastward it joined with the north-eastern branch of the Southern Mongolia basin extending through the region of Greater Hinggan into the Zeya-Selemdzha interfluve separating the Kerulen-Argun and Bureya microcontinents.

The Amur-Okhotsk basin opening into the ocean (Palaeo-Pacific) served as the further eastern continuation of this oceanic realm. The Gobi-Hinggan system of the Palaeo-Asian ocean having been transformed at the end of the Silurian into the Southern Mongolia ocean (Ruzhentsev et al., 1991), joined with the Palaeo-Pacific and another latitudinal Girin branch extending south

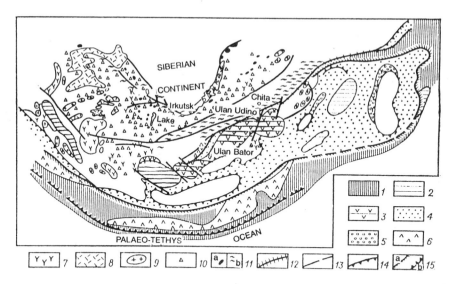

Fig. 3-6. Palaeotectonic-magmatic scheme of Sayan-Baikal-Mongolia region in the Middle Palaeozoic (after Gordienko, 1987).

1—predominantly marine complexes including ophiolite complexes; 2, 3—deepwater silt-clay flysch deposits (2—including turbidites, 3—with volcanics of mafic and intermediate composition); 4—shallow-water terrigenous and carbonate-terrigenous deposits; 5—coarse clastic continental, at places littoral marine deposits; 6—marine volcanogenic strata of calc-alkaline composition of intermediate and felsic volcanics (differentiated series); 7—continental volcanic strata of calc-alkaline composition of intermediate and felsic volcanics (differentiated series); 8—continental volcanic strata of contrasting composition; 9—calc-alkaline granites and granodiorites; 10—subalkaline granites, granosyenites and syenites; 11—alkali-gabbroid and alkali-granitoid complexes (a) and ultrametamorphic granitoids (b); 12—zone of the Mongolia-Okhotsk fault; 13—other faults; 14—boundaries of epicontinental shelf seas; 15—assumed boundary between continent and ocean (a) and probable disposition of the seismofocal Zavaritsky-Benioff zone (b).

of the Khanka microcontinent right up to commencement of the Devonian, constituting a single entity with the Bureya-Jiangxi massif.

In both these basins deepwater sedimentation continued with the formation of flysch beds along the continental slope and rise and at places with manifestation of island arc volcanism.

The southern, Sino-Korea margin of the eastern segment of the Palaeo-Asian ocean was transformed in the Silurian from a passive into an active margin of the Andean type, as suggested by calc-alkaline volcanics and granites aged 430—379 m.y. and also by the unconformable attitude of the Upper Silurian molasse on much older Palaeozoic formations (Wang Quan and Liu Xueya, 1986).

150

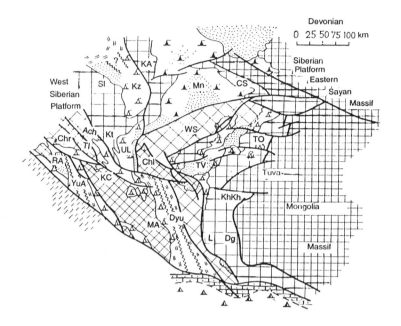

Fig. 3-7. Palaeotectonic scheme of Central Asian Caledonides in the Devonian (after A.B. Dergunov. In: Mossakovsky and Dergunov, 1983). See Fig. 1–10 for legend. Kz—Kuznetsk basin.

3.3.2.4 *West Pacific belt.* In the stage under consideration this belt could already be traced in its entire present-day stretch from Koryakia to Australia and New Zealand. In western Koryakia, an island arc, the future Udino-Murgal already existed. The eastern margin of the Bureya-Khanka microcontinent probably remained passive while an active margin of the Andean type with intermediate and felsic volcanism and shallow-water carbonate-terrigenous sedimentation and transition to a deepwater, oceanic region with hemipelagic sediments is quite distinctly developed in the region of the Japanese islands. Major changes occurred by the commencement of the Silurian in south-eastern China where a major portion of the Cathaysian geosyncline was transformed into a montane terrain and the remaining troughs were filled initially (S_1) with flyschoid-terrigenous, later (S_2) with shallow-water terrigenous and finally continental (D_1) sediments, except Qinfan trough in the south-west where marine and relatively deepwater conditions persisted until the Early Carboniferous inclusive. This trough fused in the south with the Viet-Lao geosynclinal system representing, as mentioned earlier, a branch of the Palaeo-Tethys.

According to palaeomagnetic data, Eastern Australia in this stage was disposed quite close to southern China and eastern Indochina.

The western part of the Lachlan system west of the Melbourne meridian was transformed by Benambran deformation into a montane terrain with some basins filled with shallow-water or continental clastic deposits and with manifestation of felsic volcanism, suggesting a margin of the Andean type. In the east conditions characteristic of the modern West Pacific belt clearly prevailed. Deepwater basins on an oceanic or transitional and even continental (Cas, 1983) crust continued to exist east of Melbourne (Melbourne 'trough') and in the Bathurst region in the north-east (Hill End 'trough'). Between them lay an extensive volcanic arc (up to 400 km) complicated by interarc rift troughs (Kaura and others). It formed the hypothetical Molong microcontinent (Scheibner, 1987) and should have been underlain by a seismofocal zone gently inclined westward. To the back of this arc, in fact a volcanoplutonic belt, the existence of the Wagga marginal basin in the Early Silurian is suggested, possibly with an oceanic crust. By the Middle Silurian, its formations were deformed, metamorphosed and thrust westward onto the Victoria microcontinent.

Farther east, within the New England system, formation is assumed of another volcanic arc, already ensimatic (Calliope arc; Scheibner, 1987), in the Middle Silurian. Between this arc and the Molong-Canberra arc of the Lachlan system to the east evidently lay a marginal basin with an oceanic crust.

A comparison of the distribution of deposit facies from the Late Ordovician to the Early Devonian points to gradual shoaling of the basins of the Lachlan system. This was promoted by deformation and uplift of three epochs of tectogenesis: Benambran (O/S), Kwidonian (S_1/S_2) and Bowning (S/D). At the same time they represented epochs of granitoid plutonism. The granitoids are anatectic (in the west) as well as of mantle origin (in the east).

The zone of Transantarctic mountains (Rossides) in the stage under consideration continued to survive the orogenic period of development, as pointed out by Lower Devonian molasse of the Shackleton and Pensacola mountain regions. This development concluded with the Shackleton event after which the Rossides became a constituent of the Antarctic Platform.

3.3.2.5 *East Pacific belt.* C o r d i l l e r a n   s y s t e m. Palaeotectonic conditions in this active margin of the North American continent changed little during the Silurian compared to the Ordovician. On the basis of the present position of structural elements, simple relationships are seen here (Fig. 3–8): west of the continental margin, a marginal sea with deepwater, thin carbonate-clay sediments and a volcanic arc more to the west. Fragments of this arc are preserved in the Alexander archipelago, Klamath Mountains and in northern Sierra Nevada. These island arc blocks are presently regarded, however, as exotic 'terranes' which may have occupied

152

their contemporary position after considerable displacement along the continental margin in a northerly direction: this is particularly true of the Alexander
archipelago, to a lesser extent of the Klamath Mountains and Sierra Nevada
and for many 'terranes' which have been proven as not very remote in origin
(Gray, 1986). Thus, in principle, the foregoing picture is perhaps quite close
to reality.

Some changes occurred in the Early Devonian, in the development of the
system: marginal seas shoaled, volcanism on island arcs became dormant
and they sank more extensively below sea level. This possibly points to
increasing subduction and accretion of the Palaeo-Pacific crust but for some
reason is not at all reflected on the continental margin, which remained
amagmatic as before.

A n d e a n   s y s t e m. In the stage under consideration the North-
ern Andes could be considered, as before, a probable continuation of the
Appalachian system. This is supported by the unconformable bedding of
Silurian on metamorphosed Cambro-Ordovician representing the Taconian
epoch of deformation and the Appalachian-type fauna of Venezuela and
Colombia. The marine basin in the southern half of the Andes, on the con-
trary, appears closely associated with the basin of the southern Sierras of
Buenos Aires, the Folkland (Malvinas) islands and the Cape province of
South Africa. Marine-glacial formations deposited here at the base of the
Silurian. The sediment supply distinctly manifest in the Silurian from the
Arequipa massif in Southern Peru points to the continuing intracontinental
nature of the Southern Andes geosyncline. Transition from Silurian to Devonian was gradual here, without traces of Late Caledonian diastrophism. but
with characteristics of basin shoaling (quartzites!). This, however, is a unique
exception in the background of involvement of almost the entire remaining
area of the Andes in Caledonian deformation and uplift.

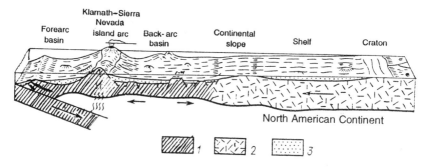

Fig. 3–8. Model of Cordilleran continental margin in the Silurian and Devonian (after F.G. Poole.
In: Frasier and Schwimmer, 1987):

1—oceanic crust; 2—continental crust; 3—deposits.

4

Middle Devonian—Early Carboniferous. Mature Stage of Development of Hercynian Mobile Belts. Inundation of Cratons

.1 REGIONAL REVIEW. CONTINENTAL CRATONS AND THEIR MARGINS

.1.1 **North American Craton**

Middle Devonian. A new wave of subsidence gradually ffected the central parts of the craton but its south-western part, i.e., the ranscontinental High, remained in the form of dry land. The West Canadian nd Illinois basins enlarged, transgression penetrated the Williston basin nd all three were joined together by straits (Fig. 4-1). The area of the lichigan and Hudson Bay basins enlarged. The partially enclosed nature f all these interior seas, their shallowness, quiescence, slightly rugged pography of the surrounding land and warm climate promoted formation them of stagnant-water conditions, high salinity of water and extensive eposition of carbonates and evaporites, including salt-bearing members the West Canadian, Williston and Michigan basins (*Stratigraphic Atlas f North and Central America,* 1975). Barrier reefs developed extensively the West Canadian basin. Terrigenous rocks were not typical in all the terior basins and the total thickness of the Middle Devonian formations onstituted a few tens of metres, less often 200 to 300 m; only the Michigan asin experienced rapid subsidence. About 800 m of carbonates as well as vaporites accumulated there.

Conditions in the marginal basins of the craton were more dynamic: andy-clay beds up to 800 to 900 m in thickness, transported from the djoining orogen, formed in the Peri-Appalachian basin; accumulation of errigenous deposits combined with formation of reef limestones in the Peri-nuitian basin; terrigenous and carbonate rocks alternated in the Peri-ordilleran basin and its area underwent change. Several sequences in the entral and northern segments of the Peri-Cordilleran basin experienced

154

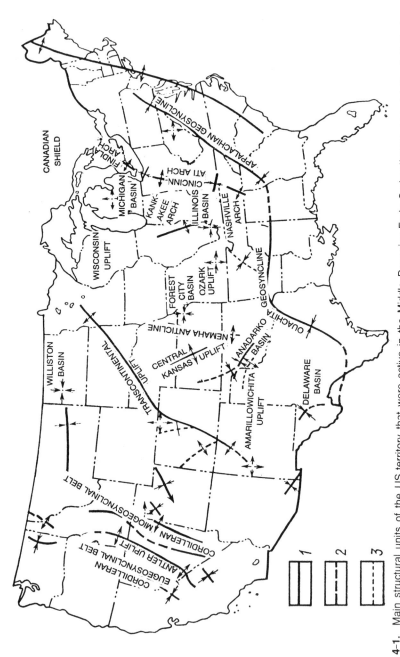

Fig. 4-1. Main structural units of the US territory that were active in the Middle Devonian-Early Carboniferous period (after L.K. Craig and K.L. Varnes. In: Frasier and Schwimmer, 1987, simplified).

1—Mississippian (C_1) structural units; 2—structural units of unknown or assumed Mississippian age; 3—structural units that arose in the post-Mississippian or Early Pennsylvanian period and continued to develop in the Pennsylvanian.

ntense subsidence while the area of this basin enlarged in the south due to subsidence and transgression in the extreme south-western part of the craton. The thickness of the Middle Devonian formations in the Peri-Cordilleran zone rose from a few metres to 500 to 600 m (in the west).

Late Devonian. Following emersion of almost the entire craton at the verge of the Middle and Late Devonian, a new transgression encompassed almost the entire Midcontinent. Uplift and regression over small areas occurred only in the northern part of the craton, in the Peri-Innuitian zone, and the sea regressed from this region. Nevertheless, the remaining basins ceased to be partially enclosed and sea almost wholly covered the area of the craton from the Peri-Cordilleran zone in the west to the Peri-Appalachian zone in the south-east. The Williston basin maintained its individuality in the distribution of facies and thickness of sedimentary formations. In this basin accumulation of carbonates and evaporites (including salts) prevailed at the beginning of the epoch. The West Canadian basin too retained its individuality with carbonate rocks (reef limestones) playing a major role in the composition of formations. In this feature, the West Canadian basin differs from the Peri-Cordilleran continental margin adjoining the basin from the west. Clay and sand deposits predominated on the continental margin. Accumulation of diverse carbonate rocks likewise prevailed in the south-western part of the craton although here, too, carbonates were gradually replaced by terrigenous rocks in the Californian segment of the Peri-Cordilleran zone, towards the Cordilleran mobile belt, while thickness increased in this direction from a few tens of metres to 600 to 700 m. Continental facies of terrigenous rocks are noticed in the proximity of small table islands.

A specific environment prevailed in the south-eastern and eastern parts of the craton. Here, on an extensive territory covering north-western Mexico and Arizona to Hudson Bay, the Upper Devonian is represented by a characteristic complex of the same type of black bituminous shales. These Chattanooga shales (and their stratigraphic equivalents) sometimes contain bands of cherts usually running into a few tens of metres in thickness (less often 200 to 300 m). These bands are replaced only in the Peri-Appalachian marginal zone by the more 'common' sandy clay beds whose thickness increases rapidly to 1.2 km or more close to the Appalachian orogen, reflecting manifestation of Acadian orogen in the Appalachians proper.

Early Carboniferous. At the end of the Devonian the sea regressed from almost all of the craton except for the Michigan and Illinois basins and Peri-Cordilleran and Peri-Appalachian margins. The Early Carboniferous epoch as a whole was regressive for the craton. Uplift affected much of the West Canadian basin and considerable portions of the Michigan and Illinois basins. The basin of Hudson Bay experienced total regression (Avcin and Koch, 1979; Burchett, 1979; Chronic, 1979; Ebanks et al., 1979;

Fay et al., 1979; Pierce, 1979; Schoon, 1979; Thompson, 1979). Further, almost all of the remaining water basins experienced only very weak subsidence and were shallow, sometimes swampy lagoons. Carbonate rocks (often reef) and occasionally cherts formed deposits in them up to tens of metres thick, reaching 200 to 300 m occasionally. However, in the Michigan and Illinois basins conditions are more contrasting. The thickness of the Lower Carboniferous reaches 500 to 1000 m in their central area, some levels comprising evaporites and fairly extensively represented by continental terrigenous facies. Rapid subsidence here was evidently wholly compensated by sedimentation. The same is true of the Williston basin whose area progressively shrank during the Early Carboniferous. The thickness at the centre of this structure goes up to 1 km and the formations contain gypsum, anhydrite and sometimes salt with a general preponderance of carbonates while continental sand facies are present in the upper part of the sequence. In the West Canadian basin (in northern Alberta) coal-bearing deposits were manifest.

Significant changes occurred in the Peri-Cordilleran marginal zone of the craton. Manifestation of Antler orogeny in the adjoining Cordilleran mobile belt extended into a significant part of the marginal zone that had stabilised earlier. Subsidence compensated by the accumulation of sandy clay and reef carbonate deposits, however, continued in the narrow Peri-Appalachian zone. Apart from the Appalachian orogen, the Innuitian and Eastern Greenland orogens represented sources of clastic material that reached the northern regions of the Midcontinent at the beginning of the Carboniferous.

Thus in the development of the North American Craton a transgressive-regressive cycle occurred in the Middle-Late Devonian and Early Carboniferous with maximum subsidence in the middle of this stage, in the Late Devonian, and commencement of regression in the Early Carboniferous. Warm climate and other palaeogeographic and palaeotectonic conditions made for predominance of carbonate deposits in the composition, extensive development of reef facies and deposition of evaporites. Ellesmere, Acadian and Antler orogenic zones developed along the northern, north-eastern and western periphery of the craton respectively, and were active sources of clastic material supply.

4.1.2 East European Craton

M i d d l e   D e v o n i a n. Much of the territory experienced extensive transgression from the east which reached the centre of the craton (Fig. 4-2). This transgression was most widespread in the Givetian stage when sea covered more than two-thirds of its area. In the Middle Devonian, the Peri-Caspian zone of subsidence was manifest positively for the first time, almost at once as a zone of accumulation of deepwater sediments comprising carbonate-clay muds. Transgression advancing from east to west enveloped

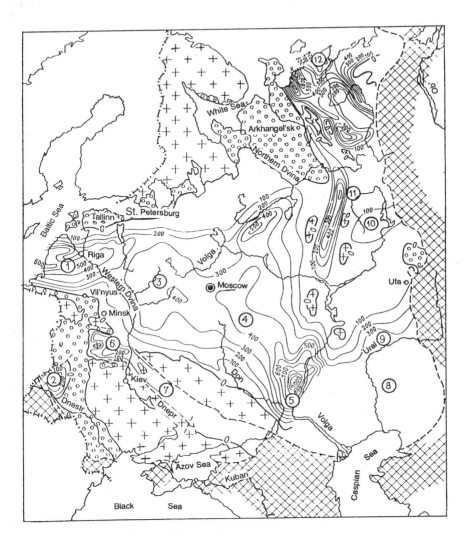

Fig. 4-2. Scheme of distribution and thickness variation of formations in the transgressive stage of the Middle-Late Palaeozoic stage of development of the East European Craton, end of Early to commencement of Late Devonian (after I.I. Bykova, V.I. Gorsky-Kruchinin et al. In: *Geological Structure of the USSR and Pattern of Location of Mineral Deposits.* Vol. 1, *Russian Craton,* 1985). See Fig. 1-2 for legend.

Structures of the Middle-Late Palaeozoic stage (circled numbers): 1—Baltic syneclise; 2—L'vov trough; 3—Moscow syneclise; 4—Ryazan-Saratov trough; 5—Don-Medveditsa trough; 6—Pripyat trough; 7—Dnepr-Donets basin; 8—Peri-Caspian syneclise; 9—Buzuluk basin; 10—Upper Kama basin; 11—Viatka aulacogen; 12—Pechora syneclise.

positive features of the topography, whose height was rather significant. This could be judged, for example, from the fossil valley formed in the Eifelian in the Saratov region and comprising river and lake deposits of various colours, their thickness exceeding 600 m. The complex of Middle Devonian sedimentary formations usually commences with a variously coloured lagoon-continental association changing into shallow-water carbonate at the end of the epoch. The Moscow basin represented a lagoon in which dolomites, gypsum and rock salt accumulated and these rocks extended in the southeast up to the Lower Volga region. Coal-bearing facies developed in the northern part of the craton where the climate was already humid. Middle Devonian formations are usually 200 to 300 m in thickness. Subsidence of the Dnepr-Donets aulacogen commenced in the Givetian stage with the formation there of a deep trough and accumulation in it of micritic limestones and marls.

The marginal zone of the craton adjoining the Mediterranean belt also experienced subsidence and organogenic carbonate and, less frequently, terrigenous formations up to 600 to 700 m in thickness were deposited here in the Polish-Lithuanian and L'vov basins. However, subsidence was most stable in the Peri-Uralian marginal zone from where waves of subsidence and transgression advanced westward onto the craton. This zone is traced from the south of Novaya Zemlya to the Southern Urals. Several fairly large islands maintained their elevated position in this zone. In the littoral zones of these islands, aluminium-iron-bearing weathering crust formed and bauxite accumulated in the periods of regression. Most of the Middle Devonian formations in this eastern marginal zone representing a carbonate shelf, are made up of carbonates (including reef limestones) and only the lower parts of the sequence contain sandstones. The thickness of formations here is quite variable, from 300 to 1000 m.

L a t e D e v o n i a n. The area of the craton increased in the northwest with the inclusion of the former British-Scandinavian orogen (including Spitsbergen) where tectonic activity had died out and the orogenic regime changed into a platform one. The general palaeotectonic environment of this epoch is distinguished by complexity and contrast of vertical movements. In many regions the marine environment changed into continental and vice versa while deepwater basins uncompensated by sedimentation arose sometimes. This made for the diversity of facies composition and thickness of Upper Devonian formations. Among them, carbonates predominated, especially in the east, and continental alluvial-lacustrine rock complexes (coal-bearing in the north) played a significant role. In the Famennian stage shoaling of many basins and a change in climatic conditions led to salinisation of water and replacement of limestones by dolomites with anhydrite inclusions.

Eruptive and intrusive magmatism was manifest on the craton in the Late Devonian. Eruptions of trachybasalts occurred in the northern Timan, in the Dnepr-Donets aulacogen and, on a small scale, on the Voronezh high and in the Viatka-Sergiev basin. Nepheline syenites of the Khibiny and Lovozero massifs, associated with subaerial trachybasalt lavas, intruded into the Baltic Shield.

Uplift preceded manifestation of volcanism in the Dnepr-Donets aulacogen. This was followed by mighty subsidence, resulting in the formation of a deep trough uncompensated by sedimentation but later transformed into a salt lagoon. The thickness of terrigenous and carbonate deposits in the littoral part of this trough reaches 1.5 km. In the Pripyat segment of it, zones of salt accumulation and of deep sea existed simultaneously.

In the Transvolga region uncompensated basins were formed in the Frasnian age and black bituminous clay sediments of the Domanik type were deposited in them. In the north-eastern part of the craton such formations filled four submeridional troughs joining in the south-east into a single relatively deep basin up to 600 km in width.

The south-western margin of the craton continued to subside and carbonate and terrigenous beds accumulated from the Moesian platform to northern Central Europe; predominantly terrigenous rocks accumulated farther westward up to Ireland.

Continental sand facies were extensively developed in the British Isles. Their thickness in some shallow basins goes up to 1 km; such a high rate of sedimentation evidently reflects the last orogenic movements.

In the Peri-Uralian marginal zone subsidence was rarely interrupted by uplift but several islands continued to exist. Late Devonian formations are represented here by carbonates and terrigenous rocks, in the west by bituminous clays and oil shales (Domanik) and by reef limestones along the outer margin of the craton (see Fig. 2-9). On the Pechora platform this was the main period of development of the Pechora-Kolva aulacogen, also framed by reefs. The thickness of formations here went up to 400 to 600 m but exceeded 1 km in Pay-Khoy.

Early Carboniferous. The most stable region of subsidence inherited from the Devonian was the Pripyat-Dnepr-Donets aulacogen which joined the Polish-Lithuanian and L'vov basins in the west. In the central and western regions of the craton the structural plan underwent a significant reorientation. The southern part of the Baltic Shield and the Lithuanian saddle experienced uplift, causing considerable regression. The south-eastern part of the Voronezh high subsided. The Tokmov arch began to form. At the time of regression, it obstructed the connection between the seas of the Volga-Uralian region and Moscow basin.

On the whole, during the Early Carboniferous a double alternation of transgression and regression occurred in the central and western regions.

Shallow-water limestones and dolomites predominated in the composition of formations and clay deposits and alluvial-lacustrine facies in the littoral zones. In the Visean epoch paralic coal-bearing formations predominated. Further, accumulation of coal on a small scale manifested in the eastern part of the continent and in its south-western part (Moesian and L'vov-Volyn basins). Enhanced salinity of water and accumulation of dolomite and gypsum have been noticed in the Peri-Caspian and Pechora basins. The thickness of the Lower Carboniferous in the interior basins of the craton was 100 to 200 m but exceeded 3 km in the Dnepr-Donets aulacogen.

As before, the marginal south-western zone of the craton bordering with the Mediterranean belt underwent stable subsidence. Predominantly carbonate formations up to 300 m in thickness in northern Poland, for example, were deposited in it. Marine bays of the lagoonal type enlarged in Britain and the thickness of the Lower Carboniferous carbonate formations is also high at places (up to 1.2 km). In northern England and Scotland subsidence was accompanied by basaltic magmatism.

In the Peri-Uralian marginal zone of the eastern part of the craton, accumulation of limestones, dolomites and, less frequently, shales continued and was interrupted by a brief regression as a result of which islands appeared. Limestones and clays are sometimes bituminous while coal bands are known in the Ufa-Solikamsk basin. In the Pay-Khoy gypsum and anhydrite were deposited at places and the thickness of the Lower Carboniferous here is quite significant, reaching 1.3 to 1.4 km.

Thus during the Middle Devonian to the Early Carboniferous the East European Craton was characterised by a comparatively unstable regime: regression was repeatedly succeeded by transgression, reorganisation of structural plan occurred with a high rate of subsidence after manifestation of alkali-basaltic volcanism in the Dnepr-Donets aulacogen, eruptions of trachybasalts occurred even in other regions and a comparatively rugged topography formed in the land regions. However, the most common tendency was that increase in transgression towards the Late Devonian changed into partial regression in the Early Carboniferous. The high rate of downwarping gave rise in many regions to deep basins uncompensated by deposition. Stable compensated subsidence has been noticed in the marginal zones of the craton (Peri-Urals, Baltic-Podolian region).

4.1.3 Siberian Craton

M i d d l e   D e v o n i a n. As before, uplift predominated on the craton. The Tunguska and Taimyr basins slightly decreased in size; the instability of their boundaries as a result of repeated regression as also the probable connection of the Tunguska basin at the end of the epoch with the Viluy basin are noteworthy. Limestone, clay and sulphate formations up to 100 to 200 m in thickness predominated among the sediments of the shallow Tunguska basin; their thickness reaches 700 m in the Noril'sk region. The salinity of ' water was also high in the Khatanga basin and Severnaya Zemlya and was manifest in massive salt accumulation. The Taimyr basin situated between them represented an open shallow sea with limestone-dolomite deposits up to 500 m in thickness.

From the beginning of the Middle Devonian major movements along faults led to subsidence of the Viluy aulacogen in which plateau basalts erupted extensively and continental red-beds and marine clay and carbonate deposits (including sulphates) were formed. The thickness of the Middle Devonian is variable here; its maximum value is known in southern Verkhoyansk (up to 1.5 km) where manifestations of basalt volcanism preceded reactivation of the tectonic regime in the Late Devonian. In the Middle Devonian the territory of the Yana-Indigirka interfluve still experienced quiescent subsidence with accumulation of shallow-water carbonate deposits.

In the south-western part of the craton the Rybinsk basin formed and was filled with continental coarse clastic rocks. These rocks represented products of erosion of the adjoining Sayan orogen.

L a t e   D e v o n i a n. The area with a proper platform regime significantly decreased as a result of its north-eastern segment (entire Verkhoyansk and Yana-Indigirka interfluve) being affected by active tectonic movements. The history of sedimentation on the craton is complex due to repeated fluctuations of sea level. The Tunguska basin experienced maximum subsidence in the middle of the Frasnian stage (Fig. 4-3). On the whole, carbonate and evaporite deposits predominated here, pointing to the partially enclosed and hypersaline nature of the sea. The Taimyr basin, joining with the Khatanga and, during maximum transgression, with the Tunguska, experienced a comparatively strong subsidence, characterised by great depth, normal salinity of water and sedimentation of clay-carbonate facies of the Domanik type in the west. Limestones and, less frequently, dolomites predominated in the east. Subsidence of the Rybinsk basin and its filling with marine and continental terrigenous rocks also continued.

The Viluy aulacogen did not experience stable subsidence, was characterised by high tectonic mobility and had almost no connection with the sea. Its characteristic feature is an abundance of basaltic lava eruptions. Faults divided the aulacogen into two basins: Ygyatty and Kempendyay. Continental sands accumulated in them, in addition to basalts, while gypsum- and salt-bearing formations accumulated in the Famennian. The total thickness of the Upper Devonian formations in the Kempendyay basin exceeds 900 m, in the Tunguska basin 100 to 200 m and in the Taimyr 500 to 600 m (over 1 km in the west).

E a r l y   C a r b o n i f e r o u s. The axial zone of the Taimyr basin experienced strong subsidence. Tectonic reactivation occurred from this epoch here and the zone became a constituent of the Verkhoyansk-Kolyma

162

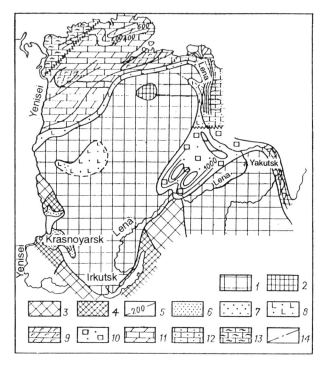

Fig. 4-3. Palaeotectonic scheme of the Siberian Craton in the Frasnian and Famennian stages (after N.S. Malich. In: *Geological Structure of the USSR and Pattern of Location of Mineral Deposits.* Vol. 4, *Siberian Craton,* 1987).

1-4—uplift regions (1—weak on platform; 2—moderate on platform; 3—weak in mobile region; 4—moderate in mobile zone); 5—isopachs, m; 6-12—formations (6—Tunguska variegated limestone-marl-sandstone; 7—Baeronov quartz-sandstone; 8—Nakokhozsk and Dyavadyakitsk sulphate-carbonate; 9—Dudinka halogenous; 10—Kygyl-Tuu tuffaceous-halagenous; 11—Kolargon and Nyukunnin limestone-dolomites; 12—variegated sandy clay); 13—bituminous-dolomite-limestone parageneration; 14—faults influencing sedimentation.

mobile region as its north-western branch. As a result, the area of the craton shrank even more. By the commencement of the Carboniferous, the Tunguska basin had experienced almost total uplift and regression and the Siberian Craton represented a vast plain bound in the south by the mountain systems of the Sayan-Baikal orogenic belt. In the central and north-western parts of the craton slight subsidence occurred and major freshwater basins existed. Sand, silts, clays and, less often, limestones were deposited in these freshwater basins to a thickness of a few tens of metres. Subsidence of the basin of the Viluy aulacogen, covered by lagoons and sometimes in connection with the sea, continued. Eruptions of basalts ceased here and limestones, dolomites and sulphates along with continental terrigenous rocks up to 500 m in thickness have been observed. In the south-western part of the craton the Kan Taseevo basin continued to subside. Its dimension increased and pebble, sand and clay beds accumulated in it, their thickness in the south-west rising to 300 to 500 m.

For the Siberian Craton as a whole, the Middle Devonian to Early Carboniferous stage was characterised primarily by a significant reduction in its area due to the north-eastern and northern regions experiencing tectonic reactivation. Elsewhere in the craton fluctuation of sea level and a general tendency towards enlargement of uplift and regression were manifest. Therefore, in the Early Carboniferous almost all of its territory was in the form of dry land. An important event was subsidence along the faults of the Viluy aulacogen accompanied by large eruptions of basalts.

4.1.4 Sino-Korea Craton

M i d d l e   D e v o n i a n. Almost the entire area of this craton experienced uplift and denudation (*Atlas of the Palaeogeography of China,* 1985). Basins with paralic conditions of sedimentation existed only within the Tarim block.

L a t e   D e v o n i a n. In addition to inherited basins on the periphery of the Tarim block, a new one arose west of the Ordos syneclise in the present-day upper reaches of the Huanghe River. This basin was wholly intracontinental. It was filled with disintegration products of the Qilianshan orogen located southward.

E a r l y   C a r b o n i f e r o u s. The area of the craton enlarged in the south-west since the orogenic development of the Qilianshan system, including the Qaidam block, was completed and transition to a platform regime took place in this vast territory. The remaining area of the craton, however, experienced only weak uplift as before, with formation of a thick weathering crust. In the region of the former Qilianshan orogenic system, two residual troughs were still preserved and coarse continental and marine sand-pebble beds, coal-bearing formations and sometimes reef limestones (in the more subsided sections in connection with the open sea) accumulated in these residual troughs. The thickness of these deposits did not exceed 200 m.

4.1.5 South China (Yangtze) Craton

M i d d l e   D e v o n i a n. As a result of the dying out of tectonic activity, much of the Cathaysian orogen became a part of the craton, slightly enlarging its area. Its northern regions experienced weak uplift and plain or undulating topography developed here (*Atlas of the Palaeogeography of China,* 1985). Fairly strong subsidence, often interrupted by brief regression, continued only in the southern marginal zone. The internal structure of this zone was complex: in the most depressed parts of the Hubei-Guizhou and

Yunnan-Guangxi basins, the thickness of the Middle Devonian formations reaches 700 m and in the southern part of the craton (within northern Viet-nam) 1500 m (Yang Shin-Pu et al., 1981). Sand-conglomerate, alluvial and littoral marine deposits were formed in the marginal parts of the South basin. Deep in the basin, they were replaced by carbonates (calcareous) and sandy-clay complexes.

L a t e   D e v o n i a n. The craton experienced subsidence at places, leading to transgression reaching the lower course of the Yangtze River. Large islands separated by basins and straits existed within the basin. Accumulation of carbonate and terrigenous deposits occurred in the basins, with carbonate deposits predominating in the Yunnan-Guangxi basin. In the northern, very shallow part of the basin, lagoonal, alluvial-deltaic and sometimes coal-bearing facies a few tens of metres thick accumulated. The thickness of the Upper Devonian rises to 500 to 900 m in the basins, reaching 1500 m in the far south.

E a r l y   C a r b o n i f e r o u s. A differentiated environment was preserved in this epoch and subsidence of isometric basins combined with stable uplift of the islands separating them. A very shallow lagoon existed in the Hubei-Guizhou basin and in the lower course of the Yangtze and organogenic limestones 10 to 30 m thick were deposited in it. The Yunnan-Guangxi and Jiangxi-Guangxi basins experienced somewhat greater subsidence; the thickness of carbonate and terrigenous rocks in them reached 700 to 1000 m. Continental facies and paralic coal-bearing formations 100 to 200 m in thickness were extensively developed in the littoral zones of these two basins.

In the Early Carboniferous continuous and stable subsidence of the southern marginal zone of the craton ceased and the Shong Lo massif was transformed into an island.

Thus the Middle Devonian to Early Carboniferous stage for the South China Craton was a period of significant increase in its total area due to the Cathaysian orogenic system completing its active development. But the palaeotectonic conditions were characterised by a comparative complexity and contrast of movements which made for variations in facies and their thickness, with a general preponderance of carbonate formations and an increase in thickness towards the south.

4.1.6 South American Craton

M i d d l e   D e v o n i a n. There was some redistribution of the areas of subsidence and uplift but sea covered a very small area while uplift prevailed on land. Clastic material originated from this uplift and was supplied to all the sedimentary basins. In the Amazon basin the subsidence zone advanced far westward and joined with the sea of the Andean mobile system as a result of transgression. However, uplift and regression manifest in the region of the strait connecting the Amazon with the Maranâo basin interrupted this connection. Sedimentary formations of the Amazon basin were represented by sandy-clay rocks while limestones developed to a small extent in the east. The total thickness of the Middle Devonian did not exceed 300 m.

In the Maranâo basin subsidence occurred within the former boundaries but its connection with the Parana basin was also interrupted as a result of uplift. The thickness of the sand deposits in the Maranâo basin reached 400 m. The Parana basin shrank markedly in size, deepened (thickness of marine sandstones up to 300 m) and acquired a stable connection with the sea of the Peri-Andean marginal zone. The latter extended in the form of a comparatively narrow strip along the entire Peru-Bolivian segment of the Andes. The thickness of the terrigenous formations depositing here increased rapidly up to 500 to 700 m from east to west.

Late Devonian. Significant intensification of uplift led to a general regression. Seas were preserved in notably reduced size only in the Amazon and Maranâo basins. Further, the connection of the former with the sea of the Andean system was interrupted and abundant clastic material wholly compensated the weak subsidence in the midcourse of the Amazon where terrigenous rocks were deposited in continental facies. The thickness here as well as in the east where marine conditions were preserved but regression continued to develop, did not exceed 150 m. As a result, even in the Famennian stage the sea regressed from the Amazon basin. In the Maranâo basin subsidence was more stable and was preserved throughout this epoch. The thickness of sandy-clay beds reaches 500 m. In the Peri-Andean marginal zone, developed as before only along the Peru-Bolivian Andes, subsidence continued and accumulation of coarse sandstones occurred in marine as well as in continental facies under conditions of abundance of clastic material brought down from the elevated parts of the craton. The thickness here is usually 500 to 600 m, sometimes reaching 1000 m.

Early Carboniferous. Subsidence and marine conditions were preserved only in the Maranâo basin whose size diminished even more. This was evidently an isolated interior basin. The thickness of alluvial and lacustrine terrigenous formations in it did not exceed 200 m. The remaining part of the craton was uplifted and remained as land.

Thus in the South American Craton the stage under consideration was a period of successive intensification of uplift, regression and complication of land topography.

4.1.7 African Craton

Middle Devonian. Subsidence intensified in the extensive territory of the Sahara basin and transgression extended south of the boundaries of the Early Devonian sea. Subsidence was significant in western

Egypt (Fabre, 1988). The Reguibat massif subsisted in the form of a major island. A small marine basin with boundaries not clearly established so far appeared in southern Ghana. These extensive seas in the northern and north-western regions of the craton were shallow; subsidence was replaced in them by uplift and regression while sandy clay deposits predominated among thin deposits (mainly tens of metres, less often up to 100 to 200 m) (see Fig. 2–3). In the littoral parts in the south (Nigeria, Chad and Sudan), alluvial-deltaic facies alternate with marine facies. The most stable subsidence was manifest in the Tindouf and Saoura-Ougarta basins (thickness of Middle Devonian up to 600 to 900 m; besides clay rocks, carbonates including reef carbonates accumulated). The zone of carbonate sedimentation in the Saoura-Ougarta basin in the north enlarged, joining the shelf of the Peri-Atlas marginal zone. Subsidence was stable in this zone too (Buggisch et al., 1979). The north-western part of Sicily, eastern part of the Apennine peninsula and western part of the Balkan peninsula (Apulian block) fall in this marginal zone. They experienced uplift and represent the northernmost salient (within present-day boundaries) of the African Craton.

Subsidence continued, as before, in the north-eastern part of the craton and only slightly enlarged in eastern and north-western Iran (Pyzhyanov et al., 1980). The seas here were very shallow and carbonate and clay deposits 200 to 400 m in thickness were deposited in them. Only the central part of the basin, where the thickness of the Middle Devonian went up to a maximum of 1200 m, stood out prominently in the intensity of subsidence.

In the Cape basin, in the extreme southern part of the craton, subsidence slowed down, interior uplift was manifest in the form of islands and continental facies played a major role in the composition of the sand formations of the Witteberg group.

L a t e D e v o n i a n. The extensive marine basin in the northern and north-western parts of the craton remained shallow. It attained maximum proportions in the Frasnian but began to decrease gradually from the Famennian as a result of intensification of uplift in the central regions of the craton. The role of coarse deposits and continental facies increased in the extensive littoral zone; their thickness was not much, rarely attaining 100 to 150 m. The Reguibat massif continued to rise while extremely stable and strong subsidence was manifest in the Saoura-Ougarta and Tindouf basins where formations 2 km or more in thickness accumulated, with clay and carbonate rocks predominating in their composition. In the Peri-Atlas zone, which likewise continued to subside, sandy-clay beds exceeding 1 km in thickness were deposited; reef limestones too were deposited in the north. The former environment was preserved in the Apulian block. At the end of the Late Devonian, almost all of the territory of the Sahara basin experienced a regression.

Subsidence enlarged in the north-eastern part of the craton and a series of basins represented the marginal zone of subsidence bordering the corresponding sector of the Mediterranean mobile belt. Transgression covered almost all the territory of Turkey, Iran, Iraq, Afghanistan and part of Syria. Seas remained shallow and even lagoon-like, as before, and carbonates and sandy-clay deposits 200 to 500 m in thickness were deposited in them. Extremely strong subsidence of the central (Iraq) basin continued and it acquired a linear (north-western strike) form. The thickness of D_3 here was about 900 m.

Development of the Cape basin in the south was completed. Subsidence was interrupted in it by uplift and it became filled with marine and continental sand formations.

E a r l y C a r b o n i f e r o u s. In the northern part of the craton (Sahara basin) subsidence renewed, following a general regression at the end of the Devonian, its magnitude being significantly less than that of the Devonian. A relatively large maximal transgression was manifest in the first half of the Viséan stage and the Sahara was no longer a single basin, but several broad basins divided by islands and extended uplifts or thresholds. Lagoonal sandy-clay and sometimes carbonate deposits formed in these basins. In the Saoura-Ougarta basin the role of carbonates was significant while the deposition of gypsum and salt points to high salinity of water. The thickness of the Lower Carboniferous in the Saoura-Ougarta and Jeffara basins exceeded 1 km. In the Late Viséan and Namurian the western and southern regions of the Sahara basin experienced uplift and alluvial-lacustrine sands, clays and at places carbonaceous deposits accumulated here under continental and lacustrine-marshy conditions. In the Air region of the south-eastern Ahaggar massif and possibly in the Jebel-Oweinat region at the confluence of the boundaries of Libya, Egypt and Sudan, glacial formations appeared in the Lower Tournaisian. In contrast to the foregoing, following regression in the Middle Viséan, a new transgression occurred in the northern Sahara at the end of the Middle Viséan and commencement of the Namurian (Serpukhov age). This new transgression was accompanied by deposition of essentially carbonates (unlike the predominantly terrigenous lower part of the Lower Carboniferous) including thick bands of limestones and evaporites (gypsum). In the west, as in the Tournaisian, shoaling is observed. This has been interpreted by Fabre (1988) as due to the influence of the Mauritanides system beginning to rise; shoaling also proceeded eastward, towards Libya.

Much of the Peri-Atlas marginal zone in this epoch experienced break-up, subsidence and sharp intensification of tectonic activity (Moroccan Meseta and High Atlas). In the region of the Moroccan Rif and in northern Algeria, shallow-water shelf conditions were preserved and accumulation of carbonate-terrigenous rocks proceeded.

In the north-eastern part of the craton enlargement of uplifts led to a regression and marine basins were preserved within Turkey, north-eastern Syria and Iran and also northern Afghanistan. The dynamic palaeotectonic environment with contrasting vertical movements made for facies variability of formations and their variable thickness. Here were deposited purely reef carbonate formations, carbonate-terrigenous, sandy-clay, continental sand and sand-conglomerate beds from a few tens of metres to 1 km (north-western Syria) in thickness. At the end of the epoch, uplift intensified and the area of sea diminished even more.

In the Cape basin in the southern part of the craton subsidence concluded with the deposition of continental sands of the 'upper horizon' of the Witteberg series. The uplift of western regions was accompanied by folding processes (western Capides) which affected all the older Palaeozoic formations. The immense territory of the craton north of the Cape basin and south of the Sahara represented a zone of moderate uplift and plain-undulating topography throughout the stage under consideration, although intracontinental basins similar to those tentatively delineated in eastern Ethiopia and Zambia may have existed in some regions.

On the African Craton as a whole, the Middle Devonian to Early Carboniferous stage (as on the South American Craton also) was a period of succession of transgressions and regressions and gradual intensification of the contrast of tectonic movements.

4.1.8 Indian Craton

Throughout the stage under consideration the entire territory of the craton evidently experienced an uplift since formations of the Middle Devonian to Early Carboniferous are not known here. Only in the north (the Himalaya and southern Tibet) did gentle subsidence continue almost incessantly (Jain et al., 1980). M i d d l e   D e v o n i a n formations were represented in this marginal zone by carbonate-terrigenous rocks up to 100 to 200 m in thickness. These are replaced by sand-beds in the littoral regions in the south. The U p p e r   D e v o n i a n is made up of only sandy-clay complexes up to 400 m in thickness and the L o w e r   C a r b o n i f e r o u s again by carbonate-terrigenous (up to 300 m) with characteristics of some regression of the basin.

4.1.9 Australian Craton

M i d d l e   D e v o n i a n. The formations of this epoch, like those of the Early Devonian, are not known on the craton; evidently its entire territory was a region of uplift and dry land.

L a t e   D e v o n i a n. Tectonic movements revived. Subsidence and transgression encompassed the Carnarvon, Canning and Bonaparte basins (Fig. 4-4). Their size was not significant but the rate of subsidence was

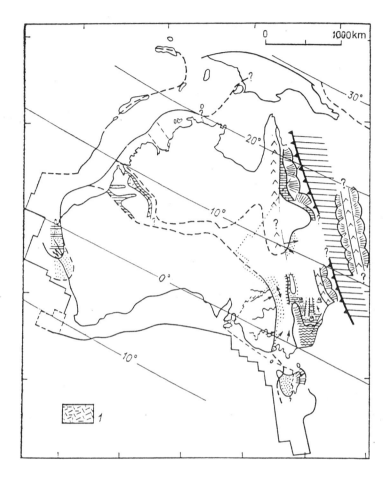

Fig. 4-4. Palaeogeographic reconstruction of Late Devonian and Early Carboniferous Australia (from Veevers, 1986). See Figs. 1-4, 2-5 and 3-3 for legend. 1—evaporites.

considerable and fairly thick (1 to 1.3 km) carbonate-terrigenous formations accumulated in them. Further, the salinity of water in the Canning basin was high (Veevers, 1986). In the Amadeus basin too extensive subsidence and sedimentation renewed. However, calcareous sandstones and conglomerates of the Pertijara formations have a distinctly continental origin and their thickness reaches 500 m. This was an interior lacustrine-alluvial basin and sediment inflow into it came mainly from the north, from the heaving Arunta massif.

E a r l y C a r b o n i f e r o u s. Subsidence continued in the Carnarvon, Canning and Bonaparte basins but conditions in them underwent change. Marine conditions were preserved only in the Carnarvon basin and about 800 m of carbonate-terrigenous formations were deposited in it. In the other two basins subsidence was evidently wholly compensated by clastic material flowing abundantly from the elevated central regions of the craton. Thus only continental sand facies up to 1 km in thickness were deposited in them. Further, in the Canning basin subsidence was concentrated in the Fitzroy graben (aulacogen) bound by faults. In the Amadeus basin territory uplift commenced at the site of the former extensive intracontinental trough. This uplift was associated with intense fold formation, resulting in complex structures, including overthrusts.

On the whole, on the Australian Craton tectonic activity intensified throughout the Middle Devonian-Early Carboniferous; the manifestation of intense fold-overthrust processes on the platform cover of the central part of the craton is interesting. The origin of such palaeotectonic conditions in a generally stable platform regime is not yet clear.

4.1.10 East Antarctic Craton

At the end of the Early, in any case by the beginning of the Middle Devonian, the Pacific margin of the East Antarctic Craton together with the Rosside orogen building it up (much flattened by that time), was affected by transgression with accumulation of predominantly sand deposits containing cold-loving Malvinokafre fauna of Brachiopoda and plant remnants. These fauna are characteristic of other Gondwana Continents (South America and South Africa). The thickness of the Devonian formations in the Pensacola mountains reaches 3100 m but is usually only a few hundred metres thick. The cover of the Antarctic Craton, termed the Beacon supergroup, commences with these formations. It has been suggested (Bradshaw, 1991) that a marginal mainland basin, similar to the South African Karroo basin, extended west of the Transantarctic Mountains, especially in the region of the Ellsworth Mountains. The thickness of the Palaeozoic in this basin reached 8 km. From the ocean side it was bound by the volcanic arc of Mary Byrd Land-northern Victoria Land (see below).

4.2 REGIONAL REVIEW. MOBILE BELTS

4.2.1 North Atlantic Belt

4.2.1.1 *Ouachita-Mexican system.* Deepwater conditions and an environment of accumulation of thin clay and siliceous silts were preserved throughout the territory of the system in the M i d d l e and L a t e D e v o n i a n. The total thickness of deposits did not exceed 100 to 150 m.

It may be recalled that such a palaeotectonic environment has been fixed in this region over an exceptionally prolonged period, at least from the beginning of the Ordovician, i.e., during three periods (about 160–180 m.y.). This is a unique example in geological history of monotypical conditions being protractedly preserved in a mobile belt. Over such a prolonged period, only 700 to 800 m of fine deposits. formed. The sequences here are incomplete, however, and there are stratigraphic breaks but no angular unconformities are known.

The first significant changes in the development of the Ouachita-Mexican system occurred in the E a r l y   C a r b o n i f e r o u s. Filling of the deep-water basin with turbidites and uplift of several of its interior blocks in the form of islands with probable volcanic eruptions commenced (Haley et al., 1979; Kier et al., 1979; Morris, 1974; Weyl, 1980). The thickness of flysch formed under continental slope and rise conditions is about 3 km. Deepwater conditions were preserved only in the Mexican part of the system in which the thickness of the Lower Carboniferous shales is 200 to 300 m.

Details of the palaeotectonic conditions in the zone of southern Sierra-Madre de Chiapas have not been clearly deciphered but it is known that submarine-volcanic, siliceous and greywacke complexes running into hundreds of metres in thickness accumulated here in the Late Devonian and Early Carboniferous. Even the old Honduras block experienced subsidence but uplift of the Yucatan block was stable.

4.2.1.2 *Appalachian system.* M i d d l e   D e v o n i a n. In the Northern Appalachians uplift became predominant and an orogenic regime was established by this stage. As a result, orogenic processes were manifest throughout the territory of the Appalachian system although residual marine troughs also existed (Merrimack etc.). The largest of these troughs was in the Northern Appalachians; it was intensely filled with coarse terrigenous deposits and in the axial zone with flysch. Here, and in other basins also, contrasting basalt-rhyolite volcanism was manifest along with accumulation of molasse. The most extensive subaerial eruptions of rhyolites occurred on Newfoundland Island. The thickness of the Middle Devonian formations is 2 to 3 km and even 4.5 km; over one-half of the sequence is made up of molasse. This suggests in particular intense uplift and formation of an extremely rugged topography. Moreover, in the Middle Devonian fold and overthrust processes continued and intensified, reflecting the growing manifestation of Acadian orogeny. In the zone of its most distinct manifestation, the interior zone of the Northern Appalachians, Acadian diastrophism was accompanied by powerful intrusive activity and regional metamorphism. Granitoids intruded throughout the system. The corresponding complex in New England is called the Oliverian.

L a t e   D e v o n i a n. High tectonic activity prevailed throughout the area of the Appalachian system. Residual marine troughs were filled with sediments, underwent uplift and subsequently regression. Intense uplift made for a high topography and molasse formed in the intermontane basins. Its maximum thickness (over 3 km) is known in the south-western part of Newfoundland Island. Here a major rift graben with subaerial manifestations of intermediate and felsic volcanism (Horton group) was formed at the end of the epoch. Coal-bearing deposits accumulated in the central sector of the system. Development of the fold-overthrust Appalachian structure concluded and intrusion of granites, especially of the New Hampshire complex of diorites, tonalites, monzonites and granites continued. The regional metamorphism of all the much older strata is associated with the plutons of this complex.

E a r l y   C a r b o n i f e r o u s. As before, an orogenic tectonic regime was manifest in the territory of the Appalachian system. Intensification of subsidence of several basins in the Southern Appalachians led to brief transgression from the adjoining Peri-Appalachian platform basin. Besides molasse, marine terrigenous sediments and sometimes reef carbonates were deposited in them. Their thickness in the Lower Carboniferous is 1 to 1.5 km. In the Northern Appalachians and on Newfoundland Island subsidence of basins led to the formation of salt lakes in which salt-bearing formations and dolomites accumulated. Andesite subaerial volcanism combined with accumulation of molasse in an intermontane basin renewed on Newfoundland Island and continued south into Nova Scotia. The total thickness of the Lower Carboniferous (Windsor group) here reaches 6 km. This basin, Fundy basin, followed the Late Devonian rift. There was gradual attenuation of fold and metamorphic processes in much of the remaining area of the system.

Thus in the Middle Devonian to Early Carboniferous, the Appalachian system developed in an orogenic regime: intense uplift prevailed, rugged topography existed, molasse formed, metamorphism and intrusive and eruptive magmatism were manifest and complex tectonic deformations developed. The intensity of this entire complex of processes, known as the Acadian epoch of orogeny, was highest at the end of the Middle to commencement of the Late Devonian.

4.2.1.3 *Eastern Greenland system*. An orogenic regime prevailed in the territory of the system throughout the stage under consideration (Harland, 1979; Henriksen, 1978). Formation of a rugged topography was accompanied by formation of molasse 2 to 2.5 km in thickness for each epoch. In this system fold and metamorphic processes and granitisation were also manifest but precise data concerning these aspects is still inadequate.

4.2.1.4 *British-Scandinavian system.* M i d d l e D e v o n i a n. The orogenic development of the system concluded. Intensity of uplift, mountain formation, fold-thrust and metamorphic processes gradually weakened. Volcanism died out and only a small burst of it has been fixed at the end of the Middle to commencement of the Late Devonian in north-eastern Scotland. The Midland Valley graben in the northern British Isles separating the northern Scotland-northern Ireland and southern Scotland uplands from the Early Devonian, was already filled with molasse by this epoch (House et al., 1978). The intrusion of granites ceased here and the graben experienced a two-sided compression. As a result, its sedimentary filling underwent rather intense folding. At this same time, the northern Scotland Highland was dissected by the Great Glen strike-slip fault and the Orcade basin formed in the north-eastern part of the Highland. The rapid subsidence of the basin was compensated by the accumulation of thick molasse of the middle Old Red (over 3 km). The main displacement along the Great Glen and the Minch fault similar to it in age and direction of movement, occurred in the Middle Devonian, i.e., in the concluding stage of Caledonian orogeny of the British segment, roughly corresponding to the Acadian orogenic epoch of the Appalachian system.

The southern part of the British Isles experienced less orogenic uplift. In the central region of the Welsh uplift fold processes had already begun in the Early Devonian but the main folding occurred there as well as in the Wales borderland belt in the Middle Devonian. Along the south-western margin of the British segment, in southern Wales, northern Devonshire and south-western Ireland, transition of continental facies of the Old Red sandstone into lagoonal-paralic and marine facies took place. Here the zone of development of the Old Red served as a margin of the Hercynian Mediterranean mobile belt that developed intensely from the Devonian time.

Filling of the intermontane basins with molasse ceased in the Scandinavian segment. These basins, unlike those in Britain, are small in size and confined only to the southern half of Norway. In the largest, Rorägen basin (graben), the thickness of the Middle Devonian goes up to 2 km (Oftedahl, 1980). A minor felsic volcanism manifested at the end of the epoch and Devonian formations were crumpled into germanotype folds; folding was very intense at places, however. An analogous palaeotectonic environment has been revealed in Spitsbergen. A complex graben situated between the Raudfjorden and Billefjorden faults completed its subsidence and filling with red continental molasse (thickness up to 1.5 km). This graben later underwent intense dislocation right up to development of a cleavage but only in its marginal parts around faults; the Middle Devonian strata preserved subhorizontal bedding in its central portion.

Commencing with the Late Devonian, the entire British-Scandinavian system (including Spitsbergen) experienced successive levelling of the topography and attenuation of tectonic and magmatic activity; it entered the platform stage of development and became a part of the East European Craton.

4.2.1.5 *Innuitian system*. M i d d l e D e v o n i a n. This system lay between the Hyperborean and North American continents and joined with the North Atlantic belt only in north-eastern Greenland. But the main events of its tectonic history conformed, as before, to the general evolution of the latter belt. Filling of the former deepwater trough with terrigenous formations was completed in the western part of the system. Turbidites were also present amongst the sediments, pointing to their formation in an environment of continental slope and rise (Trettin and Balkwill, 1979). A complex of basaltic submarine lavas (spilites), cherts and terrigenous rocks possibly formed in the northern part of Axel Heiberg Island even in the Middle Devonian. In the eastern Innuitian system an orogenic regime with manifestation of folding and metamorphism was maintained.

L a t e D e v o n i a n. A residual marine basin still prevailed in the western part of the system but the composition of its deposits changed sharply. Continental coarse clastic rocks and also coal-bearing facies played a significant role in these deposits and the overall thickness of the Upper Devonian here exceeded 3 km. All this suggests a transition of the western part of the Innuitian system to an orogenic regime. Further, orogeny evidently extended far into the south-west and affected the northern platform block of Alaska. As a result, the Innuitian orogenic system acquired a very large extent, encircling the North American Craton from north-eastern Greenland to Alaska. Much of its territory represented land with a rugged topography, active manifestation of intrusive magmatic (granite formation) and tectonic (Ellesmere folding) processes and regional metamorphism. Further, residual marine basins were still preserved in Alaska and the thickness of marine terrigenous sediments in the Upper Devonian reaches 3 km in the largest of these basins in the region of Brooks Range. But molasse, including coal-bearing, was deposited in the northern part of this basin.

E a r l y C a r b o n i f e r o u s. Orogenic development of the Innuitian system was completed. It still preserved its vast dimension but tectonic activity diminished. The residual Sverdrup basin experienced uplift and regression by the beginning of this epoch and folding manifested here as in northern Greenland. New troughs (grabens) arose in northern Axel Heiberg and Ellesmere islands and also in north-eastern Greenland. They were filled with thick molasse, partly coal-bearing, and marine terrigenous rocks. In the west, in the residual basin of Brooks Range, marine conditions of sedimentation predominated and Lisburne reef limestones formed extensively. Tectonic activity was low in Alaska.

Thus the intensity of orogenic processes rapidly rose within the Innuitian system during the period of Middle Devonian to Early Carboniferous and significantly enlarged the territory of this orogen. However, by the end of the stage under consideration, orogeny had concluded here, a stable cratonic regime was established and these regions became a constituent of the North American Craton in the Middle Carboniferous.

4.2.2 Mediterranean Belt

4.2.2.1 *West European region.* M i d d l e   D e v o n i a n. A complex structure and high tectonic activity prevailed throughout the territory of this region. The eastern part of the massif Central subsided and all other microcontinents (Bohemian and northern Armorican massif, Midgerman High, Tissia and others) heaved as before. Volcanic activity intensified significantly. Extensive submarine basaltic eruptions (sometimes spilite-keratophyre) occurred to the north and north-east of the Bohemian massif (in the Rhenohercynian zone) and eastern Tissia (in the Morava-Silesian zone). These eruptions were combined with the deposition of clayey, siliceous and greywacke beds usually not exceeding 1 km in total thickness but in some 'oval subsidences' of the Rhenohercynian zone, reaching 2 to 3 km (Belov, 1981; Behr et al., 1980; Lutzens, 1980; Ziegler, 1982). Deepwater conditions evidently prevailed in all these regions. The same is true of the Saxothuringian zone in which the Middle Devonian is represented by black shales but uplifts and a minor folding (Reiss phase) were manifest here at the end of the epoch.

Along the boundary with the East European Craton and on the eastern margin of the Bohemian massif, reef carbonate deposits formed extensively and south of the Moesian platform flysch deposits (Spassov, 1973; and Krebs, 1974). Both deposits accumulated under continental slope conditions. So did the very thick clay and sandy-clay beds of the Ardennes and southern England (Matthews et al., 1980; Ziegler, 1982). The accumulation of reef carbonates combined with terrigenous sedimentation in the vast areas of subsided margins of the Armorica and western Central massifs. Andesite-basalt calc-alkaline volcanism of the island-arc type occurred in the south-eastern part of the massif Central.

Within the Iberian peninsula, in the Cantabrian, the West Asturia and Leonese zones, a weak interrupted subsidence and deposition of carbonate-terrigenous and terrigenous formations of the shelf type to a thickness of 600 to 800 m continued under shallow-water conditions (Guy Tamain, 1978; Julivert et al., 1980). The Cantabrian zone contains Middle Devonian beds of shales which formed under more pelagic conditions. In the Central Iberian zone uplift intensified and the land mass enlarged. A deep trough was preserved to the west and south-west in the Ossa-Morena and Southern Portugal zones. Greywackes and clay formations up to 1.5 km in thickness were deposited in this trough.

The Atlas zone (Maghreb), traced from the Moroccan Meseta and High Atlas to the Kabylie massifs of the Algerian Tell, represents the extreme south-western link of the Mediterranean belt. The Middle Devonian includes greywackes and reef carbonates in the west, in the Moroccan Meseta, and sandy-clay beds with turbidites in the Atlas and Kabylie massifs. The general conditions of development in this zone resemble those of the Rhenohercynian zone. According to Boulin (1991), the oceanic crust of the Atlas basin experienced subduction under Meseta in the Early Devonian to Early Visean, causing deformation of its margin and intrusion of granites.

In the future Alpine fold belt (including Dinarides, Carpathians and Balkans besides the Alps), carbonate (shallow and deepwater) as well as terrigenous (clayey, sandy and siliceous, i.e., lydites) and volcanic (mainly diabases and porphyroids) formations were widely developed in the Middle Devonian (Belov, 1981). These formations evidently correspond to various palaeogeographic and palaeotectonic conditions. The most distinct facies differentiation is noticed in the Eastern Alps. Shelf limestones accumulated south of the Peri-Adriatic suture (i.e., on the margin of the Adriatic microcontinent) and extended into the Outer Dinarides. Through a band of reefs, they change to deeper water varieties and turbidites of continental slope and rise while shales, lydites and diabases of the bottom of the deepwater basin appear even more to the north. Various shales and lydites too predominate north of the Peri-Adriatic suture; bands of limestones are present and island arc volcanics are observed. A band of formations of this type extends eastward through the Sendrö-Chipony Mountains (northern Hungary) into the Gemerian zone of the Western Carpathians and farther into the eastern and southern Carpathians, Balkans and Dinarides. Outcrops of much older ophiolites are associated with the outcrops of these formations. Basic volcanics, diabases and spilites probably correspond to the continuing intraplate eruptions at the bottom of this basin with a crust of oceanic type while manifested felsic volcanics and coarse clastic rocks correspond to island arcs which complicate the structure of the basin. The northern flank of this ocean basin, i.e., the Palaeo-Tethys, is noticed only in the northern greywacke zone of the Eastern Alps and in northern Dobrogea from the predominance of limestones and dolomites. The axial zone of the Palaeo-Tethys coincided, according to Belov (1981), whose reconstruction is shown in Fig. 4-9, with the Vardar zone; but it could extend even more northward while the Vardar zone represents a southern marginal sea.

L a t e   D e v o n i a n. There were no major changes in the mode of tectonic development of the belt at the verge between the Middle and Late Devonian and this feature distinguishes the Mediterranean belt from the North Atlantic belt where Middle Devonian (Acadian) orogeny is so extensively manifest. The scale of submarine volcanic activity decreased and spilite-keratophyre eruptions have been fixed only in the Saxothuringian

one where deep basin conditions renewed after a brief phase of folding
nd clay deposits accumulated. Clay and greywacke beds ranging from 1
) 3 km in thickness were formed in the troughs of the Rhenohercynian
one from south-western Ireland to Poland while flysch accumulated in the
Morava-Silesian zone in the east. The Bohemian massif experienced a par-
al subsidence and the Upper Devonian was represented in its basins by
hallow-water limestones and sandstones. The Central and Armorican mas-
ifs were completely submerged but southern Armorica experienced uplift
nd remobilisation and anatexis of old complexes with the formation of
nigmatites. In the south-eastern massif Central volcanism continued but
s composition was highly felsic, i.e., andesite-rhyolite. Similar island arc
olcanic manifestations were noticed in the Corsica and Sardinia islands.

The same conditions as existed before prevailed in the northern Iberian
eninsula in the Cantabrian zone. Here carbonate and terrigenous rocks
ere deposited under shallow-water conditions. However, the trough of the
West Asturia and Leonese zones deepened and accumulation of sandstones
redominated in it. The Central Iberian island experienced uplift and erosion
s in the Middle Devonian. At the end of this epoch, folding manifested here
nd determined the structure of its central regions. On its south-western
nd western slopes flysch beds were formed in the Ossa-Morena zone.
eplacement of subsidence by uplift and formation of a small island massif
ave been assumed in the northern part of the Southern Portugal zone but
 deep trough with minor manifestation of basic volcanism was preserved
 the south.

In the Atlas (Maghreb) zone in north-western Africa, Upper Devonian
rmations are represented by sandstones, shales, limestones and some-
nes conglomerates and formation of the olistostrome type. Flysch too was
rmed in the Kabylie massifs. At the end of the Late Devonian, break-up
nd deep subsidence commenced in the south-western part of the Moroccan
eseta, uplift encompassed the Rehamna massif (Moroccan Meseta) and
tense movements causing folding and thrusting manifested in Maghreb
Beauchamp and Izert, 1987).

Like most other regions, the segment of the Mediterranean belt under
onsideration in the future Alpine fold region proper underwent no percep-
ble changes in the Late Devonian compared to the Middle Devonian. Par-
cular mention must be made of the deepening of the basin in the Eastern
nd Southern Alps, which continued into the first half of the Early Carboni-
rous.

E a r l y   C a r b o n i f e r o u s. The verge between the Devonian and
arboniferous is an extremely important one in the tectonic development of
is segment of the Mediterranean belt; it corresponded to the commence-
ent of active deformation of sedimentary and volcanic strata, uplift and
anite formation. This Bretonian epoch of orogeny commenced even in

the Late Devonian in southern Armorica and in the Moroccan Meseta and was most notably manifest in the Central European zone of uplifts. From the Early Carboniferous, upheaval encompassed the entire Armorican massif, massif Central and many minor blocks; uplift likewise intensified in the Bohemian massif. In the margins of the Central and Armorican massifs subaerial basalt-rhyolite eruptions occurred, coal-bearing deposits accumulated and granitisation, metamorphism and folding represented important processes (Autran, 1980; Paris et al., 1986; Ziegler, 1982).

Bretonian upheaval led to intense inflow of clastic material and the nature of formations on the slopes of massifs underwent notable change. These formations were represented by terrigenous deposits commonly known as Culm and comprise shallow-water-littoral coarse clastic deposits, sometimes with plant detritus and bands of organogenic limestones. They were intercalated with tuffs and lavas of felsic and intermediate composition. Culm is very close to molasse and goes up to 3 km or more in thickness. South of the massif Central (Montagne Noire), Culm deposits are replaced by flysch.

The Bretonian epoch of diastrophism did not, however, affect the northern part of the system under consideration, i.e., the Rhenohercynian zone. Shale deposits with bands of greywackes and sometimes turbidites were formed here as before while turbidites predominated in the eastern Rhenohercynian zone. A new, probably deepwater trough with deposits of black shales and bimodal volcanics extended into the Early Carboniferous in the form of an arc from the region of the Midgerman High or Polish Sudeten Mountains to Vosges (Pin et al., 1988; Ziegler, 1982). The formation of this structure is associated with rifting and extension of crust, which possibly arose as early as in the Late Devonian.

In the Iberian peninsula Bretonian diastrophism was manifest in granitisation, metamorphism and folding in the Central Iberian zone. Further, all of these processes had commenced even at the end of the Devonian. By the middle of the Early Carboniferous, they had come to an end; this zone experienced a brief subsidence and sand deposits close to Culm accumulated in it. North-east and south-west of the Central Iberian uplift, in the West Asturia and Ossa-Morena zones, flysch was formed in an environment of continental slopes. The thickness of Culm exceeded 4 km in the Southern Portugal trough to which clastic material was supplied by uplifts in Northern Portugal. Between this trough and the Ossa-Morena zone in the Early Carboniferous, there was a huge volcanic belt made up of rocks with a bimodal spilite-keratophyre association. The rich pyrite mineralisation of Rio Tinto is associated with these rocks. In the Cantabrian zone inversion occurred and its rapid subsidence over a short period led to the formation of a deepwater trough with an accumulation of lutites and micritic limestones (Julivert et al., 1980).

In the Early Carboniferous major transformation occurred in the Atlas zone. Bretonian orogeny manifested in the northern Moroccan Meseta commencing with the Late Devonian and Culm formed here in this epoch also. Culm included products of subaerial volcanism with andesite-rhyolite composition. In the north, in the region of the Moroccan Rif, reef limestones and sandstones accumulated in a shallow basin. Break-up and deep submergence in the south-western Moroccan Meseta led to the formation of a deepwater zone with submarine andesite-basalt volcanism (Beauchamp and Izert, 1987; Michard, 1976; Piqué, 1976). In the south the continental slope of this trough lay in the High Atlas region and flysch formed up to 2 km in thickness. In the east development of a shelf basin with terrigenous sedimentation has been assumed in the Kabylie massifs.

The future Alpine Europe also underwent major change during the Early Carboniferous. This change commenced in the Middle Viséan, became fully manifest in the Namurian and was lithologically apparent in the replacement of carbonate or deepwater clay sedimentation by flysch, typically represented by Hochwipfel flysch of the Carnian Alps, with olistostromes quite often associated with it. In eastern and western Serbia flysch is manifest in the Middle and Late Devonian. In the Briançonnais zone of the Western Alps, the Namurian was already represented by a continental coal-bearing formation. An essential similarity is observed between the succession of events in this part of the Mediterranean belt and the Central European Hercynides; no wonder that a similarity of Lower Carboniferous formations with Central European Culm is often remarked. But in some sequences, evidently corresponding to the central zones of basins, the nature of sedimentation right up to the end of the Early Carboniferous epoch remained identical to that of the Devonian. This is true of several areas of the Carpathian-Balkan region.

At the end of the Early Carboniferous, following Culm accumulation, the main epoch of Hercynian orogeny, i.e., Sudetan, took place. This epoch marks the transition of the entire extensive territory of the West European region to an orogenic regime. Sudetan deformations were the main ones in the formation of fold structures of the Iberian Meseta, Armorican massif, Ardennes, massif Central, Saxothuringian and Rhenohercynian zones and Moroccan Meseta. They significantly affected even the future Alpine region where not only fold but also thrust, even nappe deformations, were manifest. At the same time, processes of regional metamorphism and granitisation culminated. These processes had continued and at places even accelerated in the Late Carboniferous.

Thus the development of subsidence and differential tectonic movements came to an end in the Middle Devonian to Early Carboniferous in the West European region, which was ready for transition to an orogenic regime. The general structure exhibited a mosaic pattern. Isometric old massifs were separated by deep troughs and shallower thresholds whose dynamics of vertical movements were characterised by contrasts. Two mighty epochs of orogeny were manifest over large expanses at the verge of the Devonian and Carboniferous (Bretonian) and at the end of the Early Carboniferous (Sudetian).

4.2.2.2 *Caucasus-Pamirs region*. Outcrops of Middle Palaeozoic formations in the extensive territory from Dobrogea in the west to the Northern Pamirs in the east are very few and the tectonic history of the area in this epoch poses problems. In the Middle Devonian, within the Front Range of the Caucasus, Central Asia and the Northern Pamirs, there were large manifestations of submarine, evidently island arc volcanism (mafic to felsic) of extremely diverse composition. Conditions of deep troughs prevailed. Tuffaceous and greywacke beds, partly siliceous, and clay rocks also formed in these troughs (Belov, 1981). The thickness of the Middle Devonian in the Front Range of the Caucasus reaches 4 km. Submarine eruptions of mafic and intermediate composition continued in this epoch in western Alborz in Iran. Regions of north-eastern Turkey, southern Caucasus Minor and Kopet Dag experienced subsidence. Comparatively shallow-water terrigenous formations accumulated in these territories in basins of the shelf type. In the northern part of the region adjoining the East European Craton, accumulation of reef carbonates and shallow-water sandy-clay rocks proceeded under conditions of moderate subsidence of a passive margin. This belt included Dobrogea, Crimea, the entire Precaucasus and central Usturt.

Late Devonian. Differentiation of palaeotectonic conditions occurred in the axial zone of the system. In the northern Front Range of the Caucasus an island arc arose in which felsic subaerial volcanism was manifest. This uplift extended in the east through the central Caspian in the direction of Tuarkyr. The islands supplied abundant clastic material into deep troughs that survived from the Middle Devonian. The troughs were filled with thick flysch strata and also with the products of volcanism already of a typical island arc type, i.e., andesite-rhyolite calc-alkaline. The Upper Devonian reaches 3 km in thickness in the Front Range. By the end of the epoch here and in the Central Asian troughs, deepwater conditions no longer prevailed. Deep troughs survived only in the Northern Pamirs adjoining the mobile zones of Southern Tien Shan and deepwater clay and silica muds of small thickness accumulated there as before. The Transcaucasian massif evidently experienced uplift and metamorphism in the Late Devonian and represented an island. In the troughs of northern Turkey, southern Caucasus Minor and Alborz, the role of reef carbonates rose while sandy-clay deposits predominated, as before, in Kopet Dag. Mafic volcanism continued in Alborz.

The formation of islands in the interior zones of the system increased the flow of clastic material into the northern regions of Dobrogea, Crimea and

Precaucasus. Development of reefs almost ceased there and clastic rocks were deposited. At the end of the epoch, the Dobrogea region experienced uplift and folding, i.e., Bretonian orogeny manifested here while deformation began later elsewhere in these regions.

E a r l y   C a r b o n i f e r o u s. The island arc of the Front Range of the Caucasus subsided. Volcanism ceased and filling of the axial trough with sediments (mainly argillaceous) was completed under conditions of a continental slope inclined southward. The disappearance of islands which supplied clastic material revived extensive accumulation of carbonates in the northern part of the system, i.e., in the region of Crimea and Precaucasus. After Bretonian folding, Dobrogea experienced new subsidence and a coarse clastic formation ('karapelite') of molassoid type formed here. This formation consisted of conglomerates, greywackes and clay rocks up to 2 km in thickness. At the end of the Early Carboniferous, this zone again experienced uplift and folding (Sudetan).

The Transcaucasian massif continued to heave and subaerial felsic volcanism, metamorphism and granitisation manifested in it. Even the southern part of Caucasus Minor was affected by uplift. In the Central Asian part of the system a differentiated environment has been assumed, comprising a combination of uplifts and troughs with terrigenous sedimentation. Manifestations of submarine volcanism have been noticed in northern Iran, Afghanistan and in the Pamirs.

In the northernmost regions of the Pamirs, according to the latest data (Pai, 1992), Early Carboniferous volcanism was characteristic of a primitive island arc. In the Serpukhov age tholeiite-basalts were succeeded here by calc-alkaline basalts. South of the Kalaïhumb-Sauksay volcanic arc is a back-arc basin with an oceanic crust represented by ophiolite relicts.

The western part of the region experienced uplift in the Tournaisian and Viséan stages. Fold and fault deformations (Sudetan orogeny) commenced, marking the transition to an orogenic regime. The Caucasus Major territory was affected by particularly intense movements. Major complexes of Palaeozoic rocks here were detached and thrust northward. Major longitudinal strike-slip faults also occurred. After the epoch of nappe formation and during it, folding occurred and, as a result of all these deformations, the foundation was laid for the Palaeozoic nappe structure consisting of antiforms and synforms.

Thus development of the Caucasus-Pamirs region in the Middle Devonian to the Early Carboniferous coincides in general with that of the West European region. Submarine volcanism, rift or island-arc type, was manifest, deep troughs were filled and the sign of tectonic movements underwent change in some structures. However, the main tectonic deformation occurred later, in the Sudetan epoch of orogeny (end of Early Carboniferous) while Precarboniferous Bretonian diastrophism was manifest only in

Dobrogea. But this was adequate preparation for a general transition to an orogenic regime. The Caucasus-Pamirs region also has a distinct overall structural plan. It is generally linear and not mosaic as in Western Europe.

4.2.2.3 *Kuen Lun-Indochina region.* M i d d l e   D e v o n i a n. In the Kuen Lun system accumulation of thick terrigenous complexes continued, island arc volcanism was manifest and reef carbonates developed on the margins of island uplifts (*Phanerozoic of Siberia,* Vol. 1, *Vendian. Palaeozoic,* 1984; *Atlas of the Palaeogeography of China,* 1985). Extremely thick turbidites (up to 3 km) were formed in the Qinling system; strong uplift continued; folding, metamorphism and granitisation were manifest in the Qilianshan system including the Qaidam block; and intermontane basins were filled with molasse.

In the Yunnan-Malaya system, shelves of the old Shan and Malacca massifs were probably joined in the Middle Devonian. As a result, a single Sinoburman massif covering even eastern Sumatra island arose and shallow-water carbonate and terrigenous deposits of small thickness formed on it (Gatinsky, 1986). In the shelf of the Indosinian massif, much of which continued to heave and represented dry land, thick (up to 2 km) terrigenous strata predominated. Submarine volcanic eruptions of mafic and intermediate composition continued in the regions north of the Indosinian massif and east of the Sinoburman massif. Further, major batholiths of granite-granodiorite composition intruded into the northern part of the Indosinian massif. Their association with volcanic eruptions is evident. All of these formations represented a common volcanoplutonic belt on continental crust. East of the Shan block, andesite volcanism, probably of the island-arc type, was manifest. In Sumatra island territory the general conditions did not change compared to the Early Devonian; in the west accumulation of flysch predominated under conditions of continental slope and in the east of shelf terrigenous and carbonate deposits. The latter deposits are evidently associated structurally with the shelf of the Sinoburman massif (Gatinsky, 1986; Mishina, 1979).

L a t e   D e v o n i a n. Shallow-water Upper Devonian carbonate-terrigenous, sand deposits and intermediate volcanics which formed in a subaqueous environment are known in the Kuen Lun system. In the east, in the Qinling system, filling of a former deep trough ceased and deposition of shallow-water carbonate-terrigenous strata predominated at the end of the epoch. Later, a brief uplift occurred. The intensity of tectonic movements and magmatism weakened and the topography flattened in the Qilianshan orogen territory. However, on the whole, the tectonic regime was still orogenic and molasse continued to form.

Within the Yunnan-Malaya system, accumulation of shallow-water shelf carbonates and terrigenous rocks proceeded, as before, in the Sinoburman massif. The south-western part of the massif, however, underwent subsidence and a highly extended zone of continental slope traceable up to Sumatra Island formed here. Turbidites were deposited in this zone (Mergui, Puket etc. series) with inflow from the east from the submarine uplifts of the massif. Along the eastern border of the Sinoburman island arc andesite volcanism took place in the north while submarine eruptions of tholeiite-basalts and andesites and silica accumulation proceeded in the south within the Malacca peninsula.

The Indosinian massif experienced mighty uplift and folding in the Late Devonian. In its western margin shallow-water carbonate formations were replaced towards the west by sandy flyschoid and later by deeper water flysch (in the eastern part of northern Thailand) of continental rise.

Conditions even in the south-east of this system, i.e., on Kalimantan Island, may be satisfactorily reconstructed for this epoch. In the western part of the island clay sedimentation of small thickness occurred and submarine basaltic volcanism was manifest (Mishina, 1979).

E a r l y   C a r b o n i f e r o u s. The area of the mobile belt shrank since active development ceased in the Qilianshan system and the orogenic regime was succeeded by a platform one. This entire territory (including Qaidam block) thus became a unit of the Sino-Korea Platform. In the Kuen Lun system subsidence and heavy, almost continuous sedimentation continued. The Lower Carboniferous is represented here by terrigenous and carbonate deposits of the shelf type reaching 1 to 1.5 km in thickness. In the western part of the system ophiolites were formed and large submarine eruptions of tholeiite-basalts occurred. These basalts were replaced in the Late Viséan by volcanic products of intermediate composition; in the east andesite volcanism was preserved on the former scale. In the Qinling system, after a brief uplift, a fairly deep trough formed again and rapidly filled with turbidites, but later with shallow-water carbonate-terrigenous formations, to a total thickness exceeding 2 km. At the end of the epoch, subsidence was replaced here by uplift and weak folding. This folding preceded the transition of the eastern part of the system to a platform (quasi-platform) regime. The main tectonic deformation including nappe formation and affecting the entire complex of Palaeozoic deposits occurred in this zone slightly later, at the end of the Triassic.

The Sinoburman massif preserved its structural entity and probably shallow-water carbonate-terrigenous sedimentation of small thickness occurred throughout the entire area under conditions of slow, interrupted subsidence. In its former western extended zone of continental slope, the irregularities of submarine topography were filled with a prism of sedimentary rocks, i.e., turbidites, by the end of the Devonian and, as a result, the massif enlarged here. East of the massif break-up and extension led to the destruction of the crust and a new deepwater zone appeared with a large manifestation of submarine tholeiite-basalt volcanism.

The palaeotectonic environment in the Early Carboniferous may be reconstructed for the first time in south-eastern Malacca. Here a shallow shelf basin with carbonate (partly reef) formations and, in the west, a continental slope with accumulation of turbidites may be reconstructed. However, the nature of the boundary of the latter with the Sinoburman massif is not clear. A suture is very probable here in the present-day structure. In the southern Sinoburman massif the shelf carbonate basin in Sumatra enlarged at the expense of the continental slope and deposition of pelagic sediments occurred in the south-western part of the island from the Early Carboniferous, evidently on an oceanic crust. Lower Carboniferous shallow-water carbonates and sandstones are known in the western part of Kalimantan Island and probably deeper water shales in the east (Mishina, 1979).

The Indosinian massif experienced subsidence again and was almost wholly drowned by a shallow sea. In the northern part of the massif felsic subaerial volcanism occurred and a band of flysch was deposited along the western slope. This flysch band evidently corresponded to the continental slope of the massif and its rise.

Along the south-western part of Sumatra Island, deposition of deepwater pelagic sediments commenced in the Early Carboniferous, probably on oceanic crust. In western Kalimantan Island, Lower Carboniferous carbonates and sandstones of the shelf type are known and in the east shales, probably of deepwater origin.

Thus the eastern part of the Mediterranean belt, Kuen Lun-Indochina region, reveals a pattern of Middle Devonian-Early Carboniferous tectonic development which is common to the rest of the belt. Prolonged subsidence was succeeded by uplift at the end of the Early Carboniferous (Sudetan phase of orogeny) but the previous Precarboniferous (Bretonian) phase of uplift and folding was manifest here only in a very weak form. Like almost the entire extent of the belt, the Kuen Lun-Indochina region occupied an intercontinental position and only in the south-east did it border directly with the Palaeo-Pacific.

4.2.3 Ural-Okhotsk Belt

4.2.3.1 *Uralian system.* M i d d l e   D e v o n i a n. Following the epoch of tectonic reactivation at the end of the Early Devonian in the form of uplift, folding and intrusion of plutons of gabbro-plagiogranite formation, the Middle Devonian opened up with renewed and intense subsidence covering the whole of the Urals. In Pay-Khoy, the Polar, Northern and Central Urals as well as the Bashkirian anticlinorium, the Middle Devonian lay transgressively and with more or less sharp unconformity on formations ranging from the Lower Devonian to the Riphean with basal conglomerates or sandstones. In the Magnitogorsk and Eastern Urals troughs and in the northern continuation of the Tagil trough, submarine basaltic and andesite-basaltic volcanism was intensely manifest. Apart from lavas and tuffs, terrigenous rocks, siliceous shales, jaspers and lenses and bands of limestones deposited in these troughs. The overall thickness of the Middle Devonian in them reaches 3 to 4 km. In the Petropavlovka zone of the Central Urals, reef limestones predominated and bauxites were deposited on them. The thickness of the Middle Devonian here is 1 to 2 km. This was an extended band of barrier reefs with a deep uncompensated trough east of them. In the Polar Urals volcanism was feebly manifest and negligible in Novaya Zemlya. Pelagic phtanites and carbonaceous-clay-siliceous rocks were deposited in the Sakmara-Lemva deepwater zone, their thickness sometimes reaching 500 to 600 m but more often only a few tens of metres. Formation of these sediments occurred most probably under conditions of the East European continental rise.

Late Devonian. The intensity of volcanic activity progressively decreased and the composition of its products underwent a change. Undifferentiated diabase volcanism was replaced by differentiated diabase-quartz-albitophyre type. In the Magnitogorsk zone eruptions had totally ceased by the Famennian stage and accumulation of the greywacke-clay flyschoid Zilair series commenced. The latter was widely developed, thickest in the west in the Zilair trough proper and initially also covered the Uraltau uplift. Small uplift and deformations including nappe formation preceded the deposition of this series. Nappe formation is fixed in particular by unconformities and conglomerates at its base. A similar tectonic environment prevailed in the regions of Shchuchya trough of the Polar Urals. Here, too, at the verge of the Middle and Late Devonian, intense movements were manifest and represented by folding, deep erosion, accumulation of coarse clastic rocks about 300 m in thickness and later greywacke-clay-siliceous flyschoid strata.

All this points to some increase in uplift and the appearance of islands. At the same time, terrigenous formations, volcanics of extremely diverse composition ranging from mafic to felsic, clay-siliceous rocks, lenses and bands of reef limestones continued to accumulate in the eastern Uralian and Tagil zones. Development of the Petropavlovka zone of barrier reefs of the Central Urals was completed and carbonates were replaced here by sandy-clay formations. However, in the Sakmara-Lemva zone, deepwater conditions and comparatively thin siliceous clay sedimentation continued to prevail.

Early Carboniferous. A new phase of rejuvenation of uplift and deformation manifested at the commencement of the Tournaisian stage. Further, this time upheaval also affected the Transuralian zone. On its western periphery, in the Eastern Urals trough, paralic coal-bearing formations were deposited in the Late Tournaisian to Early Viséan stage.

These formations filled the superimposed basins (for example, Poltava-Breda) that overlay sharply unconformably much older formations up to the Ordovician inclusive. Subaerial felsic eruptions occurred here too. Upward movements at the commencement of the Carboniferous affected with the same intensity the entire Eastern Urals zone of uplift where they were accompanied by the formation of a complex of intrusives with gabbro-granite composition. A similar intrusive activity was likewise manifest in the Tagil zone.

Uplift at the commencement of the Tournaisian stage partly covered the eastern Magnitogorsk zone also but prevailed in its western region only in the Early Viséan. In the Early and Middle Viséan, here as well as in the Eastern Urals trough, coal-bearing formations accumulated. Further, clastic material came from the marginal structures of the East European Craton in the west. At the beginning of the Viséan, these structures were uplifted above sea level and dissected by the Kama-Kinel' buried river system.

In the Late Viséan and Namurian stages, the Uralian system experienced a period of diminished tectonic activity, growth of subsidence and an extensive transgression with deposition of predominantly carbonate deposits usually not exceeding 1 km in thickness. Volcanic activity renewed in the Late Tournaisian in the eastern part of the Magnitogorsk zone. It was manifest most extensively almost throughout the Early Carboniferous in the more eastern zones, i.e., Eastern Urals trough and Transuralian zone, as also in the North Sos'va region. Formation of the Valeryanov volcanic belt along the eastern boundary of the Uralian system is placed in the Viséan-Namurian stage. The composition of the products of volcanism was highly differentiated everywhere, ranging from basalts and andesites to rhyolites and trachytes, and the thickness of the Lower Carboniferous was 2 to 3 km (up to 5 km).

Conditions in the erstwhile deepwater Sakmara-Lemva zone had changed by the Viséan stage. It experienced partial uplift and in the Namurian, accumulation of large sandy-clay (flyschoid) strata had commenced.

Thus the Middle Devonian-Early Carboniferous stage corresponds to an independent cycle of tectonic development of the Uralian system from pre-Middle Devonian orogeny to the end of the Early Carboniferous; thereafter transition to a true orogenic regime took place. This cycle commenced with a wave of subsidence, interrupted by brief phases of uplift and reactivation of tectonic processes, and ended in a mightier uplift encompassing the system after the Namurian. The phase of intensification of tectonomagmatic activity at the commencement of the Carboniferous, corresponding in general to the Bretonian epoch of the Mediterranean mobile belt, was the most powerful in the stage under consideration. The conclusion of this epoch in the Urals coincided with the Intra-Viséan phase of movements of Altay and Tien Shan. The pre-Late Devonian phase was manifest weakly and followed by formation of the Zilair series. This phase corresponds somewhat to the Acadian orogeny of the Appalachians.

4.2.3.2 *Southern Tien Shan system.* Subsidence continued and accumulation of shelf carbonate and terrigenous formations prevailed in the M i d d l e D e v o n i a n within the system as a whole. Andesite-basaltic volcanism continued in the Southern Ferghana zone. In the north-western part of the system, in the Sultanuizdag Range, volcanism of mafic and intermediate composition was also manifest and greywackes and limestones were formed under conditions close to those of the formation of volcanosedimentary complexes of the same age in the Eastern Urals. Thin siliceous and carbonate sediments accumulated south of the aforesaid system in the Alay-Kokshaal zone under conditions of uncompensated subsidence and tholeiite-basalt probably as a part of ophiolites, also formed.

In the L a t e D e v o n i a n the structural plan underwent a minor reorganisation. In the northern part of the system bordering with the Kazakhstan orogen, accumulation of shallow-water sandy-clay and carbonate rocks continued. Marshy lagoons formed here and dolomites, gypsum and sometimes rock salt were deposited in them. Volcanism ceased in the south and uplift of island arcs commenced. Beds of the flyschoid type (Pushnevat and Taushan suites) up to 3 km or more in thickness accumulated on their slopes. In the Alay-Kokshaal zone these formations were replaced by deepwater clay deposits of small thickness.

In the E a r l y C a r b o n i f e r o u s subsidence prevailed in the area of the system as before and limestones, often dolomitised, predominated in its northern part (Central Tien Shan). Red sandstones, conglomerates and intercalations of gypsum were also formed in the partially enclosed lagoons. The total thickness of the Lower Carboniferous here exceeded 2 to 2.5 km. Island series separated this shallow-water basin from the deepwater zone located in the south in which thin carbonate-siliceous deposits had accumulated. In the Pamirs thick terrigenous beds were formed and submarine volcanism of mafic and intermediate composition manifested. The intensity of volcanism, renewed from commencement of the Carboniferous, accelerated throughout this epoch. In the Namurian stage submarine eruptions occurred in Central Tien Shan in the region of Kurama Range. A structural relation between Southern Tien Shan and the Uralian system is evident in the Early Carboniferous. At the end of the latter epoch, the relatively quiescent development of the Southern Tien Shan system was interrupted by major uplift and horizontal displacements, leading later to a transition to an orogenic regime and thrusting of northern structural zones on southern ones.

4.2.3.3 *Kazakhstan region.* M i d d l e D e v o n i a n. Intense upheaval prevailed here as before and the region represented an upland with highly eroded mountain massifs situated on it. The highest mountains probably lay in the Ulytau region and Atasu zone. Tectonic activity was maximal in the central parts of the region where new faults with a north-western strike appeared; centres of volcanism and formation of small basins in the form of grabens are associated with these faults. These basins were filled with molasse together with extrusive and pyroclastic rocks of diverse composition to a total thickness ranging from hundreds of metres to 1.5 km.

In the latter half of the Middle Devonian the structural plan underwent change and uplifts and troughs were redistributed. The new superposed basins represented a zone of intermontane and submontane troughs relative to the volcanic upland and clastic material from eroded volcanoes and mountains filled these new basins. These materials were continental red-beds of various composition, ranging from silts to boulder conglomerates up to 2.5 km in thickness.

Development of the Kazakhstan marginal volcanic belt girdling the Junggar-Balkhash marginal sea continued but the composition of the products of volcanism changed, with felsic lavas and tuffs predominating. Rhyolitic ignimbrites were characteristic of the northern and eastern parts of the belt. They accumulated in the volcanotectonic depressions, i.e., calderas, as well as over extensive expanses in the form of sheets of small thickness. The formation of such large masses of ignimbrite-welded tuffs points to the generation of enormous volumes of gas-saturated felsic magma and its geologically instantaneous catastrophic eruptions. Moreover, within the belt, volcanosedimentary rocks were formed and subordinate (in terms of scale) eruptions of mafic-intermediate lavas occurred; however, most rocks of the Middle Devonian are represented by felsic volcanics up to 3 to 4 km in thickness. Across the strike, the belt of volcanic rocks was replaced by molasse.

In the Kazakhstan region extensive manifestation of volcanism was associated with intrusion of granitoids comagmatic to volcanics. Massifs of feldspathic and alaskite granites (sometimes alkaline granites) formed under near-surface and low temperature conditions and were usually confined to the cores of anticlinal structures or volcanotectonic depressions.

L a t e   D e v o n i a n. In the intermontane basins accumulation of molasse continued but clastic material became finer since the zone of uplift decreased in area and its topography smoothened. Subaerial volcanism was weakly manifest and felsic extrusives predominated, as before, but such volcanism completely ceased in several graben-like basins. All the properties of the continental formations of the Middle-Upper Devonian of the Kazakhstan orogenic region as well as its subaerial volcanism suggest their essential similarity with Old Red formations of the British Isles, albeit the Kazakhstan molasse is slightly younger and the position of the main Devonian unconformities is slightly shifted upwards along the sequence.

Development of the Kazakhstan volcanic belt came to an end at the commencement of the Frasnian stage. A new reorganisation of structural plan occurred, folding manifested at places, movement intensified along faults and intrusion of granitoids continued on a small scale.

Significant changes in tectonic regime also occurred in the Famennian stage. Smoothening of the topography of the orogenic region and subsidence of extensive areas led to widespread transgression of sea into the south-east. Deposition of red molasse was replaced by accumulation of limestones and shallow-water terrigenous rocks. A regime close to a platform (quasi-platform) was established in the Kazakhstan region at that time. This is supported by the nature of bedding and composition of formations. They formed a continuous cover of sustained thickness (100 to 300 m) and facies. Lagoons with high salinity were located in the marginal parts of the marine basin.

In spite of this transgression, however, uplift still prevailed in the area of Kazakhstan and elevated land remained in Karatau, BetpakDala, the Kokchetav massif and the Chingiz-Tarbagatai zone, where accumulation of molasse in intermontane basins continued. In the Karaganda region Telbessian folding was followed by rifting in Famennian time (Veimarn and Milanovsky, 1990).

Direct palaeogeographic relations of the Kazakhstan region with the Uralian system in the west and the Ob'-Zaisan system in the north-east were established from the Famennian stage onwards.

E a r l y C a r b o n i f e r o u s. Subsidence and transgression peaked at the commencement of this epoch, in the Tournaisian stage. The shallow-water marine basin gradually extended from north to south, i.e., from Dzhezkazgan into the region of BetpakDala, southern and south-western Peri-Balkhash region and into Northern Tien Shan. Flat islands associated with block uplifts existed in the interior parts of this marine basin. Accumulation of carbonate and terrigenous formations continued in the basin while anhydrite, gypsum and salt were also deposited in the Teniz-Chu basin.

Uplift intensified and regression grew later, from the second half of the Viséan stage, and lagoonal coal-bearing facies and continental volcanic formations (mainly tuffs) of intermediate and felsic composition were developed in the eastern Kazakhstan region along with marine deposits. At the end of the Early Carboniferous, a major residual marine basin was preserved in the west within the Dzhezkazgan and Teniz-Chu basins where formation of carbonate and siliceous deposits and, at places, evaporites continued. Uplift predominated in the rest of the area, and only small lagoons as well as bays with terrigenous and, less frequently, carbonate deposits were preserved. Coal formation continued in the freshwater basins and river mouths in the east. Intrusive magmatism was negligible in the Early Carboniferous in the

Kazakhstan region while eruptive magmatism was localised in the south-eastern regions along the periphery of the Junggar-Balkhash system and was represented by comparatively small subaerial eruptions of intermediate and felsic composition.

Thus at the commencement of the stage under consideration, the Kazakhstan region represented a mobile structure with distinctly manifest signs of an orogenic tectonic regime. Strong uplift, mountain building and the associated formation of molasse, intrusive and eruptive magmatism, metamorphism and diverse tectonic deformations characterised the development of this territory until the middle of the Late Devonian. However, the intensity of all the aforesaid processes decreased sharply later, subsidence enlarged and a more quiescent (quasi-platform) regime continued almost up to the end of the stage, with some reactivation commencing from the Late Viséan.

4.2.3.4 *Altay-Northern Mongolia region.* M i d d l e D e v o n i a n. Intense, predominantly upward movements continued over the entire area of the region, leading to structural reorganisation, frequent change of composition of formations and notable development of magmatic processes (Belousov et al., 1978; Gordienko, 1987; Dergunov et al., 1980; Leont'ev et al., 1981; Tsukernik et al., 1986). An orogenic regime prevailed everywhere. Active uplifts of Gorny (High) Altay and Mongolian Altay, Kuznetsk-Alatau, Western and Eastern Sayan and Sangilen were separated by numerous intermontane basins. Among them, two groups are distinguishable: some originated on a Precambrian basement and were discordant relative to Caledonian structures while others follow the structural plan of the Caledonides. The latter are characterised by very early beginning of deposition and a low degree of unconformity between the Devonian and underlying deposits. Rybinsk, Minusa, the central part of Western Sayan, Tuva and other basins belong to the first category. In the Middle Devonian they were filled with red lagoonal-continental (occasionally marine carbonate) terrigenous formations in the form of a coarse molasse up to 2 to 4 km in thickness.

Uimen-Lebed', Anuy-Chu and Yustyd basins and the western parts of Kuznetsk and Altay-Salair basins belong to the second category. Grey-coloured marine formations, black shale, often flyschoid and bands of reef carbonates up to 2 to 3 km in thickness, developed extensively in them.

In the Eifelian stage intense volcanic activity continued within the region under study. For example, in the Minusa basin volcanic products constitute almost eighty per cent of the volume of all Eifelian formations. Eruptions occurred through volcanoes of linear and central types and were distinguished by highly variable composition. In intermontane basins volcanism was predominantly subaerial but marine in the Rudnyi (Ore) Altay zone.

At the end of the Eifelian and in the Givetian stage, volcanism decreased sharply, intrusions of granitoid massifs commenced, enlargement of sedimentation basins occurred and accumulation of continental terrigenous and marine deposits increased, especially in the Minusa and Anuy-Chu basins. New block uplifts caused rejuvenation of land topography in Kuznetsk-Alatau, eastern Tuva and Western and Eastern Sayan.

Late Devonian. The area of the Altay-Northern Mongolia orogenic region increased with inclusion of the Mongolia-Amur zone where subsidence had been replaced by uplift and an orogenic regime established by this time. On the whole, in the Late Devonian repeated alternation of uplift and subsidence occurred within intermontane basins and contrasting movements also manifested with the formation of a rugged topography in the elevated regions. In the basins, seas-lakes manifested at times with grey and variously coloured terrigenous and terrigenous-carbonate formations up to 2 km and above in thickness but formation of typical molasse complexes 3 to 4 km in thickness predominated. In the Kuznetsk basin territory and Gorny Altay a marine basin (Salair-Gubin) with carbonate sediments was preserved in the Frasnian stage and a regression occurred later, limestones were replaced by clays and Salair became a low plain. The Sayan, Eastern Tuva, Gornaya Shoria and southern Kuznetsk-Alatau uplifts renewed. Marine sedimentation was replaced by continental in the Minusa basin.

Volcanic manifestations in the Late Devonian continued to attenuate. Most intense subaerial eruptions of mafic, intermediate and felsic composition occurred in Uimen-Lebed' trough but they, too, ceased in the Famennian stage. However, by this stage extensive granitoid magmatism as well as fold deformation affected this trough. Major granitoid plutons intruded into Gorny and Mongolian Altay, Western and Eastern Sayan and Tuva. Their composition changed from moderately felsic to leucocratic, alaskite and potash granites and in Eastern Sayan and Tuva, to alkaline and subalkaline granites in the concluding stage of granite formation.

Early Carboniferous. The Altay-Northern Mongolia orogenic region was a montane country in the east and a plain, poorly rugged land in the west. The activity of tectonic movements gradually diminished everywhere. Dimension decreased and shoaling of the Kuznetsk basin occurred. Accumulation of carbonates (detrital limestones) was replaced in it by deposition of lagoonal and later continental terrigenous rocks. The area of Tuva basin, a freshwater lake, shrank. From commencement of the Viséan stage, carbonate accumulation was here replaced by tuffaceous and sand deposits and clastic material built up over time. The thickness of the Lower Carboniferous in Tuva exceeds 2 km. Minusa basins represented a system of interconnected but later disjointed lakes. Clastic material almost did not enter them and ashy masses and carbonate muds were deposited. But, at the end of the epoch, in the Namurian stage, these lake-troughs were joined by river systems which transported large amounts of coarse clastic sediments. The mountain massifs of Kuznetsk-Alatau and Western and partly Eastern Sayan represented regions of erosion. In the valleys of rivers, lakes and marshes of the Minusa and Tuva basins, plant material accumulated and coal formation proceeded. These continental formations occurred under more quiescent tectonic conditions and their thickness measured hundreds of metres.

In the Mongolia-Amur zone an orogenic regime was succeeded by an environment of extension and widespread subsidence, with the formation of a deep marine trough.

Thus the intensity of manifestation of the orogenic regime in the Altay-Northern Mongolia region gradually diminished in the Middle Devonian-Early Carboniferous stage. The high contrast of tectonic movements and mighty effusive volcanism of the Middle Devonian, along with the formation of typical molasse, gradually died out and, after an extensive Late Devonian burst of granitoid magmatism, the general mode of tectonic development of the region approached the platform type in the Early Carboniferous.

4.2.3.5 *Junggar-Balkhash system.* The size of the marine basins slightly decreased in the M i d d l e D e v o n i a n. They preserved their shallow-water nature and were filled, as before, with thick littoral-marine terrigenous deposits. The trough submerged most intensely in the region of Junggar-Alatau where, apart from sandy-clay deposits, reef carbonates were formed; the total thickness of the Middle Devonian formations reaches 4 km.

In the L a t e D e v o n i a n filling of marine basins with the products of disintegration of islands and surrounding hills of the Kazakhstan orogenic region concluded. In the north-west, limestones were deposited extensively in the Frasnian stage. Felsic and intermediate volcanism manifested in the eastern part of the system under marine conditions. Volcanic beds in the region of Junggar-Alatau reached 1.5 km in thickness. Clay-silica rocks were also deposited here.

In the E a r l y C a r b o n i f e r o u s the palaeotectonic environment underwent rapid change: volcanic islands grew and active volcanoes supplied masses of tuffaceous, lava and clastic material to residual basins. In the Viséan stage, intensification of uplift provided conditions of littoral plains where coal-bearing subcontinental formations (Karaganda basin) accumulated. In this stage, representing pre-orogenic development for the Junggar-Balkhash system, intrusion of granodiorites occurred. In the middle of the Viséan stage the Junggar-Balkhash system experienced the action of the Saourian phase of orogeny. The deformation of this phase most intensely affected the north-western part of the system between the Spassk zone of faults and the northern Peri-Balkhash region.

4.2.3.6 *Ob'-Zaisan and Gobi-Hinggan systems.* M i d d l e D e v o n i a n.
In the marginal segments of the Ob'-Zaisan system extensive shelf zones were
formed and shallow-water terrigenous formations and reef carbonates were
deposited in them. Felsic volcanism was manifest sometimes. In the axial,
deepwater part of the system, submarine volcanism was of andesite-basalt
composition and pelagic clays predominated among sedimentary rocks.

In the Gobi-Hinggan system the northern zone represented a narrow
shelf and continental slope adjoining the neighbouring Altay-Northern
Mongolia orogenic region. Carbonates and shallow-water sandy-clay rocks
accumulated on the shelf while turbidites accumulated in the south in
the region of continental slope and rise. The thickness of the Middle
Devonian was 2 to 3 km and intermediate and felsic volcanism manifested
here (Gordienko, 1987). A deepwater zone with distinct submarine uplifts
(sometimes islands) evidently lay even more to the south and clay sediments
of small thickness were deposited in it.

Uplift maintained in the eastern part of the Gobi-Hinggan system within
the boundaries of old massifs—Bureya, Kerulen etc. In the Greater Hinggan
region carbonate deposits were extensively formed and submarine andesite
(sometimes rhyolite) volcanism was manifest (*Atlas of the Palaeogeography
of China,* 1985). In the southern part of this system, along the boundary
with the Sino-Korea Craton, subsidence in shelf seas was compensated by
terrigenous-carbonate deposits while spreading continued in the axial zone
of the Southern Mongolia ocean.

Development of the Mongolia-Amur zone concluded in the Middle Devonian in northern and north-eastern Mongolia. Here conditions changed compared to the Early Devonian and deep troughs were rapidly filled with turbidites, but in the east conditions of a deepwater trough with accumulation
of spilite-silica complexes prevailed in the Amur-Okhotsk zone. The thickness of these complexes is significant, exceeding 5 km for example, on the
Shantar Islands.

L a t e D e v o n i a n. In the axial zone of the Ob'-Zaisan system deep
trough conditions evidently prevailed as before and a complex of dark-
coloured terrigenous deposits, sometimes flysch, deposited in this trough.
In the peripheral parts of the system, especially in its south-eastern segment, volcanism of andesite composition was manifest. This volcanism was
probably confined to island arcs.

Within the Gobi-Hinggan system, the former tectonic zoning was preserved. In Southern Mongolia, spreading continued (Ruzhentsev et al., 1992)
and siliceous clay deposits were formed under deepwater conditions while
sandy-clay and carbonate beds of shelf seas prevailed in the north. In the
Greater Hinggan region, Upper Devonian flysch formations are known while
volcanics of andesite-rhyolite composition of the island-arc type are known
in the Lesser Hinggan region. Subsidence continued in the southernmost

zone of the Gobi-Hinggan system bordering the Sino-Korea Craton. The thickness of terrigenous and carbonate shallow-water formations here runs into hundreds of metres.

E a r l y   C a r b o n i f e r o u s. The central zone of the Ob'-Zaisan system was divided into western and eastern parts as a result of upheaval of the Chara uplift in the form of an island. Submarine volcanoes ejecting mafic and felsic lavas existed west of this uplift and shallow-water terrigenous and carbonate deposits formed on it. In the Viséan stage volcanism ceased here, brief uplift caused regression of the sea and, by the end of the Early Carboniferous, this territory adjoined the land of the Kazakhstan orogen.

The trough east of Chara Island was very deep and subsidence was very stable. Fine muds were deposited in it in the Tournaisian stage followed by flysch-type deposits. Within the Chara uplift, volcanism was manifest in the Viséan stage and intermediate lava outpourings occurred along the faults. In the Zharma and Western Kalba zones small plutons of granodiorites intruded at the end of the Early Carboniferous and fold deformations commenced at the same time. All of these together were precursors for transition to orogenic conditions.

In the western part of the Gobi-Hinggan system, during the stage under consideration, the tempo of subsidence slowed, manifestations of intermediate and felsic volcanism enlarged and new uplifts in the form of islands arose. In the east, filling of troughs with terrigenous, volcanogenic and carbonate material was completed in the Greater and Lesser Hinggan regions while uplift intensified in the Kerulen and Bureya massifs.

A new phase of extension, break-up and deep subsidence affected the Mongolia-Amur zone. The trough formed here grew rapidly and formation of spilite-siliceous complexes was replaced by greywacke and turbidite sedimentation and island arc volcanism of intermediate and felsic composition even by the end of the Early Carboniferous. In the east, this zone was connected with the Amur-Okhotsk zone of the West Pacific mobile belt where deep trough conditions with basalt eruptions were preserved much longer, up to the end of the Carboniferous.

On the whole, the general tendencies of tectonic development throughout the Middle Devonian-Early Carboniferous in the Ob'-Zaisan and Gobi-Hinggan systems were differentiation of environmental conditions, comparative expansion of uplifts and shoaling of sedimentary basins, shift in composition of volcanic eruptions towards andesites and rhyolites, the birth of fold structures and commencement of granitisation processes.

4.2.4 West Pacific Belt

4.2.4.1 *Verkhoyansk-Kolyma region.* M i d d l e   D e v o n i a n. Further break-up and subsidence of hitherto stable continental blocks occurred in the Kolyma-Omolon interfluve region. Differentiation of tectonic movements intensified and manifestation of submarine volcanism (basalts) in the Omulevka zone as well as subaerial (alkaline basalts, andesites and rhyolites) in the Omolon massif commenced. In this massif and in the Peri-Kolyma uplift, granitoids and syenites are associated with volcanics. Intense deposition of carbonates continued in the shelf basins (sulphates too in the western Omulevka zone), sometimes exceeding a thickness of 3 km, while sandy-clay strata up to 2 km in thickness were formed in narrow deep troughs (Surmilova, 1980).

L a t e   D e v o n i a n. In the Verkhoyansk-Kolyma region reworking of old continental blocks intensified. In the Verkhoyansk system with break-up and extensive basaltic eruptions, reactivation of a tectonic regime commenced in the territory of the former north-eastern margin of the Siberian Craton. As a result, these regions became a constituent of the mobile belt. Subsequent deep subsidence led to the formation of troughs filled with sandy-clay deposits and reef carbonates up to 1.5 km in thickness. A complex differentiated environment prevailed in the Kolyma-Omolon region. In the enlarged island areas in the Omolon massif, an exceptionally mighty rhyolite and, less frequently, andesite volcanism was manifested. Clay, tuffaceous and carbonate deposits and sometimes coarse clastic red-beds deposited in the troughs surrounding these areas. Minor plutons of gabbro-granodiorites intruded into the Okhotsk massif and granitoids into Southern Verkhoyansk. In the Omulevka zone, basalts erupted under submarine conditions and terrigenous formations, at places calcareous reef facies, accumulated.

E a r l y   C a r b o n i f e r o u s. Palaeotectonic conditions levelled over much of the territory of the system. A sharp intensification of subsidence led to the formation of deepwater basins with lutitic and sometimes siliceous (in the Gizhiga zone) sedimentation over large areas. In the west, along the boundary with the Siberian Craton, an extensive shelf zone (passive margin) was preserved in which a complex of sandy-clay and carbonate rocks formed up to 1.5 km in thickness. Clastic material came from the uplifts of the craton and the prism of sedimentary rocks prograded into the east towards the deep basin.

Subsidence in the area of the Kolyma-Omolon region led to transgression and reduction in the area of islands in which extensive subaerial basaltic eruptions manifested. In the Omolon River basin felsic volcanism predominated at places while minor granitoid plutons intruded into the Peri-Kolyma uplift.

On the whole, subsidence accelerated in the Verkhoyansk-Kolyma region from the Middle Devonian to the Early Carboniferous and covered more and more new hitherto stable blocks of the earth's crust, including the margin of the Siberian Craton. In this stage destruction of continental crust, rift formation, accompanied by reactivation of tectonic movements and magmatism including basalt eruptions, prevailed over an extensive area of the region. These processes broke up the extensive passive continental margin that existed here in the Early and much of the Middle Palaeozoic and formed deep basins of the type of present-day marginal seas in the rear of island arcs of the western part of the Pacific.

4.2.4.2 *Koryak system.* M i d d l e   D e v o n i a n formations are represented in the Koryak upland by greywackes, reef limestones and products of basic submarine volcanism. Their thickness is not less than a few hundred metres. Well-dated Upper Devonian formations have not yet been identified here. In the western part of the system, Lower Carboniferous coarse clastic deposits and volcanics of intermediate and felsic composition were evidently formed under island arc conditions while deepwater conditions with siliceous and clay sedimentation probably existed in the east (Terekhova et al., 1979).

4.2.4.3 *Nippon-Sakhalin system.* M i d d l e   D e v o n i a n. Accumulation of shallow-water terrigenous and, less frequently, carbonate deposits continued in the western shelf zone of Japan (Mazarovich and Richter, 1987). In the east, under conditions of a deep basin, spilite-keratophyre submarine volcanism was manifest as in the Early Devonian and clay and siliceous muds not exceeding a few hundred metres in thickness were deposited. Submarine basaltic eruptions occurred possibly in the central parts of the Sakhalin and Hokkaido islands (Brodskaya et al., 1979). Subsidence renewed in the shelf zone north of Honshu Island after a brief uplift, erosion and folding, and shallow-water terrigenous strata up to 800 m in thickness were deposited. Angular unconformity is noticed at the base of the Middle Devonian and small basalt flows are known. Products of felsic volcanism and terrigenous rocks accumulated south of the Khanka massif under continental conditions while shallow-water carbonates were deposited towards the west.

L a t e   D e v o n i a n. The western shallow-water shelf zone with terrigenous sedimentation was preserved in Japan while the palaeotectonic conditions in the eastern regions are not clearly known. Submarine andesite-basalt volcanism prevailed here and greywacke strata were formed (island arc?). Accumulation of a spilite-siliceous complex (marginal sea?) in the central zones of Hokkaido and the Sakhalin islands has been assumed. In southern Primorie the Upper Devonian is practically not known. The Khanka massif probably experienced almost total subsidence, representing a row of islands with minor manifestation of volcanism of felsic and intermediate composition (marginal volcanic belt).

E a r l y   C a r b o n i f e r o u s. Submarine mafic and, less frequently, intermediate volcanism, accumulation of lutites, greywackes and cherts predominated and a deepwater environment prevailed, as before, in the region of the Japanese islands. The Lower Carboniferous in western Honshu Island is represented by greywackes, including turbidites, with sediment supply from west to east. In the shelf zone north of Honshu Island, uplift and folding manifested at the end of the Devonian but formation of submarine volcanic and terrigenous complexes exceeding 2 km in thickness continued in the Early Carboniferous. In the Namurian stage this region again experienced uplift and folding. In Primorie territory, east of the Khanka massif, which again represented dry land, a broad meridional zone extended with mighty manifestation of submarine basaltic volcanism and terrigenous, siliceous sedimentation. The thickness of these formations exceeded 2.5 km in northern Primorie and attained 3 km in southern regions.

Thus the Nippon-Sakhalin system, occupying a marginal-continental position, preserved a fairly uniform palaeotectonic environment and products of submarine volcanism and terrigenous deposits accumulated in the marginal seas.

4.2.4.4 *Cathaysian system.* In the M i d d l e D e v o n i a n a rugged topography still prevailed in the territory of the system and tectonic conditions evidently corresponded to an orogenic regime. The rate of uplift decreased and fairly thin fluvial deposits formed in the basins in place of molasse and manifestation of fold deformation and intrusive magmatism had ceased even by the end of the Early Devonian. In the Late Devonian tectonic development became even more quiescent. From this epoch, conditions of a platform regime prevailed here and this mobile system became a unit of the South China Platform.

4.2.4.5 *East Australian region.* M i d d l e D e v o n i a n. An important tectonic deformation occurred in the territory of the Lachlan system and formation of the basic features of its present-day structure was completed (Murray and Kirkegaard, 1978; Plumb, 1979). This Tabberabberan orogeny, signifying the total involvement of the system in an orogenic regime, commenced at the end of the Emsian and concluded at the end of the Givetian stage. Uplift and deformation were most intense on Tasmania Island where a proper montane country was formed. However, even in other regions uplift predominated, residual marine basins almost vanished and metamorphism and folding were extensively manifest. In this same stage numerous batholiths of granitoids (mainly adamellites) arose and are known all along the width of the Lachlan system, from former Melbourne trough to the Caperty uplift and in the north in the Yulo and Anakee uplifts. Such granitisation points to completion of formation here of a mature continental crust and to cratonisation of all this territory. Therefore, basins of the northern part of the system, Adavale and Cobar, in which marine and continental molasse up to 2 km in thickness formed, sometimes containing salt-bearing deposits and subaerial andesites, belong to the platform type by their structure; on the other hand, even the marginal segment of the adjoining old Australian Craton was affected by processes of uplift and granitisation, thus enlarging the area of the Lachlan orogen.

A perceptible change occurred in the development of the New England system in the Middle Devonian. It was manifest in expansion of the western shelf zone with terrigenous and reef carbonate sedimentation caused by the prograding sedimentary prism on the erstwhile continental slope. As a result of compressive forces acting from the east, the complex of these rocks underwent piling up and dislocations coinciding in time with the Tabberaberan epoch of orogeny took place. Middle Devonian granitoid magmatism, unlike in the Lachlan system, is not known in this region. This is explained by the absence of a sufficiently thick granite-metamorphic crust layer.

In the eastern zone of this system, in the Yarrol-Tamworth trough, deepwater conditions continued and clayey, siliceous and volcanic rocks were deposited. However, the composition of volcanic eruptions changed from mafic to intermediate.

In the northern part of the East Australian region in the Queensland system, filling of the Hodgkinson trough with extremely thick turbidites (over 3 km) continued. In the Brock River graben inflow of clastic material sometimes abundantly compensated subsidence and a part of the Middle Devonian sequence with a thickness of over 2.5 km is represented by continental facies.

Late Devonian. In the Lachlan orogenic system the activity of tectonic movements and volcanic activity intensified even more. This orogen again enlarged in the west due to the reactivated margin of the craton and, in the south-east, as a result of transition of the former shelf zone of the New England system to an orogenic regime. Subsidence renewed and molasse formed over extensive areas of the western parts of Queensland, New South Wales and Victoria states. Although outcrops of this molasse are at present disconnected, formerly a single zone of intracontinental basins existed here (Lamb basin), drained by river systems entering the open sea in the south-east. The Grampian basin in south-western Victoria state represented an isolated structure. In the southern and northern regions of the Lachlan orogen, Late Devonian reactivation was accompanied by extensive subaerial volcanic activity, mainly of felsic composition, and intrusion of granodiorite plutons.

The western part of the New England system experienced transition to an orogenic regime. The depth of the Yarrol-Tamworth trough decreased, andesite volcanism continued, greywackes predominated and polymictic conglomerates appeared. Volcanic centres were concentrated in a belt with its convex part facing eastward and representing a network of small islands. A deep elongated trough (rift) was formed at the back of the southern part of this volcanic arc, evidently as a result of displacement. Intense accumulation of submarine basalts and greywackes occurred in this trough.

In the Queensland system, at the verge of the Middle and Late Devonian, the Brock River graben was uplifted and its sedimentary filling experienced folding. The Hodgkinson trough also experienced uplift and deformation but subsidence here was rapidly renewed and accumulation of extremely thick shallow-water marine and continental sediments (3 to 4 km) commenced. They unconformably overlie much older formations. The general regime of development in this system was orogenic.

E a r l y   C a r b o n i f e r o u s. The boundaries of the Lachlan orogen were slightly displaced eastward since upwarping died out on much of the reactivated margin of the craton and new regions of the New England system were involved in orogeny. Accumulation of molasse had ceased over much of the area of the system even by the end of the Devonian and such areas began to experience fold and fault deformation. At the end of the Early Carboniferous, this deformation was manifest even more intensely, although at no place did it go beyond the germanotype. The orogenic epoch under study is called Kanimblan in Australian literature. A fairly large number of post-tectonic intrusives (stocks) of granitoids were formed at this time and crumpling of deposits filling the Adavale and Drummond troughs occurred. In the eastern part of the Lachlan system a marginal continental volcanic belt with eruptions of felsic and intermediate composition developed in a strip adjoining New England. On the whole, intensity of manifestation of orogenic processes had diminished perceptibly within the system by the end of the Early Carboniferous.

In the New England system, the Yarrol-Tamworth trough shoaled and was already filled with coarse clastic continental sediments (molasse) by the end of the Viséan stage. However, subsidence, as before, went ahead or compensated sedimentation in the central and eastern regions of the system where flysch of continental rise (a few kilometres of dark shales with bands of turbidites) began to accumulate above thin deepwater (oceanic) deposits. But at the end of the Viséan and in the Namurian, shoaling became perceptible and bands of conglomerates were manifest in the deposits. A deep trough formed in the southern part of the system in the Late Devonian now developed very rapidly and was filled with marine terrigenous formations and continental molasse even by the Early Carboniferous.

In the Queensland system conditions of the orogenic regime were maintained. Filling of the Hodgkinson trough with molasse to 3 to 4 km in thickness was completed and fold and fault deformation of the Kanimblan epoch of orogeny took place. As a result, the entire territory of this system experienced uplift, sediments were crumpled into germanotype folds, at places metamorphosed and overthrust on the Georgetown massif of the craton margin. The Queensland system evidently extended into the central zone of New Guinea Island.

On the whole, the Middle Devonian-Early Carboniferous stage of evolution of the East Australian mobile region was marked by the culmination of orogenic processes. Intensity of uplift and manifestation of magmatism and metamorphism successively increased and, following this stage, the largest in the Palaeozoic areas experienced cratonisation, which accreted the region of older continental crust. In the Late Devonian epoch of orogeny marginal zones of the Australian Craton were thereby affected. As in the much older stages, migration of zones with high tectonic activity in an easterly direction has been distinctly expressed here. A noteworthy feature is the matching of the period of the Tabberabberan and Kanimblan epochs of diastrophism and the Acadian and Sudetan phases of orogeny respectively, well known in Europe and (or) North America.

4.2.4.6 *West Antarctic region.* In the Middle Devonian-Early Carboniferous the marginal basin extending through the Ellsworth Mountains underwent erosion and west of it, encompassing Mary Byrd Land and northern Victoria Land, a magmatic arc extended with eruptions of felsic and intermediate volcanics (age 374 to 323 m.y.) and intrusions of I-granites (age 380–325 and 370 to 295 m.y.). The volcanics contain bands of sandstones with Middle-Late Devonian flora. The rocks of the arc experienced intense deformation associated with Borggrevink orogeny.

The magmatic arc Mary Byrd-Victoria represents a continuation of the same-aged magmatism of Tasmania, Australia and New Zealand (Laird, 1991). This arc continues in the opposite direction into the Antarctic peninsula and probably into the Southern Andes.

4.2.5 East Pacific Belt

4.2.5.1 *Cordilleran system.* Middle Devonian. Deepwater conditions prevailed within much of the system with thin lutite deposition (Churkin et al., 1980). The thickness of clay rocks exceeded 1 km only in the north, in Richardson trough, which was already filled with sediments. In the island arc region of the Klamath Mountains, reef limestones, terrigenous deposits and products of subaerial volcanism of felsic composition over 1.5 km in total thickness were deposited.

On the inside of this arc, volcanoclastics were replaced downward along the slope by turbidites and near the bottom by siliceous shales (Watkins and Flory, 1986). In the island arcs of the Alexander archipelago and Vancouver Island, carbonates, terrigenous and tuffaceous rocks were also deposited and volcanism was manifest; andesite-basalts predominated in the composition of volcanic products and bands of siliceous rocks were also formed here.

L a t e   D e v o n i a n. In the northern part of the system, after the deep Richardson trough was filled, accumulation of pelagic clays was replaced in it by deposition of shallow-water sandy-clay rocks in an environment of an extensive shelf basin. In the region of the Alexander archipelago, manifestation of volcanism of mafic and felsic composition continued and was close in nature to an island-arc type. Reef limestones and greywackes were deposited here to a thickness of up to 2 km. Volcanism of the island-arc type continued on the former scale in the region of the Klamath Mountains and submarine mafic volcanism in the zone east of Vancouver Island.

In the southern part of the system, at the end of the Late Devonian, intense uplift commenced (in central Nevada and western Utah) as a result of commencement of the Antler epoch of orogeny. These processes led to the formation of submarine mountains and of an elongated zone of thick flysch sedimentation, corresponding to continental slope and rise. In the west, between this zone and the island arcs, a deepwater zone (marginal sea) was preserved in which muds were deposited to a small thickness.

In the eastern part of the central segment of the system, uplift manifested at the end of the Devonian and a major island massif arose. In the eastern zone of British Columbia intense regional metamorphism and formation of granite-gneiss domes of the Shuswap complex occurred.

E a r l y   C a r b o n i f e r o u s. Major changes occurred in the tectonic development of the Cordilleran system. Orogenic processes of the Antler epoch of orogeny encompassed the entire eastern zone of the system from California to Alaska. Uplift predominated here and folding and metamorphism were vigorously manifest. Ancient metamorphic complexes lay in the cores of arched and bloc uplifts (Ancompaghre, Colorado-Sierra-Grande etc.). But on the whole, the dynamics of tectonic movements was contrasting and major residual troughs developed and were filled with carbonate-terrigenous and flysch (up to 3 km) formations (Larson and Langenheim, 1979; Saul et al., 1979; Skipp et al., 1979). Angular unconformities are common everywhere at the base of these Lower Carboniferous strata.

In the western zone of the Cordilleran system, a deepwater trough in its southern part, separating the volcanic arc of Klamath from the mainland was filled with flysch. Further, within the arc itself, felsic volcanism continued. In the central segment of the zone deepwater conditions with thin sedimentation were preserved while submarine-volcanic complexes were formed in the north, evidently from the Early Carboniferous, in the regions of southern and south-eastern Alaska (Dutro, 1979). Deepwater spilite-siliceous and shale formations probably lie here on an oceanic crust. Their thickness reaches 4 km or more. In the region of the Alexander archipelago volcanism of the island-arc type continued and greywackes and reef limestones accumulated.

The protracted Early-Middle Palaeozoic stage of development of the Cordilleran system was thus completed. A characteristic feature of its devel opment was the existence of an extended deepwater zone with thin muddy siliceous sedimentation. The complex processes of Antler orogeny mani fested at the end of the Devonian and in the Early Carboniferous caused closure of the Richardson trough and led to a mighty uplift of the eastern flank of the Cordilleran system, which resulted in the filling of a significant portion of the deep troughs adjoining the western margin of the North American Craton with terrigenous formations (mainly flysch). Based on the prevailing position of volcanic arcs of the Alexander archipelago, Vancouver Island and Klamath Mountains, it may be suggested that they separated the deepwater zone of marginal seas from the open ocean (Palaeo-Pacific). But, according to several researchers, these structures represent terranes and their former disposition relative to the Cordilleran system may have been different.

4.2.5.2 *Andean system.* M i d d l e D e v o n i a n. In the Ecuador-Colombian segment subsidence continued within its eastern part while maintaining the general prevalence of uplift and orogenic regime conditions (Zeil, 1979). Predominantly terrigenous formations deposited here in shallow marine basins joined with the Amazon basin of the South American Craton Adequate information is not available for the central zone. Maintenance of conditions for uplift with manifestation of folding and metamorphic processes has been suggested.

In the Middle Devonian vigorous filling of the central deep trough of the Peru-Bolivian segment with flysch continued, the main source of material being, as before, the Arequipa massif. Uplift of the latter slowed down some-what and its marginal area was covered by sea; however, intense erosion of the rest of the massif supplied an adequate quantity of clastic material. The thickness of the Middle Devonian in the axial zone of the trough exceeds 2 km, decreasing to 500 to 1000 m on its flanks.

In the Chile-Argentinian segment Middle Devonian formations are almost unknown and uplift evidently predominated in the whole of its territory and conditions of an orogenic regime were maintained (Miller, 1987). A series of molasse up to 2 km in thickness formed in the western part of the northern regions of Chile. In the Middle Devonian a new phase of tectonic defor-mation manifested in the Sierra Pampeanas massif. A marine trough with terrigenous sedimentation existed only in the far south of the segment. It was distinctly inherited from the preceding stage of development but its nature and boundaries are not yet sufficiently clear. Tectonic deformation occurred in the Patagonian zone, representing an elevated land. It was accompanied by intrusion of granitoids in the northern part of the Patagonia-Deseado mas-sif. These processes concluded the orogenic development of the Patagonian one whose territory, together with the Patagonia-Deseado massif, later formed a part of the young platform which accreted the old craton.

Late Devonian. Intensification of uplift in the eastern zone of the Ecuador-Colombian segment caused regression. Marine basins ceased to exist and Devonian formations underwent fold deformation of moderate intensity. Molasse was deposited to an insignificant thickness in a small residual trough. Evidently granitoids intruded in this epoch into the region of Cordillera-de-Merida.

During the Late Devonian the axial trough of the Peru-Bolivian segment was wholly filled with flysch about 1.5 km in thickness and its development was completed with inversion and folding, as suggested by angular unconformities observable everywhere at the base of the Lower Carboniferous (Dalmayrac et al., 1980). Further, uplift extended from the periphery of the system to its centre. However, folding is not discernible everywhere and some eastern regions of the system apparently did not undergo folding before the end of the Mesozoic since Cretaceous formations in them overlie the Devonian only with disconformity. In the marginal portions of these Late Devonian troughs, marine terrigenous facies alternate with continental facies. The area of uplift enlarged in the Arequipa massif.

Small marine troughs with terrigenous (sometimes carbonate) deposits were preserved in the southern part of the Chile-Argentinian segment. But in the main part of its area conditions of an orogenic regime were manifest as before: uplift predominated, molasse accumulated in intermontane troughs and folding and metamorphic processes evidently continued (Hervé et al., 1981).

Early Carboniferous. Formations of this age are almost unknown in the Ecuador-Colombian segment. There is likewise no data on the manifestation of magmatism and metamorphism. It may be assumed that uplift prevailed and an orogenic regime, albeit of a very weak nature, was manifest.

Within the Peru-Bolivian segment, after folding and general uplift at the end of the Devonian fixing transition to an orogenic regime, subsidence renewed at several places (Castanos and Rodrigo, 1980). Basins (intermontane) were filled with marine and continental formations (molasse) including coal-bearing facies. They overlie much older formations everywhere with angular unconformity. In the south-eastern part of the system, glacial material played a significant role in the composition of formations, the amount of this material decreasing northward. In a small zone north of the Arequipa massif, subaerial volcanism of felsic and intermediate composition took place.

In the Chile-Argentinian segment accumulation of terrigenous and carbonate formations continued In the south while moderate uplift predominated in other regions; lacustrine and fluvioglacial formations accumulated only at

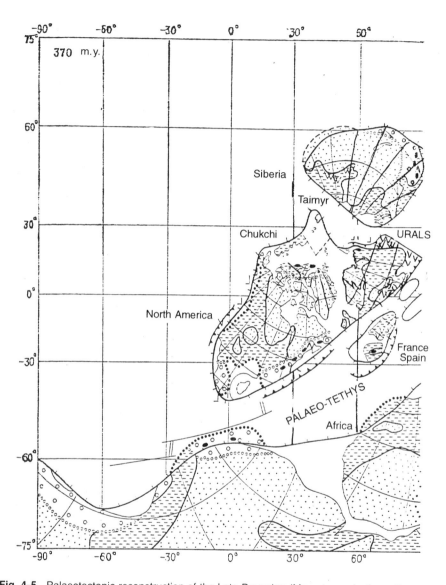

Fig. 4-5. Palaeotectonic reconstruction of the Late Devonian (Mercator projection with centre

The configuration and relative positions of some continental blocks suggested for the Late part of the South American continent, the Colombia-Ecuador orogen has been given no place. has been significantly shrunk in area. The marine sedimentary basins of South America be attributable to an inadequate study of the Devonian formations of Africa. The continental due to its proximity to Kazakhstan, no place has been allowed between the two for the Tarim Late Devonian.

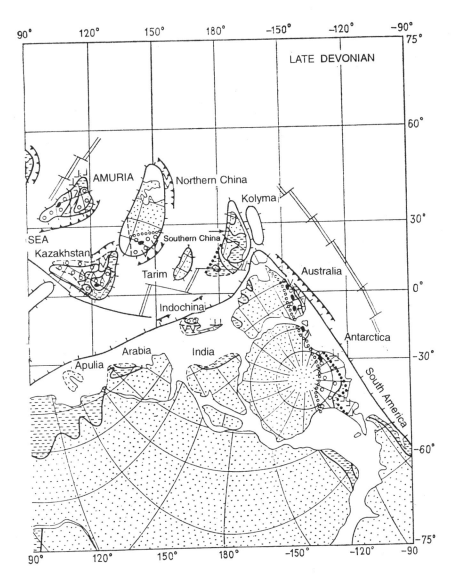

at 0° N lat. and 90° E long.). See Fig. 1–7 for legend.

Devonian by L.P. Zonenshain and coauthors give rise to some doubts. In the north-western
The north-western margin of the African continent, compared to the Early Devonian scheme,
(Amazon and Maranâo) are depicted without continuation into Africa. These anomalies may
block of Northern China, compared to the Early Devonian scheme, is diminished westward and,
and Qaidam blocks which, according to geological data, were joined with Northern China in the

206

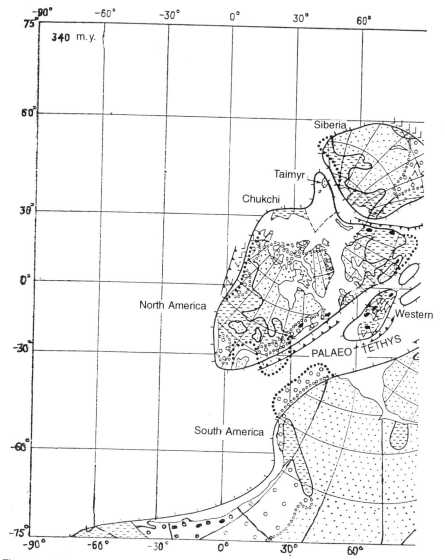

Fig. 4-6. Palaeotectonic reconstruction of the Early Carboniferous (Mercator projection with

The arrangement and forms of continental blocks as reconstructed by L.P. Zonenshain and for much older epochs. These are: inaccuracy of the profiles of blocks of South America, transformation of the contours of Northern China, which vary from one scheme to another. the East European Craton.

some places. Further, the age of the latter is not always clear. Processes of granitisation evidently commenced in the Early Carboniferous. They

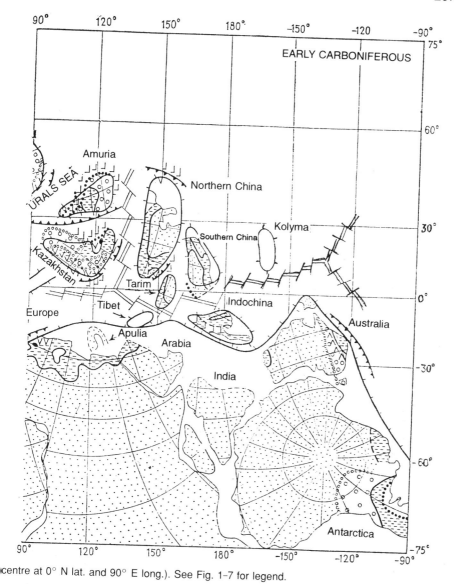

centre at 0° N lat. and 90° E long.). See Fig. 1-7 for legend.

coauthors give rise to several queries, a repetition of those expressed in relation to schemes
Kazakhstan, Australia and Antarctica; dubious displacement of Kolyma and Amuria; and
The Peri-Caspian basin has not been taken into consideration in the south-eastern part of

vere most significant in the Cordillera Coastal (Chile) where formation of
the Coastal batholith commenced and in the Pampeanas massif also.

On the whole, throughout the Middle Devonian-Early Carboniferous stage total transition of the Andean system to an orogenic regime was completed. Later than elsewhere, at the end of the Devonian, this occurred in the Peru-Bolivian segment where orogeny was accompanied by intense folding. In the Patagonian zone the orogenic regime had been replaced by a platform one already at the end of the Middle Devonian.

4.3 PALAEOTECTONIC ANALYSIS

By the commencement of this stage, North America and Eastern Europe were joined into a single continent, Laurussia or Euramerica, as a result of the final disappearance of the Iapetus ocean that separated them before (Fig. 4-5). The Palaeo-Asian (Ural-Okhotsk) ocean underwent certain changes and shrank in size; spreading in it evidently ceased and the process of reduction of expanse covered by oceanic crust commenced. A similar process characterised development of the Innuitian basin which entered its concluding phase; as a result Pearya and the entire Hyperborea approached North America. The Palaeo-Tethys experienced changes and opened extensively into the west as well as east into the Palaeo-Pacific (Panthalassa). As before, Gondwana represented a vast supercontinent which was almost wholly displaced into the Southern Hemisphere where it reached its highest latitudes (Fig. 4-6).

4.3.1 Continental Cratons

4.3.1.1 *Gondwana.* No major changes whatsoever occurred in the tectonic regime and structural plan of the supercontinent. As before, the major regions of subsidence were northern parts of South America turned towards the Palaeo-Tethys, Africa (Sahara), Arabia-Iran-Afghanistan and north-western Australia. But the developmental tendencies of these Gondwana units were not the same: uplifts grew in South America, Africa, Asia Minor and Antarctica; in Australia, on the other hand, the Carnarvon (regenerated), Fitzroy, Bonaparte and Amadeus aulacogens developed and accumulated several kilometres of deposits in the Late Devonian-Early Carboniferous, which were continental in Amadeus. On the Sahara Platform, probably under the influence of deformation in northern Maghreb, differentiation intensified, creating relative (absolute already by the Early Carboniferous) uplifts and basins; in addition to Tindouf and Saoura-Ougarta, as pointed out before, these basins included the more eastern and less deep Murzuk and Kufra. The thickness of formations here is measured in kilometres. Thus in Africa and Australia, the same tendency towards break-up of continental crust and rifting is expressed, so distinctly manifest in the northern group of continental cratons.

Since the sedimentation zone on the northern margin of Gondwana, especially in Asia Minor and Australia, lay in the low latitudes, carbonates are present among the deposits and sandy-clay formations and evaporites (sulphates) are also encountered. Against this background, the appearance of glacial formations in the Tournaisian of the southern Sahara stands out as an anomaly.

4.3.1.2 *Laurussia (Euramerica).* On both the platforms constituting Laurussia, the stage under consideration commenced with transgressions that continued by and large almost throughout the entire stage, attaining maximum in the Late Devonian. The disposition in the arid zone of the Northern Hemisphere made for extensive development of carbonates, mostly dolomites, and the presence of evaporites. A vast region of salt accumulation existed in the Middle Devonian in the north-western North American Craton, up to the Williston syneclise in the south; it lasted there until the Early Carboniferous inclusive. On the Russian Platform the main zone of salt accumulation was the Pripyat-Dnepr-Donets aulacogen regenerated in the Middle Devonian. Rifting was generally characteristic of the development of the East European Craton in the Middle-Late Devonian. It extended onto the much younger Timan-Pechora Platform and was accompanied at several places by alkali-basaltic volcanism. Both cratons are characterised by the accumulation of black shale formations (Chattanooga, Chester and Domanik) in basins uncompensated by sedimentation but their tectonic background is disputable.

4.3.1.3 *Siberia.* Development of this continent, including Verkhoyansk and the southern half of the Taimyr peninsula, apart from the present-day platform, proceeded in general similar to development of the Laurussian platforms, especially, the East European Craton. Devonian rifting manifested in the Siberia and Viluy and Yenisei-Khatanga aulacogens and rifts Kharaulakh, Sette-Daban and of the north-eastern margin of the continent arose or regenerated. This process here, too, was accompanied by basaltic volcanism. Formation of most of the diamondiferous kimberlite pipes in the north-eastern part of the platform falls in the Devonian. The Tunguska syneclise represented, as before, the main region of sedimentation.

Drastic change in development of the north-eastern, Verkhoyansk margin of the platform occurred in the middle of the Viséan stage. General subsidence sharply intensified here and shelf carbonate formation of moderate thickness was replaced by far thicker terrigenous formations of the outer shelf and continental slope. The appearance of olistostromes points out that subsidence occurred, at least in part, along fault scarps, evidently of the listric type, but fold deformation too is noticed in Sette-Daban.

4.3.1.4 *North China and Korea.* The single Sino-Korea continent, including the Tarim block, formed at the commencement of the Devonian, remained as land and an erosion zone in the Middle Devonian. Deposits of littoral alluvial plain accumulated only on the margins of Tarim. Troughs with continental sedimentation arose in the Late Devonian in the regions of Qilianshan and Alashan but the former was covered in the Early Carboniferous by shallow sea which separated Tarim-Qaidam land from Sino-Korea proper. Thus the Qilianshan basin was regenerated but already as a part of a craton.

4.3.1.5 *South China (Yangtze) Craton.* During the stage under consideration gradual decrease in elevated land areas formed by the Caledonian orogeny in south-eastern China took place. This was reflected in reduction of development of red, coarse clastic formations, which disappeared altogether in the Early Carboniferous. Areas of sedimentation enlarged simultaneously due to the formation of extensive lacustrine-alluvial plains in the Early Carboniferous. These plains were at times flooded by sea, with an accumulation of paralic coal measures. In the rest of the area, mainly along the western and northern margins of the craton and in its southern half, a shallow sea with carbonate sedimentation was preserved. Its relatively deeper portions and submarine slopes of the continent in the west and south are characterised by siliceous-carbonate or siliceous-clay sedimentation (Domanik homologues?).

A common rhythm of transgression and regression was characteristic of almost the entire series of Laurasian cratons (Ross and Ross, 1985). Transgressions grew up to the verge of the Frasnian and Famennian, then were replaced by regressive tendencies. In their background, however, partial transgressions in the Early Tournaisian and Late Viséan-Early Namurian (Serpukhov stage) were manifest.

4.3.2 Mobile Belts

4.3.2.1 *North Atlantic belt.* This belt continued to survive the orogenic stage of development but the intensity of upwarping gradually attenuated. In the Middle Devonian intense granite formation and regional metamorphism commencing in the Silurian-Early Devonian continued within this belt. The end of the Middle Devonian marked the peak of deformation; this stage of orogeny is known as the Acadian stage on Newfoundland Island and in the Northern Appalachians.

It has been suggested that the eastern zone of subduction disappeared and only the western zone inclined under the Avalonian microcontinent with a volcanic belt overlying it was preserved in the Middle Devonian in the Northern Appalachians. As a result, the eastern accretionary wedge overlay the western (Fig. 4-7). Here, too, the Iapetus was closed in the Late

Devonian as Avalonia collided with the margin of the North American continent. Many researchers assume a similar course of events for the Southern Appalachians while some suggest a more complex history (Higgins, 1972). Some basins-grabens filled with thick continental molasse of the Old Red sandstone type developed against a background of uplifts. In the Middle Devonian they existed everywhere; in the Late Devonian and Early Carboniferous, everywhere except Spitsbergen and Scandinavia. Subaerial volcanism was manifest in the Appalachians in the Middle-Late Devonian and Early Carboniferous—bimodal in the Middle Devonian and andesite-rhyolite at the end of the Devonian and Carboniferous. Volcanism accompanied accumulation of deposits in the graben of the Midland Valley of Scotland; it was of the calc-alkaline type in the epoch of Old Red sandstone, being comagmatic in the Devonian with so-called young granites of the northern and southern uplands of Scotland and alkaline, rift type in the Early Carboniferous. Molasse of the Lower-Middle Devonian is moderately folded by pre-Late Devonian deformation of the Svalbard phase in Spitsbergen (Svalbard) and Scandinavia. These deformations were also reflected in the attitude of the Old Red of the British Isles while the Upper Devonian (in spite of its molassoid character) and Lower Carboniferous already constituted the base of the sedimentary cover of the Epicaledonian Platform. In the Northern (Maritime) Appalachians, mountain formation concluded in the Late Devonian and at the end of this epoch, in Nova Scotia and on Newfoundland Island, the major Fundy rift basin appeared with continental clastic sediments and intermediate and felsic volcanics. In the second half of the Early Carboniferous a more extensive syneclise was formed at its base into which the sea penetrated.

In the Southern Appalachians, Acadian diastrophism affected only the interior zones; the outer zone of the Ridge and Valley Province continued quiescent subsidence in this stage too. The Ouachita-Mexican system did not at all undergo compressive deformation in this epoch and deepwater sedimentation continued within it during the Devonian. In principle, after the Acadian events, this system was transformed from a southern continuation of the Iapetus into the western continuation of the Palaeo-Tethys. During the Early Carboniferous, clastic and pyroclastic material began entering the deepwater basin of northern Ouachita in ever-increasing quantities from the south-east; this material was brought by turbidity currents from an assumed volcanic arc that arose as early as the Silurian (see Section 3.2.2.1).

This stage was critical for the Innuitian system: gradual approach of the Hyperborean and North American continents progressed with the elimination of ocean area between them and its involvement in orogeny. This process commenced from the east (see Section 3.3.2.1), from the junction with the Eastern Greenland Caledonides: clastic material arising from their erosion (and possibly erosion of the Spitsbergen Caledonides too) gradually filled the Early Palaeozoic deep basin. However, in the Middle Devonian, a

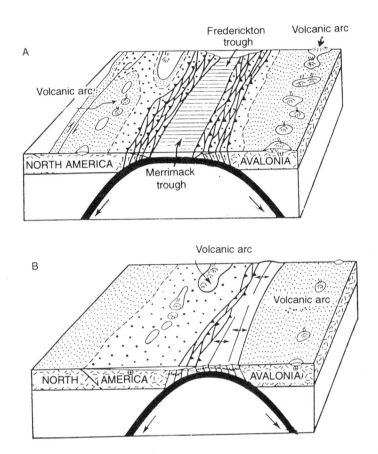

Fig. 4-7. Model of Acadian orogeny (after D. Bradly. In: Frasier and Schwimmer, 1987, simplified).

A—Palaeotectonic relationships in Late Silurian: Iapetus ocean basin was surrounded on two sides by subduction zones inclined on opposite sides and the accretionary features associated with them; sparse dots depict the shallow marine deposits; B—commencement of collision in Early Devonian; accretion prism associated with the Avalonian block covered the North American prism along an overthrust.

deepwater sequence still accumulated in the northern part of Axel Heiberg Island where additionally submarine basaltic eruptions occurred. Even more to the north, in the Early Devonian, the Pearya island arc existed. It was underlain by a seismofocal zone inclined northward. In the Middle Devonian this arc adjoined the northern continent and in the Late Devonian the slope of subduction zone changed to the opposite direction (Fig. 4–8). In the Late Devonian the entire belt from Peary Land in Greenland to northern

Alaska had already entered the orogenic stage of its development; this is the so-called Ellesmere orogeny, only slightly later than the Acadian. Its first phase, which is the most important for the axial zone of the orogen, falls in the Late Devonian; it was accompanied by intrusion of granites and regional metamorphism; the conclusive and main phase for the southern periphery of the orogen manifested already in the Early Carboniferous. But in the Late Devonian two residual basins still existed—marine in the region of Brooks Range and paralic in the Canadian Arctic archipelago; sediments of considerable thickness accumulated in both basins. The former continued to exist in the Early Carboniferous. In this same epoch, the Innuitian orogen was complicated by rift grabens in the northern parts of Axel Heiberg and Ellesmere islands and in north-eastern Greenland.

4.3.2.2. *Mediterranean belt.* W e s t e r n   p a r t. Unlike the preceding belt, this one only now entered the period of its most active development.

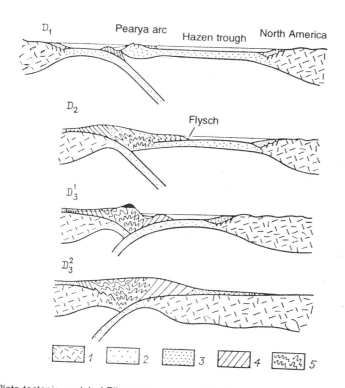

Fig. 4-8. Plate-tectonic model of Ellesmere orogeny (after Frasier and Schwimmer, 1987).

1—continental crust; 2—oceanic crust; 3—sediments; 4—accretionary wedges; 5—the same, which have experienced folding.

Two contrary tendencies coexisted in this development. The Central European microcontinent, evidently under the influence of subduction from the south, was overlain by a volcanic arc whose activity was distinctly manifest in the region of the French massif Central. In the north, Saxothuringian and Rhenohercynian basins continued to subside at the rear of this arc, suggesting extension conditions accompanied by bimodal, diabase-keratophyre volcanism. In Devon and Cornwall there are signs of development of listric faults. The limited dimension of these basins, separated by the Midgerman High, promoted their filling with thick black shales. In the Early Carboniferous intensification of the Moldanubian island arc uplift led initially to shoaling of the Saxothuringian and then of the Rhenohercynian basin and accumulation of the flyschoid (molassoid in south) Culm formation. In the middle of the Viséan volcanism ceased evidently as a result of change of extension into compressive conditions. This process developed from south to north and reached the northern boundary of the Rhenohercynian zone by the end of the Early Carboniferous. There it began to form the foredeep known as the 'European Coal Channel'. Corresponding to this direction of migration of compressive deformation, fold-overthrust structures acquired a distinct northern vergence.

Events in the region of the Hercynides of the Iberian peninsula and Maghreb (Morocco) developed in a similar manner (Julivert et al., 1980). Here migration of flysch formations, deformation and metamorphism occurred on both sides of the axis of the orogen but are most distinctly manifest on its south-western flank. Accumulation of Culm in the Southern Portugal zone (equivalent to Rhenohercynian) commenced in the Late Viséan and progressed up to the Namurian. A characteristic feature of the zone is the heavy bimodal rift volcanism of the Pyrite Belt in the 'preflysch period' i.e., in the Tournaisian and Early Viséan. In the Moroccan Meseta, Late Hercynian deformation likewise commenced in the Middle Viséan and was accompanied by intrusion of granites.

Development of the part of the Hercynian belt of Europe (and Asia Minor), now included in the Alpine belt, occurred synchronously by and large with the rest of the area. Here uplift and deformation are discernible even before the Late Devonian and intense movements commencing from the Middle Viséan, especially formation of nappes in the Eastern Alps, Carpathians, Balkans and the Front Range of Caucasus Major. Formation of Lower Carboniferous and at places of Upper Devonian are similar Middle European Culm in type, i.e., they likewise represent a shallow water variety of flysch formation with a predominance of clastic rock along with lutites, silicites and limestones; bimodal volcanism is also characteristic feature. Formations of the Middle and partly Upper Devonian preceding Culm are somewhat more diverse in composition and formation conditions. Being developed in some zones in a very quiescent tectonic regime (microcontinents?), these formations are represented by limestones. At other places, island arc volcanics are distinct while at still other places, siliceous-terrigenous deposits, evidently of marginal seas, are seen. A.A. Belov (1981) undertook to reconstruct the Devonian palaeotectonics of the future Alpine belt in the section through the Dinarides, southern Carpathians and Balkanides (Fig. 4-9). The profile does not show subduction zones. The main one evidently runs along the boundary of the Vardar zone, representing the axial zone of the Palaeo-Tethys in the Early and Middle Palaeozoic, and the Serbo-Macedonian massif; it was inclined to the north-east. In Caucasus Major subduction also proceeded from the south but nappes in the Front Range moved northward, i.e., obduction onto the northern continent occurred. Following these deformations, potash granites intruded. Potash plutonism occurred somewhat earlier in the eastern Carpathians, later in the Balkans and Dobrogea and in the Western Alps (here, even in the Middle Carboniferous).

After the Acadian diastrophism, the northern part of the Iapetus entered the platform stage of development and, from the Late Devonian, its southern part became the western continuation of the Palaeo-Tethys (see Section 3.3.2.1). To this belong the Southern Appalachians and Ouachita-Mexican system. In the Southern Appalachians interior regions of the Blue Ridge and Piedmont, after the Acadian events, experienced uplift, causing inflow of considerable masses of clastic material into the basin of the outer zone of the Ridge and Valley Province which continued to subside. This occurred at the end of the Middle and in the Late Devonian. In the Early Carboniferous, the supply of clastic material diminished, pointing to a reduction in rate of uplift of the interior zones. The Ouachita-Mexican system did not experience Acadian compressive deformation and subsidence uncompensated by accumulation of thin clay and siliceous deposits continued within this system in the Middle-Late Devonian and commencement of the Early Carboniferous. It continued in Mexico, throughout the Early Carboniferous. In the Ouachita-Marathon region turbidites appeared at the end of the Early Carboniferous, pointing to the commencement of collision between Euramerica and Gondwana (South America) in this segment. Sierra Madre del Sur zone with manifestation of submarine mafic volcanism and accumulation of cherts and greywackes in the Middle Devonian-Early Carboniferous belongs more to the Pacific belt.

E a s t e r n   p a r t. The structural plan of this part of the belt and the general character of its development in the Middle-Late Devonian changed little compared to the preceding stage. The Qilianshan system, the connecting link between the Ural-Okhotsk and Mediterranean belts, ended its orogenic development, evidently associated with collision between the Tarim and Sino-Korea continents. In the Early Carboniferous a platform regime was

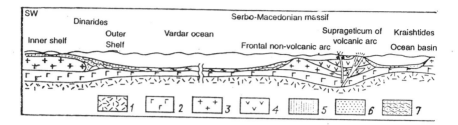

Fig. 4-9. Palaeotectonic profile of the Moesian platform-Dinarides in the Devonian period (after

1—upper mantle; 2—'basalt' layer and the third layer of the ocean floor; 3—granite-metamorphic marginal and interior seas; 7—formations of slopes of continents and island arcs; 8—layers 2 10—granites; 11—plagiogranites; 12—folds; 13—volcanoes.

established in the Qilianshan system and it became an integral part of the Sino-Korea continent, which enlarged with the inclusion of Tarim.

The Kuen Lun-Qinling basin was filled with clastic material, including turbidites, and experienced general shoaling. This, however, did not affect the Pamir-Badahshan continuation of this basin. Island arc volcanism manifested in Kuen Lun. The Yunnan-Malaya system extending between the Sinoburman and Indosinian microcontinents was set off from the former by a volcanic arc, evidently overlying a seismofocal zone that was inclined under the microcontinent to the west. Sinoburma was set off from Gondwana in the Late Devonian by a narrow but deep rift trough. In the Yunnan-Malaya basin fine clastic deposits not compensating its subsidence accumulated, probably under deepwater conditions. In the Viet-Lao branch of the Palaeo-Tethys, in the Shongda trough, sedimentation proceeded more intensely, alternating with basalt eruptions. At the verge of the Devonian and Carboniferous, in both systems, Viet-Lao and partly Yunnan-Malaya, deformations and uplift occurred but the axial zone of the latter system, traceable commencing from the Late Devonian to the western part of Kalimantan, subsided continuously. Felsic subaerial volcanism associated evidently with the seismofocal zone inclined under the massif was manifest in the northern part of the Indosinian massif.

Important changes occurred in the Early Carboniferous in the extreme north-west of the segment of the Mediterranean belt under consideration, i.e., Badahshan-Pamirs and Kuen Lun. Here in the Tournaisian-Early Viséan a second (?) opening of a basin with an oceanic type crust occurred, which is documented with ophiolites; in the Late Viséan-Namurian these were overlain by island arc volcanics. In Qinling an axial deepwater trough was regenerated in the Early Carboniferous. The total thickness of the Devonian and Lower Carboniferous in western Qinling exceeds 12 km.

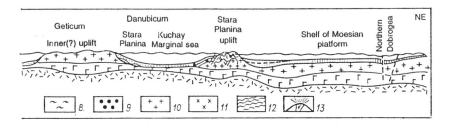

| Geticum | | Danubicum | | | Stara Planina uplift | | | | NE |
| Inner(?) uplift | | Stara Planina | Kuchay Marginal sea | | | Shelf of Moesian platform | | Northern Dobrogea | |

| $\sim\sim$ | 8. | $\bullet\bullet\bullet$ | 9 | $+\ +$ | 10 | $\times\ \times$ | 11 | $\approx$ | 12 | $\nearrow$ | 13 |

Belov, 1981).

layer; 4—island arc complexes; 5—shelf formations of epicontinental seas; 6—formations of
and 3 of the oceanic crust; 9—volcanic and sedimentary complexes of orogenic basins;

4.3.2.3 *Ural-Okhotsk belt.* The Urals basin of the Palaeo-Asian ocean
in the D_3-C_1 stage experienced three phases of development. In the first
(Givetian-Frasnian) and also in the second (Tournaisian-Early Viséan), active
volcanic arcs existed and marginal seas in their rear. Arcs were associated
with subduction directed westward, according to some (*History of Development of Urals Palaeo-ocean,* 1984), and eastward according to others
(Ivanov et al., 1986). They fall within the present-day Magnitogorsk zone
but the much younger arc was displaced eastward compared to the older.
In the interval between these two island arc phases a phase of general
compression and piling up manifested in the Famennian. With it was associated formation of the flyschoid terrigenous Zilair formation (D_3fm-C_1t) whose
deposits covered for the first time the carbonate beds on the western slope
(in the same manner as at the end of the Middle and Late Devonian in the
outer zone of the Appalachians). Volcanism ceased in the Middle Viséan
and shallow-water carbonate deposits accumulated over the entire area in
the Late Viséan-Namurian and during the Bashkirian stage of the Middle
Carboniferous, pointing to temporary cessation of subduction. An exception in this background is the formation of a new volcanic arc at the same
time (Valeryanov) in the eastern, Kazakhstan continental boundary of the
Urals basin. Thus a jump of the seismofocal zone to the east most probably
occurred while maintaining the same eastern dip.

In the Late Devonian-Early Carboniferous, Central Tien Shan represented the southern passive margin of the Kazakhstan-Northern Tien Shan
microcontinent with accumulation of shallow-water carbonates replacing the
Lower-Middle Devonian continental molasse. In Southern Tien Shan formation of the Turkestan palaeo-ocean basin was completed or neared completion, a volcanic arc developed and accumulation of carbonates, including
reefogenic, continued on its southern margin. The opening of a new basin
with an oceanic crust even more southward, in the northern part of the

218

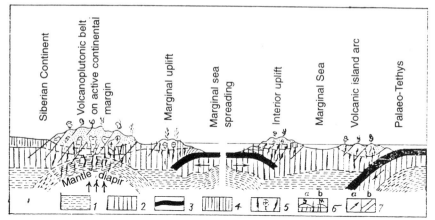

Fig. 4-10. Schematic reconstruction of the Sayan-Baikal-Mongolia region in the period of formation of the Devonian volcanoplutonic belt on the active margin of the Siberian continent and island arcs in the Southern Mongolia system (after Gordienko, 1987).

1—asthenosphere; 2—lithosphere; 3—oceanic plates of the Palaeo-Tethys and marginal seas; 4—blocked continental plates and microplates; 5—heat, fluid and magmatic melt flow; 6—volcanic arcs on transitional (a) and continental (b) crust; 7—direction of stress and movement of plates and also fluid flow in the mantle diapir (a) and fault dislocations in shear zones (b).

northern Pamirs in the Early Carboniferous, represented a significant event (see Section 4.3.2.2.).

In the Kazakhstan microcontinent the eastern (in present-day co-ordinates) volcanoplutonic belt completed development in the Middle Devonian while the Junggar-Balkhash basin simultaneously continued to subside. The Karaganda zone between the two represented an accretionary wedge along the subduction zone, having already experienced intense deformation in the Middle Devonian. In the Late Devonian-Early Carboniferous the Kazakhstan area experienced smoothening and was partially affected by transgression with deposition of carbonates similar to those of Central Tien Shan (Sarysu-Teniz watershed). The Junggar-Balkhash basin continued to subside with accumulation of deepwater sandy-clay deposits and eruptions of mafic volcanics. Further, the southern part of the basin experienced shoaling in the Late Devonian (*Atlas of the Palaeogeography of China*, 1985).

The Chingiz-Tarbagatai volcanic arc separated the Junggar-Balkhash basin from the Ob'-Zaisan, which continued into Southern and Inner Mongolia and Dunbei, then extended further with two branches—Greater Hinggan and Southern Mongolia-Girin—at its juncture with the Amur-Okhotsk and Sikhote-Alin zones of the West Pacific belt. These branches were separated by a microcontinent including the Hinggan-Bureya and Khanka massifs called Amuria (shown as Sunliao land in the Chinese atlas) by L.P. Zonenshain and coauthors (1987).

The south-eastern and southern salients of the Siberian continent (which enlarged considerably in size due to accretion of the Altay-Sayan-Baikal and Northern Mongolia regions) were fringed by a subduction zone above which a volcanoplutonic belt formed (Fig. 4-10). This belt covered the territory of Rudnyi Altay in the north-west. Behind this belt, in Mongolia and western Transbaikalia, the Hangay-Hentey-Dauria marginal sea continued to exist and, like the Junggar-Balkhash basin, filled with thick terrigenous sediments; its ophiolites outcrop along the Onon River in Transbaikalia (G.S. Gusev, pers. comm.).

4.3.2.4. *West Pacific belt.* The Verkhoyansk-Kolyma region west of Chersky range right up to the Middle Viséan still represented the north-eastern passive margin of the Siberian continent with a predominance of shelf carbonate sedimentation but experienced rifting on an ever-increasing scale with submarine eruptions of highly alkaline basalts. By the middle of the Viséan age this process led to commencement of a generally intense subsidence of the entire Verkhoyansk-Kolyma margin of Siberia along listric faults and accumulation of a thick terrigenous Verkhoyansk complex, also promoted by the change of arid to humid climate. Unconformities noticed at places at the base of the Verkhoyansk complex evidently represent extension unconformities; at some places (Kharaulakh and Sette-Daban) these are accompanied by olistostromes formed most probably along the scarps of listric fault planes.

At some unknown distance from the north-eastern margin of Siberia lay the Kolyma-Omolon microcontinent on which, too, volcanism was manifest but in the form of subaerial eruptions of felsic lavas of the Kedon series, possibly associated with subduction from the side of the Palaeo-Pacific. The main water area of the latter was set off at the end of the Devonian-commencement of the Carboniferous from the Okhotsk margin of Siberia and the Kolyma-Omolon microcontinent by the Udino-Murgal volcanic arc developed above a seismofocal zone inclined to the north-west. Somewhat later, in the Viséan-Namurian, a new volcanic arc, Talovka-Mayna, arose in the east, evidently as a result of the 'jumping' of this zone eastward.

More to the south, in the region of Sikhote-Alin' and the Japanese islands, ophiolites and deepwater silica-shale (central Sikhote-Alin') as well as shallow-water sandy shale (Japan) or carbonate of the seamount type (coastal zone of Sikhote-Alin') deposits and manifestations of mafic (Sikhote-Alin') or intermediate (Japan) volcanics are known from the Middle Devonian-Early Carboniferous. The development of a thick series of subaerial, mainly felsic volcanics of the Middle Devonian along the eastern periphery of the Khanka massif points to the existence there of a marginal volcanic belt of the Andean type. At the verge of the Devonian and Carboniferous, a weak phase of deformation and uplift was noticed in the region of the Kitakami Mountains (northern Honshu Island).

Even more southward, in south-eastern China, in the Caledonian fold system of Cathaysia, the orogenic stage of development was completed in the Middle Devonian. A considerable part of the region experienced active uplift and another part represented a trough filled with red molasse. In the Late Devonian and Early Carboniferous intensity of uplift clearly diminished and paralic coal-bearing formations came to replace molasse and accumulated on the south-eastern periphery of the epicontinental marine basin of the Yangtze Platform.

In the Middle Devonian the Lachlan system in Eastern Australia entered the main epoch of its deformation and mountain building, the Tabberabberan epoch. As a result, a sharp change in the nature of sedimentation occurred: accumulation of molasse proceeded to a thickness of up to 4 km or more over an extensive area from Kanmantoo zone in the west almost up to the Sydney meridian in the east within New South Wales. The zone of uplift and erosion lay in the south, in Victoria, where granitoid plutonism manifested actively. On the coast south of Sydney a narrow rift zone arose with manifestation of bimodal volcanism. Tabberabberan diastrophism also affected the western part of the New England system, i.e., the marginal sea that separated the Lachlan system from the latter closed and the Calliope volcanic arc was transformed into a marginal volcanic belt of the Andean type. East of it lay shelf formations (forearc trough) and the slope of the deepwater ocean basin. The cause of the Middle Devonian diastrophism has not been established; it has been assumed (Degeling et al., 1986; Scheibner, 1985) that it resulted from the collision of a hypothetical microplate, probably oceanic, or even microcontinental, with the margin of the Australian (i.e., Gondwana) continent. The orogeny extended from south to north and from west to east, commencing at the end of the Early Devonian and terminating in the Late Devonian. On Tasmania Island it was manifest in folding between 430 and 400 m.y. and intrusion of granites aged 395 to 345 m.y. These Middle and Late Devonian events also manifested in the north, in Thomson zone. Only in the extreme north-east did flysch accumulation still proceed in the Late Devonian in the Hodgkinson zone. In the rest of the area of the Lachlan system development of intermontane molasse troughs continued in the Late Devonian and Early Carboniferous; lava eruptions of mafic or bimodal composition occurred in them. The end of this orogenic stage was marked by the Kanimblan epoch of diastrophism (Viséan/Namurian); after it, a platform regime was established on much of the area of the system. In the New England system, Kanimblan diastrophism manifested weakly (in the west). But glacial material brought in from the eastern (probably elevated) land appeared in Namurian formations in its western part; very similar material is known in the Amadeus aulacogen. This is one of the earliest manifestations of Late Palaeozoic glaciation of Gondwana (Veevers and Powell, 1987) but information has recently been published about traces of glaciation in the Tournaisian and Viséan formations of North Africa.

A magmatic arc developed in the West Antarctic system in the period under consideration. The resultant formations soon experienced intense deformation and metamorphism. Probably, at this same time, there was accretion of terranes of Mary Byrd Land and northern Victoria Land to the Antarctic continent which already included the Rossides in the west.

4.3.2.5 *East Pacific belt.* Development of the northern portion of the belt pertaining to the North American Cordillera proceeded almost up to the end of the Devonian, conforming to the same scenario already described, i.e., continental margin, marginal sea and volcanic arc are delineated. The outer

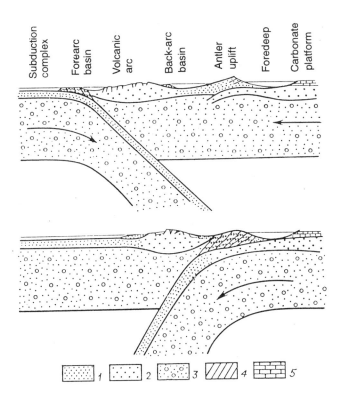

Fig. 4-11. Alternative plate-tectonic models of Antler orogen (after Frasier and Schwimmer, 1987).

1—oceanic lithosphere; 2—continental lithosphere; 3—asthenosphere; 4—accretionary wedge; 5—carbonate platform.

shelf was covered by a carbonate platform with a barrier reef in the rear (in the Canadian segment); slope and rise were made up of finely layered limestones, radiolarites with participation of micritic carbonates, graptolite shales and turbidites accumulated in the marginal sea and a volcanic arc was formed of diverse volcanics, subvolcanic intrusives, greywackes and reef limestones. This environment underwent sharp change at the very end of the Devonian with the onset of Antler orogeny, which continued almost throughout the Early Carboniferous. In this epoch formations of the interior eugeosynclinal zone of the Cordillera were displaced by roughly 100 km to the east onto the shelf of the continental margin. Two versions of the geodynamic mechanism of this event have been suggested (Fig. 4–11): according to one, obduction of marginal sea formations occurred on the margin of the continent while the back-arc basin enlarged. The second, collision of an island arc with the continent, is preferred (Frasier and Schwimmer, 1987). The main zone of manifestation of Antler orogeny is the Cordillera of the USA; in Canada it is weakly manifest. Tectonic deformation was accompanied by metamorphism to amphibolite facies but intrusive magmatism is negligible. Tectonic piling up led to the formation of a non-volcanic island arc before which small foredeeps filled with terrigenous flysch arose in the east, in Nevada and Idaho.

Development of the Andes in the stage under consideration was distinguished by high complexity with a general increase in upwarping and proceeded, as before, on the continental crust with the exception of the western zone of the Northern Andes. In the Northern Andes uplift and evidently folding and metamorphism manifested almost continuously in the central zone and, after the Middle Devonian, in the eastern zone also where a shallow marine basin existed until then. A relatively deep basin in the Central Andes separating the Arequipa microcontinent from the mainland continued to survive even in the Late Devonian, having been filled with flysch. Before the Carboniferous it dried up, however, and its filling experienced fairly intense folding. In the Southern Andes, including Sierra Pampeanas, uplift and fold deformation manifested even from the Middle Devonian, having been accompanied by accumulation of molasse, but a small marine basin still existed in the south until the Early Carboniferous inclusive. The Patagonian zone completed its orogenic development in the Middle Devonian but from the Late Devonian together with the Patagonia-Deseado massif became a constituent of the young platform, accreting the South American Craton from the south. This transition was accompanied in the massif by intrusion of granitoids. Granitoid plutonism in the Late Devonian was also manifest in the Northern Andes and Sierra-de-Merida and in the Early Carboniferous in Sierra Pampeanas and Cordillera Coastal of Chile. Subaerial volcanism of intermediate-felsic composition was noticed in this stage only in a small section of the Central Andes.

5

Middle-Late Carboniferous—Early Triassic. End of Development of the Hercynian and Birth of Cimmerian Mobile Belts

5.1 REGIONAL REVIEW. CONTINENTAL CRATONS AND THEIR MARGINS

5.1.1 North American Craton

Middle-Late Carboniferous. The territory of the craton enlarged northward due to Innuitian orogen completing its development. Subsidence of the extensive Sverdrup basin with its central portion filled with carbonate-terrigenous and salt-bearing deposits commenced. The littoral regions of the basin were filled with coarse sandy deposits. The thickness of formations is high, more than 2 km at the centre of the basin. Plateau basalts are known on the northern slope of this basin. A prominent feature of the rest of the craton is regression in the West Canadian basin, of which only a small shallow Peace River basin remained in the west. This basin maintained contact with the seas of the Cordilleran belt and was filled with sandy deposits and conglomerates in marine and continental facies. Another remnant is the Williston basin. It shrank markedly in size and lost contact with the Cordilleran troughs; concomitantly its contacts with the extensive Midcontinent sea covering the entire southern portion of the craton from Cordillera to the Appalachians enlarged (Avcin and Koch, 1979; Burchett, 1979; Chronic, 1979; Ebanks et al., 1979; Fay et al., 1979; Larson and Langenheim, 1979; Schoon, 1979; Smith and Gilmour, 1979; Thompson, 1979).

Fusion of the Michigan-Illinois and Peri-Appalachian basins occurred in the eastern part of the Midcontinent which, according to Klein (1987), was caused by tangential stress exerted by the growing Appalachian orogen. This area represented a littoral plain bound in the east by mountains of the Appalachians and in the south of Ouachita orogen. These were periodically inundated by the sea advancing from the western regions of the Midcontinent and the regions of the Great Plains. A typical paralic coal-bearing formation was deposited under these conditions. Eustatic fluctuations of ocean level associated with variations in the area of continental glaciation of Gondwana (Veevers and Powell, 1987) made for a distinct cyclic structure of this formation, some cycles traced on an immense area, from Kansas to Pennsylvania. The Illinois and Kansas basins experienced maximum subsidence up to 750 m. The La Salle swell arose in the northern part of the Illinois basin.

Formations of Middle-Late Carboniferous gradually became wholly marine west of the Mississippi. These formations are still preserved west of the Nemaha ridge (arising at the end of the Early to commencement of the Middle Carboniferous) and in the Williston basin together with the underlying Lower Carboniferous formations of substantially carbonate composition. Further, Late Pennsylvanian transgression covered the Transcontinental uplift, which long remained a barrier between the seas of the western and eastern parts of the craton. Close to the Rocky Mountains front and in the Anadarko and Ardmore basins, which experienced intense subsidence (thickness 4 to 5 km) at this time, carbonate sediments were replaced by purely terrigenous, erosion products from uplifts of adjoining orogenic systems. There were some small partially enclosed seas (Paradox basin etc.) in this western part of the Midcontinent and the south-eastern Rocky Mountains. Their water salinity was high and halite was deposited in them. These small seas were separated by block uplifts of the 'Ancient Rockies' and Wichita zone.

E a r l y   P e r m i a n. Subsidence of the Sverdrup basin in the northern part of the craton continued but its rate diminished and conditions became more quiescent in the littoral regions. This large basin was filled predominantly with carbonates up to 1 km in thickness, which included evaporites in its southern part. At the end of the epoch, this territory experienced a brief uplift almost everywhere, leading to an interruption in sedimentation.

In the Peace River basin in the north-western part of the craton, subsidence was even more quiescent (thickness up to 300 m); predominantly carbonates, including bands and lenses of cherts, were also deposited here.

In the southern part of the craton, the sea area shrank significantly as a result of uplift and regression in the eastern part of the Midcontinent and the Peri-Appalachian zone. In the latter only a small basin filled with coarse continental deposits was preserved. However, subsidence still continued in the western part of the Midcontinent in the region between mountain massifs of the 'Ancient Rockies' and the Nemaha ridge with the subsidence centre in the Western Texas basin and a bay in the south-western part of the Williston basin. Seas were partially sealed, contained lagoon-type deposits (carbonate and terrigenous) and were highly saline, with deposition of gypsum and anhydrites. On the whole, carbonates predominated but the large facies variation points to high tectonic activity as in the Late Carboniferous.

This is particularly true of the zone of 'Ancient Rockies' in which the thickness of Lower Permian formations reached 2 km in spite of a reduction in tempo of subsidence.

The Western Texas basin, situated in the south-western corner of the craton and opening into the ocean in the south, represented a particularly interesting structure of Early Permian and early Late Permian epochs. This basin was divided by an inner uplift of the Central Platform into two subbasins: Delaware and Midland (see Fig. 4-1). On the periphery of the basin and the Central Platform, shelf carbonates accumulated and large barrier reefs were formed along their margins in the second half of the Early Permian and first half of the Late Permian. Accumulation of thin dark, bituminous, thinly laminated limestones, siltstones and shales proceeded in the subbasins during this period (Fig. 5-1).

L a t e   P e r m i a n. By the beginning of the Late Permian, the North American Craton underwent some changes in contour. Uplift continued in the south-west, in the Ancient Rockies. In the north-east, in eastern Greenland, the orogenic regime was replaced by a platform one and this zone, on the contrary, added to the area of the craton. The same is true of the Newfoundland and Northern Appalachian territories, which had likewise completed active tectonic development and became a part of the craton.

Within the Cordilleran margin of the continent, mainly in Idaho, Montana and Wyoming, a relatively deepwater basin was formed at the commencement of the epoch. The basin is well known for its deposits of bedded phosphorites—products of upwelling—of the Phosphoria formation. Apart from phosphorites, it consists of fine dark clastic sediments rich in organic matter. These sediments contain bands of limestones and cherts and measure up to 500 m in thickness.

In the Sverdrup basin in the northern part of the craton, following a brief pre-Late Permian uplift, quiescent subsidence renewed and the basin was filled with thin clay deposits in the north, lime-clay deposits in the south and littoral, partly alluvial-deltaic terrigenous formations in the marginal portions. Their thickness does not exceed 800 m. In the north-eastern part of the craton the former Eastern Greenland mobile system was affected by subsidence and a carbonate-terrigenous complex of small thickness and evaporites deposited here in a lagoonal sea.

Subsidence of the small Peace River basin continued in the western part of the craton. In this partially sealed bay, joining probably with troughs of the Cordilleran system, subsidence was replaced by uplift in the second half of the epoch, with the thickness of Lower Permian formations not exceeding 200 m.

Uplift and regression had developed further in the Midcontinent region by this period and seas survived only in the form of almost closed, greatly

226

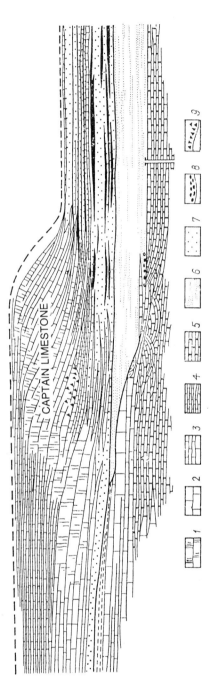

Fig. 5-1. Profile of exposed beds in southern Guadalupe Mountains, showing stratigraphic relationship between Captain limestones and other units (after King, 1978, simplified).

1—massive limestones some hundreds of feet thick; 2—thick limestone beds; 3—thin light grey limestones; 4—thin dark grey limestones; 5—thin black limestones; 6—coarse-grained thick sandstones; 7—fine-grained thin sandstones; 8—conglomerates; 9—dolomite and sand breccia (in Captain limestones).

reduced lagoons of the Western Texas and Williston basins. High water salinity and warm climate led to the deposition of gypsum and anhydrites in addition to clay, red sandstones and carbonate rocks. The thickness of formations here rarely exceeds 200 m but in the Western Texas basin subsidence was very intense and the thickness of the Upper Permian reached 3 km.

E a r l y   T r i a s s i c. The territory of the craton enlarged in the south as a result of completion of the orogenic development of the Southern Appalachians and the Ouachita-Mexican system. These regions became parts of the platform.

On the craton as a whole, uplift predominated and the area of marine basins shrank even more since the Western Texas experienced .uplift and regression and the Williston basin experienced regression. In fact, weak subsidence still continued in the Williston basin but the Lower Triassic here is represented only by lacustrine-alluvial formations running into a few tens of metres in thickness. Evaporites also were deposited in a large lake in the northern part of Williston basin. However, in the south-western part of the craton, the Anadarko and Ardmore basins filled almost exclusively with carbonate rocks to a thickness of up to 1.5 km and still continued to subside.

The Peace River basin in the north-western part of the platform also continued to subside but carbonate sedimentation in it was almost wholly replaced by terrigenous sedimentation. The thickness of the Lower Triassic in this basin reaches 400 m.

The Sverdrup basin in the northern part of the craton subsided continuously and, in its central part, fine clay sediments were deposited, as before, up to 1.7 km in thickness. Sand deposits predominated on the slopes of the basin. Eastwardly this basin extended to north-eastern Greenland and later evidently joined with the Eastern Greenland basin which, compared to the Late Permian epoch, experienced a partial regression.

Thus during the rather prolonged stage under consideration, the area of the North American Craton enlarged significantly in the north and south as a result of the disappearance of the Innuitian and in part the Appalachian orogens. A general tendency towards regression was preserved and regression manifested distinctly even at the end of the preceding stage. Uplift enlarged progressively and the area of marine basins in the Midcontinent area decreased, this process advancing from east to west. Protracted subsidence of the major Sverdrup and Western Texas basins and small Peace River and Williston basins continued.

5.1.2 East European Craton

M i d d l e - L a t e   C a r b o n i f e r o u s. The general contours of the craton did not change compared to the Early Carboniferous but its structural plan underwent some reorganisation (Fig. 5-2). The area of subsidence in the Volga-Uralian region enlarged, Tokmov arch flattened considerably and its position could be determined only from the change of facies. The dimensions of the Voronezh uplift, which joined with the Belorussian—Peri-Baltic one increased. The Moesian block and Peri-Carpathian zones too were affected by uplifts.

In the eastern part of the craton subsidence developed gradually. At the beginning of the Middle Carboniferous, in the Bashkirian epoch, sedimentation occurred only in the eastern and southern regions and the Dnepr-Donets basin. It extended into the Moscow syneclise in the second half of the epoch. Further enlargement of the regions of subsidence and reduction in the relief of the basin bottom commenced in the Moscovian stage and continued up to the end of the Carboniferous.

The Volga-Uralian basin was shallow and limestones and dolomites were deposited almost everywhere in it while variegated and red-coloured silts and clays accumulated in the marginal parts. Terrigenous material was transported from the Ukrainian and Voronezh uplifts disposed in a humid tropical zone. The total thickness of formations is 200 to 400 m, occasionally exceeding 600 to 700 m. In the Late Carboniferous the aridity of climate rose sharply and evaporites (gypsum and anhydrite) expanded widely in the middle Volga and western Peri-Timan region.

Filling of the Dnepr-Donets aulacogen was mainly completed in the Early Carboniferous and, at the end of this epoch, consequent to cessation of activity of boundary faults, it was transformed into a trough. The boundaries of the trough enlarged in this epoch and it was gradually transformed further into an ordinary platform basin, i.e., the Ukrainian syneclise. However, the tempo of subsidence was still high and the thickness of the Middle and Late Carboniferous formations exceeded 3.5 km. Terrigenous rocks predominated in their composition and paralic coal-bearing formations played a significant role.

A narrow strip of intensely subsided basins extended from the Dnepr-Donets trough westward towards the boundary of the craton and later up to the British Isles (Ramsbottom, 1979). These basins were filled with comparatively thick (up to 2.5 km) terrigenous and coal-bearing complexes and form the so-called 'European Coal Channel'.

In the British Isles per se, subsidence was succeeded by uplift at the end of the Late Carboniferous. Sedimentary formations of Devonian and Carboniferous troughs experienced fold deformation. This folding, however, did not expand everywhere and many folds lay mainly above or around faults.

Early Permian. In the eastern part of the craton, the extensive Volga-Uralian basin continued to subside and to shrink steadily. Its size was maximum in the Asselian stage (Schwagerina beds) when the sea reached almost up to the Moscow region and covered the south-eastern part of the Voronezh uplift. The shallowness of the basin and its high salinity are characteristic features which caused the deposition here of dolomitic silts, less

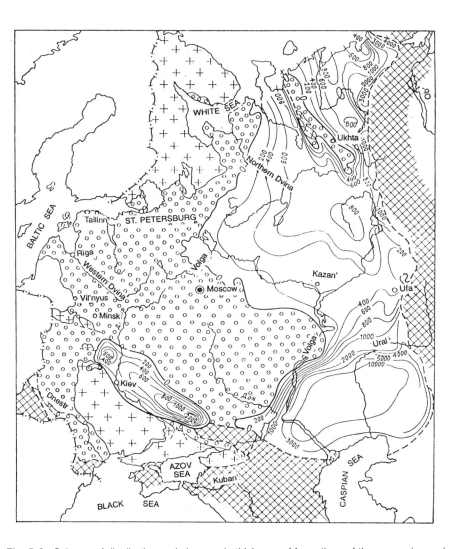

Fig. 5-2. Scheme of distribution and changes in thickness of formations of the regressive and emersive stages of the Middle-Late Palaeozoic stage of development of the East European Craton—Late Carboniferous-Early Triassic (after N.N. Forsch and A.A. Sultanaev, In: *Geological Structure of the USSR and Pattern of Location of Mineral Deposits.* Vol. 1, *Russian Craton*, 1985). See Fig. 1-2 for legend.

frequently gypsum, and sometimes salts. Salinity was maximal in the second half of the Kungurian stage. Small quantities of terrigenous material were deposited in the littoral zones but carbonates predominated. The thickness of the Lower Permian usually did not exceed 350 m. The zone of shallow sea in the east ended in a zone of reef formation, which represented low islands, shoals and atolls. East of the reef strip the depth of the sea already increased sharply within the Uralian foredeep. At the end of the Early Permian, uplift led to a regression in the western part of the Volga-Uralian basin.

Development of the Peri-Caspian basin proceeded in a specific manner. At the beginning of the epoch it was joined with the Volga-Uralian basin as well as with the seas of the Mediterranean belt in the south and was characterised by conditions of uncompensated subsidence. In the Kungurian stage the Peri-Caspian basin completely lost contact with the southern seas of the Palaeo-Tethys as a result of the formation of an orogen in the south. This sea-gulf became rapidly salinised and salts filled the immense expanse of the basin in a short time. Further, the deposition of salts proceeded on considerable depths, at least initially.

The Dnepr-Donets basin was joined with the Peri-Caspian basin at the beginning of the epoch and formation of terrigenous and carbonate deposits continued in it. In the Artinskian age inversion commenced in the eastern part of the basin, contact with the Peri-Caspian Sea was interrupted and evaporites (sometimes salts) were deposited in the isolated lake. At the same time, fold processes commenced but their development became widespread in the Kungurian age. Thus development of the fold structure of the Donets range was associated with the Saalian phase of Hercynian orogeny.

The manifestation of Early Permian volcanism of intermediate and felsic composition in the area of the Moesian block and the Peri-Carpathian zone deserves attention (Belov, 1981). It points to reactivation of the tectonic regime in these regions as a result of orogeny in the adjoining Mediterranean belt.

The Middle-Late Carboniferous basins of northern Europe ('Coal Channel') had experienced regression and weak folding by commencement of the Permian but later were again affected by subsidence. In the broad flat basins, situated even on their northern periphery, accumulation of the Rotliegende commenced. It represents a red molassoid formation due mainly to supply from the south, from the Hercynian mountain system of the Mediterranean belt and, in the British Isles territory, also from its reactivated Caledonian and Precambrian northern frame. At the end of the Early Permian, an extensive isolated salt-bearing basin originated in the central part of this newly formed North European megasyneclise. This occurred evidently due to intensification of subsidence and commencement of the invasion of sea waters from the north from the 'Scandic Sea' (H. Stille, 1948). The thickness of the Lower Permian in the most subsided part of the basin reaches 1 km.

L a t e   P e r m i a n. At the beginning of the epoch, almost the entire eastern part of the craton represented an undulating plain and a small basin survived only in the extreme east in the Peri-Urals. Later, increasingly larger areas were affected by subsidence and the Late Permian Volga-Uralian basin increasingly extended westward. In the Kazanian age it encompassed as much area as in the Asselian age of the Early Permian. The Ural Mountains lay east of this immense basin. Rivers flowing from the Urals transported enormous quantities of terrigenous material and this, together with the warm climatic conditions, made for all the characteristic features of sedimentation. Its main feature was the successive change of terrigenous formations from coarse to fine and their replacement by carbonate and halogenic deposits from east to west. The thickness of the Upper Permian exceeds 700 m in the east but is only a few tens of metres in the west. The process of subsidence of the western part of the Volga-Uralian basin was interrupted by regressions (for example, at the commencement of the Tartarian age).

In the north-eastern part of the craton the Pechora basin was set off from the Volga-Uralian basin by uplift of the Timan range. Coal-bearing terrigenous formations accumulated in it under conditions of desalinised lagoons and marshy banks.

All the characteristic features of development of the Volga-Uralian basin mentioned above are also characteristic of the Peri-Caspian basin. After its filling in the Early Permian, Ufa and partly Kazanian ages with salt-bearing deposits, the same shallow sea remained there and the slow subsidence of the basin was compensated by deposition of a variegated molassoid terrigenous formation of moderate thickness. The thickness of the Upper Permian exceeds 900 m only in the south-eastern part of the basin.

In the western part of the Dnepr-Donets basin moderate subsidence continued and uplift renewed in its south-eastern part at the end of the Early Permian. As a result, this basin was again joined with the Peri-Caspian but was desalinised and shallow-water terrigenous deposits (partly in subcontinental facies) were deposited in it. The material for these deposits came from hilly elevations in the central regions of the craton.

Within the Peri-Carpathian zone, commencement of the Late Permian was characterised by intensified subsidence and transgression, which penetrated into the Transcarpathian region in the form of a narrow bay through a lagoon of the Moldavian basin. Thin limestone-dolomite deposits with gypsum accumulated there. At the end of the Late Permian, uplift had already developed in this region and total regression occurred.

In the region of the Moesian block manifestation of volcanism ceased while intense subsidence was wholly compensated by the deposition of thick (up to 2 km) terrigenous beds of molasse supplied from the adjoining mountain systems of the Mediterranean belt. As before, the tectonic regime here was close to orogenic.

In the western part of the East European Craton, formation of a major Middle European saline basin was completed (Ziegler, 1982). Already by the commencement of the epoch, it appeared as the main structure of western and central Europe. The Vistula-Dnestr zone became its eastern flank and the Baltic and Podlias-Brest basins its bays. Subsidence extended into the northern part of the North Sea but the Scotland-Norway-Denmark basin formed here was set off from the main part of the Central European basin by a sublatitudinal uplift represented in the contemporary structure as the Northumberland salient in the west and the Southern Danish anteclise in the east.

The main part of the sedimentary filling of the Central European basin in the Late Permian comprises the thick (about 1 km) Zechstein evaporite series. This was deposited under conditions of continuous subsidence and alternation of normal marine conditions with conditions of high salinity and gypsum-salt accumulation. Bituminous shales, dolomites, limestones, gypsum and salt beds were deposited here and terrigenous sandstones in the littoral zones. At the end of the epoch, subsidence slowed down in the Peri-Baltic region and the shoreline receded far to the west.

Early Triassic. Completion of orogenic development in the West European part of the Mediterranean belt led to a change there of an orogenic regime into a platform one. As a result, this entire extensive territory adjoined the East European Craton whose area was thus enlarged by one-third. Uplift affecting the eastern part of the craton at the end of the Late Permian intensified and led to a total regression in the Volga-Uralian basin. Only a broad flat depression filled with lacustrine-swampy and alluvial deposits up to 200 m in thickness remained of this basin. At the end of the epoch, this depression was wholly affected by uplift.

Regression occurred also in the Pechora basin where, too, continental terrigenous deposits up to 400 m in thickness were formed. The low montane system of the Urals orogen represented the source for these clastic beds.

The Peri-Caspian basin acquired a partially closed, lagoon-continental nature in the Early Triassic. Its subsidence nearly ceased, its size shrank and filling with fine terrigenous and, less frequently, carbonate formations in continental and marine facies was completed.

The Dnepr-Donets basin became a fairly large elongated lake. Subsidence here maintained the former moderate tempo and persistence. The thickness of terrigenous and partly carbonate formations went up to 300 to 400 m.

Unlike the eastern part of the craton, in the western part development of the Central European basin proceeded generally almost within the former boundaries and scale. The contact of this basin with the northern 'Scandic Sea' was interrupted but contacts were established with marine basins of the Mediterranean belt. Zechstein formations in the basins were conformably replaced by a 'German Triassic' complex and its Lower Triassic section was represented by 'Buntsandstein'—a complex of variegated lagoonal-continental formations with evaporites. In this epoch also the internal structure of the Central European basin became more complicated. In the east, the Pomorie-Kuyava basin, containing Lower Triassic formations up to 1.3 km in thickness, became separated. A similar thickness is characteristic of the German basin. The North Danish basin was also isolated. Trains of alluvial-deltaic deposits were formed in the broad littoral zones of the Central European basin.

Thus during the Middle-Late Carboniferous to the Early Triassic development of the western and eastern parts of the European Craton proceeded independent of each other. In the east the extensive Volga-Uralian basin underwent a complete transgressive-regressive cycle with some variations in tempo of subsidence and general upheaval at the end of the Early Triassic. The association of tectonic development of this basin with evolution of the Urals orogen is wholly evident. In the eastern part of the craton the main stage of development of long-surviving structures, such as the Pechora and Peri-Caspian basins, and also of the Dnepr-Donets basin which inherited the position of the aulacogen, was completed by the Early Triassic.

In the western part of the craton subsidence grew successively from the Permian as a result of commencement of formation of the extremely large Central European basin. Development of this basin continued even into the succeeding Mesozoic history of the craton. If the formation of the 'Coal Channel' of Europe in the Middle-Late Carboniferous was positively associated with the manifestation of Hercynian orogeny in Central Europe, subsidence and development of the Central European basin are most likely attributable to other reasons since transition of Hercynian orogenic structures from the beginning of the Triassic to a platform regime of development was reflected in its evolution.

The last episode is the formation of the young West European platform adjoining the old East European Craton.

5.1.3 Siberian Craton

Middle-Late Carboniferous. Completion of the orogenic development of much of the Altay-Sayan region by the Middle Carboniferous with transition to a platform regime, enlarged the area of the south-western part of the craton. Simultaneously a thorough change occurred in the territory of the craton regarding the nature of sedimentation: marine sediments of the arid climatic type were replaced by continental coal-bearing deposits, forming a limnic coal-bearing formation. The Siberian Craton represented a lacustrine-alluvial plain on which carbonaceous clays, silts, sands and sometimes coal (in Kan Taseevo basin) were deposited. The main region of accumulation of these formations was the Tunguska basin which enlarged in size as the slopes of the Anabar and Central Siberian (Nepa-Botuoba) uplift were affected by subsidence. In some periods this interior basin was in contact with the Taimyr, Verkhoyansk and Ob'-Zaisan marine basins. The thickness of subcontinental formations measured a few tens of metres. Thicker Middle-Upper Carboniferous deposits accumulated in the Peri-Verkhoyansk zone (up to 2 km) and along the periphery of the Taimyr mobile zone (up to 700 m).

E a r l y   P e r m i a n. The character of a vast plain within which huge lakes and swamps arose at times was preserved on the craton. In these sealed basins, mainly in the central part of the craton (in Tunguska basin), quartz sands and other terrigenous (sometimes coal-bearing) rocks not exceeding a few tens of metres in thickness were deposited. The area of such basins decreased compared to the Late Carboniferous. The elevated territory of the former Yenisei-Baikal orogen was the main source of clastic material.

Stronger subsidence and marine basins developed only in the Peri-Verkhoyansk zone and on the Taimyr peninsula. The thickness of sandy clay formations here reached a few hundred metres.

L a t e   P e r m i a n. At the beginning of the epoch, conditions of extensive continental low plains continued to exist on the craton (Fig. 5–3). Large, shallow, freshwater lake reservoirs periodically arose here, then gave way to swampy lowlands with abundant peat formation. Sometimes upheaval led to drying of almost the entire craton; flowing rivers cut beds and filled them with alluvial formations. In the western and central parts of the craton, which generally experienced maximum subsidence, conditions were created for the accumulation of thicker continental sedimentary and coal-bearing strata. Marine deposits were formed only in the Peri-Verkhoyansk zone and in the eastern part of the Taimyr peninsula.

A new important phase set in at the end of the Late Permian in the development of the Siberian Craton: its central (Tunguska basin), northern and north-eastern regions became an arena of unusually intense trappean (flood basalt) magmatism. This commenced with a burst of explosive activity, succeeded by lava outpourings. The centres of these eruptions were confined to zones of high permeability along the periphery of the Tunguska basin and to similar zones intersecting this basin latitudinally. Intrusive bodies of dolerites and gabbro-dolerites were also formed in these same zones. Magmatic activity in the north affected Taimyr and, in the east, the Peri-Verkhoyansk zone. A particularly thick sheet (over 500 m) of Upper Permian volcanic rocks accumulated in the Noril'sk region.

E a r l y   T r i a s s i c. Magmatic activity on the craton reached culmination in the first half of this epoch. As before, the most intense eruptions occurred in the north-west, in the Turukhansk-Noril'sk region. The thickness

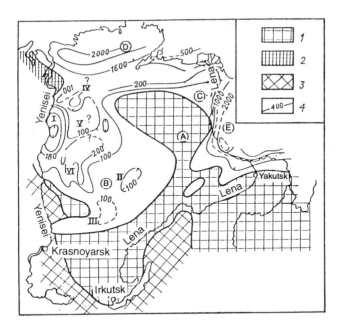

Fig. 5-3. Palaeotectonic scheme of the Siberian Craton in the Late Permian (after N.S. Malich. In: *Geological Structure of the USSR and Pattern of Location of Mineral Deposits. Vol. 4, Siberian Craton*, 1987)'.

1 to 3—zones of uplift (1—weak on platform, 2—moderate on platform, 3—weak in fold zone); 4—isopachs, m.

Structural-formation regions (circled letters): A—Anabar-Angara, B—Tunguska; C—Khatanga-Olenek; D—Taimyr; E—Verkhoyansk-Taimyr. I to III—basins (I—lower Tunguska, II—central Tunguska, III—Angara-Muya); IV to VI—uplifts (IV—Khantay-Rybnik, V—Tembenchia; VI—Uchamin).

of lava sheets and pyroclastic layers here exceeds 1 km. In the eastern (Maimecha-Kotuy region) and south-western frame of the Tunguska basin, trap association coexisted with ultrabasic-alkaline and kimberlite intrusive formations. Pipes and dykes of kimberlites intruded into the Anabar uplift as well.

In the Peri-Verkhoyansk zone subsidence intensified (thickness of marine and continental Lower Triassic formations exceeds 1 km). The marine basin enlarged even onto the northern part of the platform in the Yenisei-Khatanga region. Terrigenous strata were deposited here under marine and sometimes continental conditions. These strata represented erosion products of the northern part of the Taimyr mobile zone, which experienced mighty uplift, folding and mountain building in the Early Triassic.

Thus, at the commencement of the stage under consideration, an extremely quiescent tectonic style of development was established on the

Siberian Craton. An environment of flat swampy lowland was long preserved throughout almost the whole of its territory. A sharp change set in in the Late Permian when trappean magmatism began to manifest in the craton's northern margin. This pointed to major changes in the state of the asthenosphere.

5.1.4 Sino-Korea Craton

M i d d l e - L a t e   C a r b o n i f e r o u s. At the beginning of the Middle Carboniferous, subsidence again affected the northern China and Korean territories and the sea returned to these regions (*Atlas of the Palaeogeography of China*, 1985). Bauxites and ferruginous ores, representing erosion products of the weathering crust, formed over a protracted continental period under conditions of a peneplain and humid climate, were often deposited at the base of the transgressive Middle Carboniferous. However, subsidence in the northern part of the Sino-Korea Craton was extremely slow and the sea exceptionally shallow. In the phase of negative eustatic fluctuations, the sea receded beyond the boundaries of this part of the craton and transformed into a swampy plain, an arena for peat accumulation. As a result, paralic coal-bearing formations a few hundred metres in thickness accumulated in the Middle-Late Carboniferous in the basins of northern China and Korea. The main regions of erosion lay along the periphery of the central basin—Alashan, Inner Mongolian range, Qinling and Dabieshan—and also some interior uplifted blocks.

E a r l y   P e r m i a n. A very brief tendency towards uplift and emersion was manifest at the end of the Carboniferous (Zu-qi Zhang, 1984). As a result, marine sedimentation ceased and renewed subsidence was compensated by deposition of continental coal-bearing deposits, i.e., limnic coal-bearing formations, reaching a few hundred metres in thickness. The general environment here very much resembled that of the area of the Siberian Craton established in the Late Carboniferous.

In the eastern part of the lagoon-continental basin in the western Tarim platform block, eruptions of basalts also commenced at the end of the Early Permian.

L a t e   P e r m i a n. Tectonic activity intensified. In the composition of formations filling the extensive flat basins, red clastic rocks predominated while coal accumulation almost ceased except for basins in the Korean peninsula. The thickness of formations increased to more than 1.5 km in some regions.

E a r l y   T r i a s s i c. The territories of northern China and the Korean peninsula experienced mainly uplift while quiescent subsidence continued only in the central regions. In the extensive Ordos interior basin predominantly continental redbeds up to 350 m in thickness were deposited.

Sometimes subsidence intensified, sea-water intruded and carbonates were deposited in shallow lagoons.

Thus after some intensification of subsidence and marine transgression in the Middle-Late Carboniferous, during the Permian and the Early Triassic only continental sedimentation with a gradually decreasing rate proceeded on the Sino-Korea Craton, suggesting uplift of its territory.

5.1.5 South China (Yangtze) Craton

During this period the South China Craton continued to experience stable subsidence, which was compensated by the accumulation of shallow-water carbonate deposits. Subsidence of part of the Jiannan uplift enlarged the boundaries of the basin. The thickness of sediments here is significant; it exceeded 1 km in Sychuang, Guizhou and eastern Yunnan.

E a r l y P e r m i a n. Following a brief uplift, subsidence again manifested on the craton, being wider and more intense than in the Middle-Late Carboniferous. In a vast basin submerging almost the entire uplift, accumulation of carbonate rocks predominated, their thickness in the Sychuang and Hubei-Guizhou basins reaching 600 to 700 m. Limestones were replaced by terrigenous deposits only along the periphery and by coal-bearing deposits also in the south-east. At the end of the Early Permian, uplift commenced within the Sikan-Yunnan range in the western part of the craton and block movements occurred along the faults. At this time eruptions of flood basalts commenced in western Yunnan and the adjoining regions of Sychuang and Guizhou.

L a t e P e r m i a n. Carbonate deposits in the southern part of the craton were replaced by paralic coal-bearing formation and the area of the basin gradually decreased. However, a new wave of subsidence and transgression followed, with deposition of limestones at the end of the Permian in southern China.

Within the major Sikan-Yunnan (Kam-Dian) uplift, manifestations of trappean magmatism continued and enlarged considerably in area and intensity. The thickness of basalt sheets here reaches 300 to 400 m. Eruptions of basalt in the eastern part of the Bachu-Hetiang interior basin of the Tarim platform block intensified. In the Late Permian this basin was filled with only continental formations.

E a r l y T r i a s s i c. The conditions in southern China were altogether different from those prevailing in the north. Here eruptions of basalts ceased and the stable tendency towards intense subsidence noticed at the end of the Late Permian was maintained. Carbonate deposits predominated in this marine basin. Intense supply of clastic material came exclusively from the Sikan-Yunnan uplift.

Absolute predominance of submergence with the accumulation of essentially shallow-water carbonates was characteristic of the South China Craton throughout the Late Palaeozoic and the Early Triassic. Eruptions of flood basalts commenced in the south-western part of the craton in the Early Permian and attained utmost development in the Late Permian.

5.1.6 South American Craton

M i d d l e - L a t e   C a r b o n i f e r o u s. Major changes occurred in the tectonic development of the craton from the beginning of the Middle Carboniferous and vast expanses experienced subsidence. This was most extensive and strong in the Amazon basin, especially in its easternmost part (Szatmari et al., 1979). Limestones, shales and sandstones with beds of gypsum, anhydrites and rock salt were deposited here. Their thickness in the marginal zones of the basin ranged from 300 to 400 m, in the central zones 1300 to 1500 m, with the maximum occurring in the east at 2000 m. In the east this basin fused with the Maranâo, also a newly formed one, and filled with a similar complex of rocks but to a smaller thickness (up to 400 m).

The extensive Parana basin, also evidently joined with the Amazon narrow trough, was filled mainly with terrigenous deposits in continental facies. In this same part of the craton subsidence commenced in the Late Carboniferous and was preceded by the development of glaciation. For this reason glacial formations (Itarare formation and others) lie at the base of the sedimentary filling of the basin and are overlain by marine and lacustrine-swampy terrigenous complexes including coal-bearing. The Parana basin deepened eastward and sediment thickness there reached 1 km.

Late Carboniferous glacial formations are extremely widespread and are encountered from the Falkland (Malvinas) islands to northern Bolivia. Their thickness constitutes tens of metres (but over 500 m in the northern Parana basin). Palaeogeographic analysis has shown that the main centre of glaciation lay south-east of the Parana basin and that there were also other centres of glaciation in the territory of the craton.

Westward, the Parana basin was connected with the Chaco narrow trough whose southern continuation was the Sierra Australes of the Buenos Aires trough (Amos, 1981). Massive beds (1 to 1.5 km in thickness) of marine-glacial deposits and also continental conglomerates and sandstones were formed in these deep troughs, possibly representing a major rift.

Formation of a series of elongated basins (Paganzo, Rio Blanco, Calingasta-Upsalata and others) commenced along the western boundary of the craton with the Andean orogen. They bordered mountain massifs covered by ice in the west and were filled with coarse clastic continental, including glacial formations.

E a r l y   P e r m i a n. The phase of intense subsidence affecting the territory of the craton in the Middle-Late Carboniferous was of short duration and upwarping again predominated in the Early Permian. These upward movements extended over a considerably larger area than in the Carboniferous. As a result, the area of sedimentary basins shrank substantially and they became disjointed. The former Amazon basin represented land at the beginning of the Permian and about 800 m of continental terrigenous sediments accumulated only in a small central basin. The Maranâo basin shrank in size, shoaled, and alluvial-lacustrine, lagoonal-continental and partly littoral-marine formations were formed in it.

Subsidence of the Parana basin continued south of these regions. The basin consisted of isolated or interconnected post-glacial basins periodically representing regions of coal accumulation. Bands of conglomerates point to brief bursts of tectonic activity. Sometimes marine waters penetrated here but the environment of a low swampy plain predominated and the thickness of formations did not exceed 200 to 300 m.

The Chaco trough experienced uplift even before commencement of the Permian but subsidence of this narrow graben continued in the region of the Sierra Australes of Buenos Aires and it was filled with marine terrigenous formations of considerable thickness (up to 1.2 km). Subsidence of the Central Patagonian basin continued uninterruptedly in the south; marine and continental conditions of sedimentation alternated in it. Compared to the northern regions of the craton, the mobility of its southern part was much higher.

In the northern part of the Peri-Andean marginal zone marine basins with carbonate-terrigenous and sometimes evaporite deposits were preserved but their area and rate of subsidence diminished. Continental marginal basins in the southern part of Peri-Andean zone had wholly filled with sediments by the commencement of the Permian.

L a t e   P e r m i a n. Subsidence of all the Early Permian basins, i.e., Maranâo, Parana and Central Patagonian, ceased by the beginning of this epoch and uplift prevailed on the craton during the Late Permian. Upper Permian formations are not known on the craton except for the Sierra Australes of Buenos Aires and the Falkland (Malvinas) islands where filling of the Late Palaeozoic grabens was completed, not with marine but continental sandy-clay formations. At the end of the Late Permian, or perhaps slightly later (in the Early Triassic), folding was manifest in these grabens.

E a r l y   T r i a s s i c. uplift predominated on the vast territory of the craton and if some sediments were deposited in individual regions, they occurred in continental facies and on a very small scale. Continental basins of this type existed evidently in the Peri-Andean zone of the territory of Peru. The southern part of the craton, including the Patagonia-Deseado massif, underwent tectonic reactivation in the Early Triassic. Orogenic processes, i.e., mighty subaerial volcanism, intrusion of plutons and fault and fold deformation, extended here from the Andes. As a result, the platform regime was replaced by an orogenic one and these regions became a part of the Andean mobile belt.

Thus the Middle Carboniferous-Early Triassic stage of development of the South American Craton is characterised by a strong but comparatively brief wave of subsidence and transgression in the Middle-Late Carboniferous and later by prolonged quiescent upheaval. Late Carboniferous glaciation was extremely extensive.

5.1.7 African Craton

Middle-Late Carboniferous. The tendency towards intensification of uplift noticed at the end of the Early Carboniferous persisted in the northern part of the craton (Fabre, 1988). The western and southern parts of the Sahara basin experienced a regression in the Middle Namurian while coal-bearing beds were deposited in the residual basins under continental and lacustrine-swampy conditions (see Fig. 2-3). Ougarta basin (aulacogen) experienced inversion while sedimentation still continued in the adjoining Reggane and Bechar basins in the Middle Carboniferous. During the last minor Middle Carboniferous (Bashkirian-Moscovian) transgression in the Palaeozoic, marine and lagoonal carbonate-terrigenous rocks were deposited in the eastern part of the Sahara basin (Nedjari, 1981). In the Late Carboniferous a marine bay reaching the Murzuk basin in the south still existed in Libya.

Manifestations of felsic and alkaline volcanism—rhyolites, trachytes and tuffs—are known in the Jebel Oweinat region. Annular dykes of granites intruded here while dykes and thick dolerite sills intruded into the southern Reguibat and Ahaggar uplifts.

In the north-eastern part of the craton uplift continuing in southern Arabia also affected the northern regions. Small marine basins were preserved only in eastern Turkey, Syria, northern Iran and central Afghanistan. Shallow-water terrigenous and carbonate beds up to 200 to 300 m in thickness were deposited in them.

The tectonic development of central and southern Africa differed significantly. In the Middle Carboniferous these regions represented a zone of moderate uplift, but in the Late Carboniferous perceptible subsidence and comparatively heavy sedimentation commenced here for the first time in the Palaeozoic. Sedimentation was also associated with glaciation. At the end of the Carboniferous, almost the entire southern half of the African Craton was covered with glaciers (Fig. 5-4). The most prominent centres of glaciation were the Transvaal and Zimbabwe uplifts and the Kalahari and Karroo basins. The main direction of glacier movement was from north to south. In the eastern part of the Karroo basin ice movement in Natal was from the east, from the region of the present-day Indian Ocean. This suggests the proximity of some ancient mainland massif, probably Antarctica. Accumulation of glacial and fluvio-glacial formations (Dwyka series) proceeded in isolated tectonic depressions and graben-like troughs in which their thickness went up to 300 m, at places 500 m. In the southern part of the Karroo basin marine bands are noticed among the tillites while a comparatively deep intracontinental marine basin with an accumulation of formations similar in genesis to present-day sulphurous muds of the Black Sea was formed at the end of the Carboniferous.

The formation of systems of graben-like troughs—Limpopo, Zambezi, Nyasa-Rukwa, Luangwa, Ruhuhu etc.—in the southern part of the craton at the end of the Carboniferous is explained by the reactivation of old faults. This was the beginning of a prolonged process of rift formation leading to the disintegration of Gondwana in the Mesozoic.

E a r l y   P e r m i a n. Uplift predominated in the northern part of the craton, a total regression occurred and Lower Permian formations are not known. At the same time, subsidence and transgression commenced in the north-east and affected the vast territories of Iran and the eastern Arabian peninsula (Davoudzadeh and Weber-Diefenbach, 1987). Carbonate deposits with bands of clay muds up to 500 m in thickness predominated in shallow seas. In the southern regions of Arabia there are signs of an accumulation of glacial formations at the beginning of the Permian (or at the end of the Carboniferous; McClure, 1980).

The tectonic development of the central and southern parts of the craton inherited the features of the Late Carboniferous but the intensity of movements was higher and subsidence of the system of grabens and basins continued. The most intense subsidence occurred in the Karroo basin. This formerly deep marine basin was filled with sediments and became shallow; the thickness of terrigenous, sandy-clay deposits of the Ekka series formed here in the Early Permian reached 3 km in the extreme south, in the Cape province, where they were represented by turbidites. In the basins situated more to the north sedimentation conditions were continental and coal-bearing beds were formed (see Fig. 5-4). The latter arose at the site of major lakes. The presence of coal has been established and its seams are encountered in almost all the Early Permian sedimentary basins of the central and southern parts of the continent.

The system of rift grabens that arose in the Late Carboniferous developed further in the Early Permian. Judging from the thickness of sediments, the Zambezi (1.6 km) and Ruhuhu (1 km) grabens subsided most. Coal-bearing strata are present everywhere among formations filling the grabens.

L a t e   P e r m i a n. In the northern part of the craton, as before, uplift predominated and Upper Permian formations are not known. On the contrary, in the north-east subsidence enlarged even more and sea covered

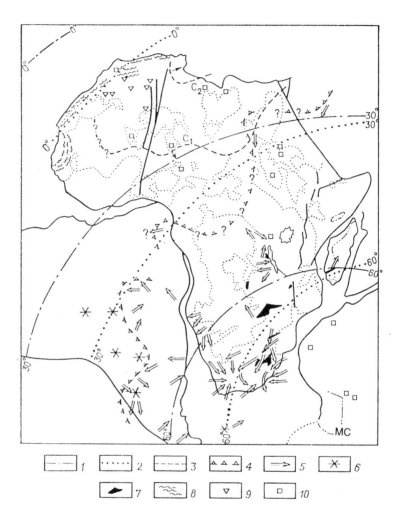

Fig. 5-4. Africa from the Early Carboniferous to end of the Permian, from 360 to 230 m.y. (after Fabre, 1988).

1—palaeolatitudes in Middle Carboniferous, about 325 m.y.; 2—palaeolatitudes at end of Carboniferous to commencement of Permian after Irwing, about 280 m.y.; 3—boundaries of transgressions in Early Carboniferous and at end of Middle Carboniferous; 4—boundaries of maximum distribution of glaciers (continental and local) at end of Carboniferous but especially in Early Permian; 5—direction of ice flow; 6—glacial formations of Middle or possibly Late Carboniferous; 7—major coal basins of Permian age under exploitation; 8—structural directions of Variscan fold systems; 9—zones of establishment of K-Ar equilibrium or paragenesis of anchizone (pyrophyllite etc.) of Permian-Carboniferous age; 10—volcanism or granite intrusions of age corresponding to it. MC—pole in Middle Carboniferous.

much of the Near and Middle East. Carbonates predominated in the composition of sediments. In the southern and western parts of this basin with a rather complex internal structure, gypsum and anhydrite were also deposited as a result of the high salinity of waters. The thickness of the Upper Permian is usually 300 to 500 m.

Rifting in the central and southern regions of the craton attenuated gradually and grabens were filled with continental lacustrine-alluvial formations. As before, the Karroo basin subsided most intensely and lower and middle parts of the Beaufort series up to 3 km in thickness were deposited in it in the Late Permian. Sedimentation conditions remained predominantly continental and were replaced by shallow marine conditions only occasionally. The great thickness of these formations suggests the maintenance of a high tempo of subsidence but the continental nature of deposits in almost all the basins and grabens underscores the fact that subsidence proceeded against a background of general uplift and was only relative.

Outcrops of the Upper Permian marine and continental formations on the coasts of Kenya, Tanzania and Mozambique and in the western half of Madagascar mark the flanks of a major basin with an axis roughly coinciding with that of the actual Mozambique channel. This structure arose at the beginning of the Late Permian at the site of a series of narrow grabens and its subsidence sometimes led to the transgression of sea from the north.

Early Triassic. In the northern part of the craton prolonged erosion and quiescent tectonic conditions led to the development of broad lowlands where extremely weak subsidence was compensated by the accumulation of lacustrine-alluvial facies. Here a transition from prolonged upheaval to subsidence has been noted.

Subsidence continued in the north-east and almost the entire territory of the Near and Middle East was covered by a shallow marine basin. Variegated shales were deposited in Arabia (Farooq, 1986) while carbonates and sometimes evaporites predominated in the north within Turkey, Syria, Iraq, Iran and Afghanistan. However, the rate of subsidence was not high and the thickness usually 100 to 200 m (at places up to 300 m).

In the central and southern parts of the craton subsidence of almost all the grabens had been completed even at the beginning of the Triassic. Lacustrine-alluvial sand and shales were still deposited to a thickness of 100 to 150 m only in the Zambezi graben. Subsidence of the Karroo basin slowed down, the basin shrank markedly and sediments accumulated here under continental conditions. A similar environment prevailed in the Congo basin. Only the eastern flank of a broad basin survived in the territory of the Mozambique channel while its western flank experienced uplift. Alternation of marine and continental terrigenous Lower Triassic formations, whose outcrops are known in western Madagascar, suggests the penetration of sea-water here. The thickness of the Lower Triassic here reaches hundreds of metres. In the extreme south of the continent a minor folding was manifest in the Early Triassic in the Cape province of the Republic of South Africa.

Even at the end of the Permian, mainly from the Triassic, subsidence commenced of the highly extended East African sedimentary basin. It occupied the territory from Ethiopia to northern Mozambique. This was an interior basin in the form of a swampy lowland. Sea penetrated into it only in the extreme east. The thickness of continental sedimentary rocks deposited here reaches 300 to 400 m.

Thus a significant difference was manifest in the tectonic regime and nature of development of the northern, north-eastern and southern parts of the craton during the Middle Carboniferous-Early Triassic, as in the Middle Devonian-Early Carboniferous. While in the preceding stage the northern half (Sahara basin) experienced subsidence predominantly and the southern half, except for the Cape province, uplift, the roles were now reversed: There was stable uplift in the northern part of the craton while numerous basins and grabens filled with sediments of the Karroo complex formed in the southern part. It may be recalled that these sediments were deposited mainly under continental conditions, i.e., submergence of basins and grabens occurred under conditions of general uplift. At the end of the stage, the difference in development of the northern and southern parts of the craton levelled. In the north-east, within the Near and Middle East, tectonic development followed a different style. Here subsidence and transgression grew successively in time and space (from the margin deep into the interior of the craton).

5.1.8 Indian Craton

M i d d l e - L a t e C a r b o n i f e r o u s. Middle Carboniferous formations in the territory of the craton are known only in the marginal zone of the Lesser Himalaya in the north where subsidence and accumulation of littoral and marine terrigenous formations continued. The major part of the craton, however, represented a stable uplift in a state of relative tectonic quiescence.

In the Late Carboniferous the craton continued to heave quite intensely and was involved in continental glaciation (Crowell, 1983). It has been suggested that the Aravalli range, Kaimur and Vindhyan ranges in the central part of India and the Deccan plateau and Eastern Ghats, i.e., zones characterised by prolonged tendencies to predominant upheaval, represented the centres of glaciation. However, in the central part of the craton subsidence of a system of basins located along the zones of major ancient faults commenced for the first time in the Palaeozoic. These basins were later transformed into grabens and underwent prolonged development. The largest of them, the Narmada-Son, extended in a sublatitudinal direction through the entire craton. Three other grabens branched out from this basin in a south-easterly direction—Godavari, Mahanadi and Damodar. The Narmada-Son basin arose at the site of an ancient Early Proterozoic trough and the Godavari basin at the site of a Riphean aulacogen. The basins were filled with fluvial, swampy, lagoonal and glacial deposits of the Gondwana complex, their thickness sometimes exceeding 300 m.

In the northern part of the craton a similar complex of deposits as well as tillites formed in the pericratonic zone in the Late Carboniferous. Further, flood basaltic volcanism manifested here over extensive areas (Panjal series of basalts). This volcanism was evidently associated with the break-up of the northern margin of the craton and destruction of the continental crust here (Jain et al., 1980; Stöcklin, 1980), leading ultimately to the formation of the Neo-Tethys ocean.

E a r l y P e r m i a n. Zones of subsidence enlarged and transgression affected the northern and north-western parts of the craton. In the new basins, i.e., Sind-Punjab and Peri-Himalayan, carbonate and terrigenous deposits up to 300 m in thickness were formed. Subsidence of the system of basins-grabens continued in the central region of the craton and terrigenous rocks accumulated in marine and lacustrine-swampy facies. Warming and humid climate promoted the growth of abundant vegetation and deposition of coal-bearing beds in these basins. The thickness of the Lower Permian exceeds 1 km in the basins.

A major transformation occurred from the beginning of the Permian in the marginal zone of the northern part of the craton (Vergunov and Pryalukhina, 1980; Shvol'man, 1980; *Atlas of the Palaeogeography of China*, 1985; Kapoor and Shah, 1979; Stöcklin, 1980). It succeeded eruptions of flood basalts and was manifest in deep subsidence, associated evidently with commencement of formation of the Neo-Tethys.

L a t e P e r m i a n. The tendency to quiescent subsidence was maintained in the northern half of the craton. There, carbonate-terrigenous sediments were deposited in a shallow sea, as before, and the area of low islands, Aravalli, Lahore etc., probably diminished. At the end of the epoch, subsidence slowed down and was succeeded by uplift which caused a partial regression.

Grabens in the central part of the craton continued to subside. In the Mahanadi graben beds of continental and marine sandy-clay sediments, conglomerates and brown iron ores were formed to a thickness reaching 300 m. In the Godavari graben, proximate in composition deposits attained a thickness of 1800 m; their thickness and composition were similar in the Narmada-Son and Damodar grabens but, additionally, coal-bearing deposits accumulated here. In the extreme north-eastern part of the craton subcontinental coal-bearing deposits were formed in the Late Permian. This suggests the formation here of a zone of subsidence in the continuation of the

Damodar graben. At the end of the epoch, subsidence of grabens slowed down and they lost contact with the marine basin.

E a r l y   T r i a s s i c. Uplift again predominated on the craton. The sea regressed from a major part of its northern basins and only small pericontinental troughs remained. In these troughs weak subsidence was compensated by deposition of shallow-water shelf sediments. Uplift affected the eastern part of the craton. Late Permian subsidence here represented a brief episode.

The growth of grabens ceased. Their identity as a single system was disturbed and now they represented individual small basins-lakes filled with mud and sand. The significant thickness of these formations (up to 1 km in the Narmada-Son graben) suggests an even greater rate of subsidence along boundary faults.

Thus during the stage under consideration, the Indian Craton experienced considerable subsidence for the first time in the Palaeozoic. In its northern part it was continuous but in the central part localised in the graben system. The cycle of subsidence commenced in the Late Carboniferous, ended in uplift by commencement of the Triassic and gradually died out during the Early Triassic epoch.

5.1.9 Australian Craton

M i d d l e - L a t e   C a r b o n i f e r o u s. At the end of the Early Carboniferous subsidence of almost all the peripheral sedimentary basins of the craton ceased and uplift predominated in this epoch. A small basin possibly survived only within the Fitzroy graben and continental sands of the Harris formation were deposited in the Middle Carboniferous to a thickness of about 200 m.

Yet one more region of probable continental sedimentation lay in the central part of the craton in Officer basin. The thickness of sediments here did not exceed 200 to 300 m and formation of glacial deposits began at the end of the Carboniferous.

The overall area of the craton increased in its south-eastern part as a result of change of the orogenic regime into a platform one over much of the territory of the Lachlan fold system.

E a r l y   P e r m i a n. After comparatively quiescent Middle-Late Carboniferous development, tectonic activity again intensified at the beginning of the Permian. Western regions of the craton experienced subsidence and in the Sakmarian age transgression affected the Fitzroy graben and Carnarvon basin and part of Tasmania Island in the south-east (Fig. 5–5). Marine terrigenous and occasionally carbonate deposits are interstratified in these regions with continental deposits and their thickness is quite high. The Fitzroy graben (aulacogen) experienced particularly intense subsidence and the thickness of the Lower Permian in it reaches 4 km.

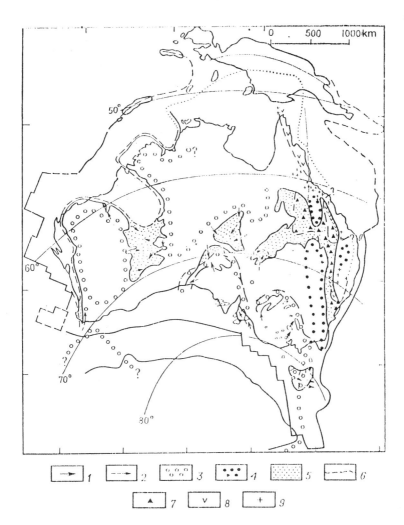

Fig. 5-5. Palaeogeography of Australia circa the verge between the Carboniferous and the Permian, i.e., 300–280 m.y. (from Veevers, 1986).

1—glacial transport; 2—transport of sediments after glaciation; 3—ice sheet; 4—mountain glaciation; 5—continental deposits; 6—coastline; 7—volcanic deposits; 8—volcanoes; 9—plutons.

However, the main events in the Early Permian period were continental glaciation and later formation of coal-bearing deposits. Glaciation in the Sakmarian age affected uplifted ancient massifs of the craton (Pilbara, Musgrave etc.), lowlands and, possibly, a part of the shelf. For this reason glacial deposits accumulated in almost all the sedimentary basins of the craton, including western parts of New Guinea and Tasmania islands. In the central part, in Arkaranga, Pedirka and Cooper basins, glacial deposits are of smaller thickness (up to 300 to 400 m) than on the western part of the craton and much finer in composition, often comprising fluvioglacial and periglacial facies. It has been suggested that glaciation proceeded in two stages. The first, much more intense, was the Sakmarian and the second, with local distribution, the Kungurian.

The first coal-bearing strata manifested among Permian formations almost simultaneous with glacial activity in the Sakmarian age in the Galilee basin among terrigenous rocks of the Reid Dome group. Later, paralic formations with coal beds were formed on the western part of the craton in Perth basin. Their thickness here was 1.8 km.

L a t e   P e r m i a n. Subsidence of the craton's sedimentary basins had practically ceased, except for the Carnarvon basin and the Fitzroy graben, by the early Late Permian. In the Carnarvon basin former conditions and exclusively marine terrigenous and carbonate sedimentation prevailed. The thickness of the Upper Permian here was 800 m. At the end of the epoch this basin experienced uplift and regression and its sedimentary filling was somewhat dislocated. The protracted and stable preceding subsidence of the Carnarvon basin and its southern continuation, Perth basin, is sometimes associated by researchers with the formation of an intracontinental rift, which continued at intervals from the Early Carboniferous to the end of the Jurassic. In present-day structure these basins are set off from Precambrian formations of the craton by a series of parallel faults and from the ocean by the horst uplifts Rockhampton and Naturalist.

In the Fitzroy graben the supply of clastic material abundantly compensated subsidence and continental coal-bearing Liveringa formations exceeding 400 m in thickness instead of marine formations began to be deposited.

Renewal of subsidence in the Cobar, Cooper and Murray basins and a new period of coal accumulation are associated with the Kazanian and Tartarian ages. The thickness of coal-bearing formations in the basins is not high (300–400 m).

E a r l y   T r i a s s i c. Uplift predominated on the craton and marine deposits of this age are not known except in the Fitzroy graben into which sea-water penetrated occasionally but continental conditions and formation of alluvial-deltaic deposits 200 to 300 m in thickness generally prevailed. According to structural disposition, duration and rate of subsidence in the Late Palaeozoic, the Fitzroy graben represents a typical rift-aulacogen. It is bound on both sides by faults with the most active movements along them having occurred in the Carboniferous and Permian. Further, the maximum amplitude of displacements attained 6 km. The sedimentary filling of the graben experienced almost no fold deformations and only a series of very simple echelon folds is noticed.

Weak subsidence continued in the continental interior basins of the eastern part of the craton (Cobar and Cooper), which were filled with lacustrine-alluvial sediments of small thickness.

On the whole, the Australian Craton experienced a stable tendency towards uplift throughout the Middle Carboniferous-Early Triassic and there were no marine basins here except the Fitzroy graben and the Carnarvon and Perth basins. Uplift promoted development of continental glaciation in the Early Permian.

5.1.10 East Antarctic Craton

In the M i d d l e C a r b o n i f e r o u s the territory of eastern Antarctica evidently represented an uplift with a rugged topography but was affected in the L a t e C a r b o n i f e r o u s by continental glaciation, traces of which are known in the Transantarctic, Ellsworth, Pensacola and other mountain ranges. Glacial formations in eastern Antarctica unconformably overlie much older ones, often with signs of exaration. The region extending from the Weddell Sea to the Transantarctic mountains through the Ellsworth Mountains was probably a marginal trough of the craton adjoining the Early Hercynian western Antarctica mobile system.

In the E a r l y P e r m i a n extensive subsidence continued in these same regions and the Beacon complex of continental formations accumulated. Tillites have been observed amongst them only in the region of Princess Astrid Coast while coal-bearing strata are present in all other basins. Sea-water penetrated into some of these basins for a brief period and deposited black shales. Subsidence of graben-like basins (rifts) in the regions of Prince Charles Mountains and Victoria Land commenced in this epoch. These were filled with similar continental and coal-bearing formations; partly marine sandstones were also deposited in the former.

In the L a t e P e r m i a n the general style of tectonic development of the craton remained as before. All of the Early Permian basins and grabens continued to subside and were filled with continental sand (including coal-bearing) formations.

Significant changes occurred at the end of the Permian and Early Triassic. Subsidence had by and large ceased in all the Permian basins and uplift predominated in the territory of the craton. The only region in which Lower Triassic formations are known is Victoria Land. Subsidence of the graben continued here and it was filled with lacustrine-alluvial deposits not exceeding 150 m in thickness.

5.2 REGIONAL REVIEW. MOBILE BELTS

5.2.1 North Atlantic Belt

5.2.1.1 *Ouachita-Mexican system.* M i d d l e - L a t e C a r b o n i -
f e r o u s. In the Ouachita zone and north-eastern part of the Mexican
zone filling of the former deep trough with thick (over 3 km) flysch was
completed in the Middle Carboniferous (Morris, 1974). Uplift of islands
intensified later, marine troughs gradually shoaled, flysch was replaced by
molasse and, in the Late Carboniferous, sedimentation proceeded under
continental conditions in the isolated Black Warrior, Arkoma, Fort Worth
and Kerr-Val-Verde troughs. These troughs lay between intensely uplifted
mountain massifs. Parallel with the accumulation of molasse to a thickness of
up to 3 km, deformation of sedimentary strata of the Ouachita zone occurred
and initially affected its central part. At the end of the Late Carboniferous,
these processes shifted westward into the Marathon Mountains. Thus, from
the beginning of the Late Carboniferous an orogenic tectonic regime was
established in this part of the system.

Western regions lagged behind in development. Unlike the Ouachita
zone, filling of the former deep trough with flysch commenced only in the
Middle Carboniferous and the high tempo of subsidence made for stability
of the system until the end of the Carboniferous. The thickness of flysch
beds exceeded 2 km. In the western part of the trough adjoining the Central
American massif, a complex of carbonate-terrigenous deposits was formed
under shelf conditions.

In the zone of Sierra Madre del Sur in the southern part of the system (Oaxaca massif), there probably existed an extensive shallow-water
shelf with terrigenous deposits. Submarine basaltic volcanism manifested
in the south and greywackes accumulated under unknown palaeotectonic
conditions (Weyl, 1980).

E a r l y P e r m i a n. In the first half of the Early Permian, deformation
continued in the Marathon zone and in the north-eastern Mexican zone. Formation of the fold-overthrust structures of Ouachita had already concluded
by commencement of the Permian. A sharply asymmetric structure with a
northern vergence was formed here. In the west, in Oklahoma, this structure had an isoclinal-imbricate character with large overthrusts and nappes.
The topography in this region evidently smoothened and accumulation of
molasse had ceased by the end of the Early Permian.

In the western Mexican zone filling of the trough with flysch concluded
only at the end of the Early Permian. The thickness of the Lower Permian
here reaches 2 km. Minor eruptions of lavas of andesite-basalt composition
occurred in the south-eastern part of this zone. In the Sierra Madre del Sur
zone, Lower Permian formations are known only in the north, where they are

represented by shallow-water terrigenous and carbonate sediments (Oaxaca massif).

L a t e P e r m i a n. In the Ouachita and Marathon zones the orogenic regime of development was finally succeeded by a platform one. In the Mexican zone, on the contrary, uplift sharply intensified and an orogenic regime was established everywhere. The former flysch trough experienced inversion and folding.

At this time, deformation affected even the rest of the territory of the zone but ceased by the end of the epoch and transition to conditions of a platform regime occurred here also. A small residual trough with accumulation of marine terrigenous formations developed at the beginning of the Late Permian in the eastern Mexican zone. The southernmost part of the system, Sierra Madre del Sur, also probably experienced a predominant uplift. Upper Permian formations are almost unknown here except in the far eastern region.

In the E a r l y T r i a s s i c a platform regime prevailed throughout the territory of the Ouachita-Mexican system and it became a part of the North American Platform.

Thus during the Late Palaeozoic the mobile system of Ouachita-Mexico underwent major transformation, its active development concluded and a mobile tectonic regime was replaced by a stable platform one. In the Middle Carboniferous to the Early Permian deep troughs were filled with flysch beds and experienced inversion and folding. Commencing from the Late Carboniferous, deformation and uplift extended from Ouachita zone west and south-west and by commencement of the Late Permian affected the entire system; however, both processes ceased everywhere at the verge between the Permian and the Triassic.

5.2.1.2 *Appalachian system.* M i d d l e - L a t e C a r b o n i f e r o u s. Manifestations of Middle-Late Devonian Acadian tectogeny including folding, metamorphism and granitisation did not lead to total consolidation of the Appalachian system. Mountain building in the expanding geoanticlinal zones continued in the Middle and Late Carboniferous and subsidence regenerated evidently along faults in the central and western zones of the system of intermontane and submontane Peri-Appalachian troughs. Extremely thick red molasse accumulated in these troughs. Its thickness in New England, for example, exceeded 4 km and the strata are coal-bearing here. Middle-Upper Carboniferous formations of the Appalachian foredeep represent a typical coal-bearing paralic molasse up to 3 km in thickness. It has been suggested that in these epochs, as in the preceding stage, intrusive magmatic activity continued almost uninterruptedly in the central zones of the system and metamorphism at the end of the Carboniferous.

E a r l y   P e r m i a n. All the zones of the system remained stably uplifted but topography smoothened, contrast of movements along faults ·decreased and development of most of the intermontane troughs was compieted. Molasse up to 2 km in thickness still formed only in the northern part of New England. The Appalachian foredeep experienced uplift and only the small Boston basin remained of it. In this basin the thickness of the Lower Permian (Dunkard group) did not exceed 500 m. Fold and fault deformation continued to develop in the Southern Appalachians but gradually died out by the end of the epoch. Probably the last of the manifestations of magmatic intrusive activity, which steadily decreased from maximum at the beginning of the epoch, were intrusions of binary alkaline granites into New England at the end of the Early Permian. These granites intruded into the Lower Carboniferous formations of Worster trough.

L a t e   P e r m i a n. Tectonic activity had diminished even more by this time. Although the territory of the system probably represented low hills, Upper Permian formations, products of the erosion of these hills, are not known. The nature of the dislocations of the Upper Palaeozoic formations in the western part of the system (Ridge and Valley province) indicates that folding and thrusting manifested here in the Late Permian. The same occurred in this epoch in the inner zone of the Appalachians since Carboniferous as well as Lower Permian formations were affected by folding and thrusting at places in the former intermontane troughs of the Northern Appalachians and experienced cleavage and metamorphism here and there. These were manifestations of the Alleghanian epoch of orogeny in which the orogenic development of the Appalachian system was completed.

E a r l y   T r i a s s i c. Some significant tectonic activity was preserved only in the central sector of the Appalachian system in the Central and Northern Appalachians. Radiometric data points to continuation here of metamorphism and possibly the last echoes of fold processes. A platform regime was established in the rest of the territory.

Thus during the Middle Carboniferous-Permian formation of the present-day structure of the Appalachians was completed, this being the result of manifestation of several epochs of orogeny from the Middle Palaeozoic. During this stage the intensity of orogenic processes gradually diminished. A noteworthy point is their extremely long duration compared, for example, to the Ouachita-Mexican system, and migration along the strike of the Appalachian system from north to south. An orogenic regime continued in the Appalachian system for about 180 m.y.

5.2.1.3 *Eastern Greenland system.* In the M i d d l e - L a t e   C a r b o n i f e r o u s   and   E a r l y   P e r m i a n   conditions of an orogenic regime prevailed throughout the territory of the system. Intense uplift made for development of a montane topography. Subsidence of intermontane basins occurred along faults and molasse was deposited in them: over 2.5 km in the Middle-Late Carboniferous and about 2 km in the Early Permian. In the Late Carboniferous the eastern part of the system experienced subsidence and brief transgression and shallow-water carbonate-terrigenous deposits accumulated here. Evidently, manifestation of folding-metamorphic processes continued in this system. At the end of the Early Permian, tectonic activity in the Eastern Greenland system became extinct and the orogenic regime was replaced by a platform one. From the Late Permian, the Eastern Greenland system was already an integral part of the North American Platform.

5.2.2 Mediterranean Belt

5.2.2.1 *West European region.* M i d d l e - L a t e C a r b o n i - f e r o u s. Orogenic development of the region commenced from the Sudetan phase of orogeny manifested at the end of the Early and beginning of the Middle Carboniferous. Sudetan deformation completed transformation of the region into a fold region and, simultaneous with it, processes of regional metamorphism and granitisation attained culmination. In the scale of granitisation the West European region surpasses most other Hercynian mobile·regions of the world.

In the Middle Carboniferous almost all the regions of intense Palaeozoic subsidence experienced uplift except for shallow residual basins. A rugged montane topography arose and major intermontane troughs were formed. These troughs, the Asturia and Saar-Hesse, lay on ancient crystalline massifs. Shallow interior basins were associated with subsidence along faults, often transverse. All these troughs and basins were filled with limnic coal-bearing Silesian molasse. The thickness of the formations in them was significant, reaching 5 km. In the Austuria and Saar-Hesse troughs and in the basins along the periphery of the Bohemian massif, marine conditions were still preserved at the commencement of the Middle Carboniferous and paralic coal-bearing deposits were formed.

In the south-western Iberian peninsula (in the Southern Portugal zone) basin development with flysch sedimentation was completed. Uplift in general affected the Atlas (Maghreb) zone in northern Africa. Tectonic deformation continued here while sedimentation was concentrated only in the small residual basins of El Ziliga and Djerada where limnic coal-bearing molasse was formed. Middle Carboniferous tectonic movements were accompanied in this zone by intrusion of sills and dykes and eruption of basalts.

In the Late Carboniferous renewal of tectonic movements occurred, i.e., the Asturian epoch of Hercynian orogeny (at the end of the Westphalian and commencement of the Stephanian). In this phase intense fold-overthrust deformation renewed in several regions, the fold-overthrust structure of the Subvariscan zone of foredeeps developed and naffe-imbricate structures of the western and southern margins of the Asturia basin were formed. The Hercynian Pyrenees system was basically formed in this same epoch. In the Southern Portugal zone, which experienced inversion, Asturian deformations affected the entire sequence of formations (up to the Moscovian stage inclusive) and led to the formation of compressed folds overturned towards the south-west and complicated by fault and flow cleavage. Small nappes were formed here somewhat later. Extremely sharp deformation of the Palaeozoic complex of the Atlas (Maghreb) zone occurred in this phase.

The latest major plutons of Hercynian granitoids—Cornubian batholith in south-western England, Brocken and Ramberg plutons in Harz, Erzgebirge batholith in Saxony, annular granitoid plutons in the Atlas zone and others—formed in the Asturian epoch. Granitisation continued at places until the beginning of the Permian.

One of the concluding episodes of Hercynian orogeny, i.e., formation of major shear faults, commenced also in the Asturian stage: initially, sinistral sublatitudinal, particularly in the South Armorican zone of shearing and an assumed shear of the Iberian Meseta relative to the rest of Mesoeuropa and later meridional, also sinistral, like the coal 'furrows' of the massif Central, Boskowitz and Blanitz furrows of the Bohemian massif. Diagonal shears (north-western and north-eastern) were formed in the concluding stage. These were distinctly manifest in the Mesetan, Armorican, Central and Bohemian massifs, i.e., predominantly in the interior zones of the orogenic system that experienced maximum consolidation and shallower depth of asthenosphere as a result of high thermal flux. Grabens and semigrabens (the latter are particularly typical) filled with coarse clastic coalbearing molasse arose along these faults. Facies distribution of sediments over the area and their vertical succession indicate occasional synsedimentary movements along faults.

E a r l y   P e r m i a n. Tectonic and magmatic activity gradually died out and their most intense manifestation was confined to the middle of the epoch, i.e., Saalian epoch of orogeny. A sharp tectonic and orographic ruggedness with numerous intermontane troughs and folded mountain systems in between of west-north-west Armorican orientation in the west and east-north-east Variscan (or Erzgebirge) orientation in the east, existed at this time throughout the territory of West Europe. Troughs and basins were filled with Rotliegende molasse formed by the erosion products of the Hercynian mountain systems. This thickness of the Rotliegende usually does not exceed 1 to 1.5 km but reaches 3 km in the deepest trough, the Saar-Hesse.

Powerful subaerial volcanic eruptions occurred in the orogenic region. Their activity was maximum at the beginning of the Permian and attenuated at the end of the Early Permian. These volcanics belonged to bimodal basalt-rhyolite series and their thickness ran into hundreds of metres.

The palaeotectonic importance of Saalian movements compared to Asturian or Sudetan was already sharply subordinate: with them was associated the last germanotype deformation of the molasse filling of foredeeps and intermontane troughs and interior basins. But at places, for example in the Subvariscan zone of foredeeps, small nappes and fold structures were still formed.

The formation of Hercynian intrusive granitoid complexes was essentially completed by the beginning of the Permian. However, intrusion of the granite plutons Brocken and Ramberg in Harz continued even in the Early Permian period.

Marine carbonate-terrigenous formations of the Upper Carboniferous and Permian in the south-eastern part of the region under consideration (Outer Dinarides, northern Hungary, southern Alps, Tuscany and Sicily) represented formations of epicontinental seas penetrating from time to time from the Caucasus-Pamirs region in the east. Further, these seas did not extend west of the Apennine peninsula and Tunisia. Sedimentation proceeded in this region almost uninterruptedly and, unlike in the rest of West Europe, no Hercynian phases of folding whatsoever manifested here (Belov, 1981). This, however, does not pertain to the northern part of the Southern Alps adjoining the Peri-Adriatic suture and Inner Dinarides.

Late Permian. Orogenic development of the West European region was completed and there were already some features of a platform regime at the end of the epoch. Hercynian mountain systems were intensely flattened by erosion, especially in the northern part of the region, which caused the Zechstein basin to spread southward at the end of the Permian into the region of Hercynian folding. But within the future Alpine mobile belt in the south, uplift in the Late Permian was still markedly intense and maintained a montane topography. As before, molasse-like sediments continued to accumulate in the intermontane troughs, predominantly in a continental environment. In the Early Triassic, however, tectonic activity died out even here and the orogenic regime was replaced everywhere by a platform one. Only the Pennine trough remained unaffected by uplift and its Triassic formations were associated with gradual transition from the Upper Permian.

In the Early Triassic predominantly quiescent conditions of extensive shallow marine basins of the platform type prevailed in the region of the future Alps and western Carpathians. The resultant unconformities at the base of the Triassic formations are manifest differently in different regions but in general are not sharp.

Thus the epoch from the Middle Carboniferous to the end of the Permian for the West European mobile region was a period of manifestation of an orogenic regime in a very distinct and full form. The exceptionally extensive granitisation in the Carboniferous and at the beginning of the Permian, powerful burst of continental bimodal volcanism in the Early Permian, mountain formation with deposition of thick molasse and intense tectonic deformation, characterise this region as a typical Hercynian orogenic structure. The intensity of tectonic deformation progressively died out from the Sudetan (Early-Middle Carboniferous) epoch of orogeny to the Asturian (Late Carboniferous) and Saalian (Early Permian) epochs.

5.2.2.2 *Caucasus-Pamirs region.* M i d d l e - L a t e C a r b o n i - f e r o u s. From the beginning of the Middle Carboniferous almost the entire territory of the region from Dobrogea the northern Pamirs experienced an orogenic stage of development and represented a montane region with a rugged topography and accumulation of molasse in graben-like intermontane basins. In the Middle and Late Carboniferous grey coal-bearing molasse was usually deposited to a thickness not exceeding 2 km. Mighty processes of thrusting and folding manifest in Dobrogea and Caucasus Major even in the Early Carboniferous migrated north, east and north-east in the Middle Carboniferous and affected vast regions of the Precaucasus, northern Afghanistan, northern Pamirs and probably the Turan Platform. These processes led to the formation of fold-nappe systems and corresponded in time to the Sudetan epoch of folding (the last of its phases).

After this important Hercynian epoch of folding in northern Afghanistan (Badahshan) and the southern part of the northern Pamirs, subsidence and, possibly, spreading continued and thick marine terrigenous strata of variable facies composition accumulated. Steady subsidence occurred even in the zone of the southern slope of Caucasus Major. The Middle-Upper Carboniferous part of the Disi series developed here is represented by marine terrigenous and to a lesser extent by carbonate rocks 300 to 400 m in thickness.

In the rest of the region uplift predominated and magmatic manifestations continued. The latter are reflected in the extensive formation of granitoid plutons, initially plagiogranites (for example, Obihumbou complex of northern Afghanistan) and later soda and potash granites. In this respect development of the Caucasus-Pamirs region is similar to the West European region.

E a r l y P e r m i a n. Mountain building within the region intensified, as revealed by the extensive development of Lower Permian molasse, much coarser in composition than the Middle-Upper Carboniferous, red-coloured and arid-continental, with a thickness up to 3 km.

Two types of structures in which these molasse accumulated can be identified. The first is represented by grabens of comparatively small dimensions laid on Hercynian folded complexes. Such grabens are characteristic of the Precaucasus and northern Caucasus but are also known in the Transcaspian region (Mangyshlak and Tuarkyr). The molasse in such grabens is usually very coarse in composition. The second type of structures is represented by large flat basins laid on blocks of pre-Hercynian basement. Later, in the Mesozoic, these basins were succeeded by syneclises and were most widespread on the Turan Platform (northern Ustyurt, southern Mangyshlak and eastern Turkmenistan).

Tectonic conditions underwent a change in the western part of northern Afghanistan and major uplifts and intermontane basins filled with molasse formed here. The Transcaucasian massif remained an uplifted montane region, as before, and uplift affected Alborz. Volcanic processes synchronous in time and very similar to the Early Permian volcanism of the West European region were widely manifest. Subaerial eruptions of basalt-andesite-rhyolite composition occurred. These formations together with molasse filled intermontane basins and became part of a very large marginal volcanic belt that extended from West Europe into central and eastern parts of Asia.

L a t e P e r m i a n. Volcanic eruptions ceased. By the end of the Permian manifestation of orogenic processes had died out as a whole, although subsidence was replaced by intense uplift in the northern Pamirs in this same period and a montane topography formed here with molasse accumulating up to 3 km in thickness. In other regions the size and number of intermontane basins diminished but they were filled, as before, with red molasse. Within the Caucasus-Pamirs region, the scale of granite formation and metamorphism decreased but in its eastern part, as a result of subduction from the south, intrusions of large granitic batholiths appeared in the northern foothills of Hindukush, in western Afghanistan along the Herirud suture and the Mazar complex in Kalaïhumb-Sauksay zone of the northern Pamirs. Tectonic activity thus migrated from west to east within this region.

As before, continuous subsidence and marine conditions were preserved in the zone of the southern slope of Caucasus Major. The Permian part of the Disi series includes the Laila suite (1 km) of sandstones and the Chelshura suite (1.4 km) of rhythmically alternating shales and sandstones with bands of arenaceous limestones and conglomerates.

At the end of the Permian, the orogenic regime in the Caucasus-Pamirs region changed into a platform one, excluding only the zone of the southern part of Caucasus Major, Kopet Dag and the southern part of the northern and also central Pamirs where high mobility and intense subsidence continued into the Mesozoic era.

Thus during the Middle-Late Carboniferous and Permian, the Caucasus-Pamirs region was characterised by orogenic development as distinctly and as typically manifest as in the West European region. Montane topography was formed everywhere here and products of its disintegration—thick molasse—accumulated in the basins; intrusive granitoid and eruptive subaerial magmatism were widely manifest and complex tectonic deformations occurred. Inadequate exposure of much of the region does not permit a detailed study of its Palaeozoic tectonic history but, nevertheless, reveals a lag of the main tectonic processes compared to the West European region and their migration in an easterly direction.

5.2.2.3 *Kuen Lun-Indochina region.* M i d d l e - L a t e   C a r b o n i - f e r o u s. In the western part of the Kuen Lun system deepwater trough conditions were preserved throughout this period. Terrigenous and sometimes carbonate rocks were deposited in it and volcanic eruptions continued. However, the composition of volcanic products changed and became more felsic with andesites and dacites predominating. Palaeotectonic conditions are not yet clearly understood for the eastern segment of this system. Here, in the south, fairly thick terrigenous-carbonate strata were formed and manifestations of andesite volcanism are known; in the north, in the region of Arkatagh and Bokalytagh mountains, thin shallow-water limestones and sandstones were deposited under quiescent conditions.

Within the Yunnan-Malaya system, the Sinoburman massif maintained stability. This massif was covered by shallow sea in which carbonate-terrigenous sediments were deposited (Gatinsky, 1986; Metcalfe, 1984). Bands of continental conglomerates and sandstones are of subordinate importance. In the Sibumasu block, under conditions of a continental margin, glacial-marine diamictites were extensively deposited. This confirms the affinity of the Sinoburman massif to Gondwana. In many regions of the massif tectonic deformations manifested and granites intruded during the verge of the Early-Middle Carboniferous. An elongated flysch zone extending to the eastern part of Sumatra Island existed in the south-eastern part of the massif on the Malacca peninsula. Turbidites are known even to the west of this island. It has been suggested that these turbidite deposits were formed in some interior troughs bound by faults. At the end of the Late Carboniferous, leucocratic rare metal granites intruded into the eastern margin of the Sinoburman massif.

The Indosinian massif maintained its structural entity and again experienced subsidence after Early Carboniferous uplift and minor manifestation of tectonic deformation (in the northern part of the massif). On almost all its surface carbonate and terrigenous deposits, including continental sandstones and conglomerates, were deposited in shallow shelf basins. Angular unconformities are often established at their base. Along the western margin of the massif, the existence of a continental slope with turbidites is assumed.

The sequences of the eastern part of the Sinoburman massif consisting of greywackes, siliceous and clay schists and micritic limestones contain tholeiite and high alumina basalts and mafic subvolcanic formations proximate to rocks of island arc complexes formed on an oceanic crust. An island arc evidently developed in the region of northern Thailand but its evolution in the Carboniferous did not proceed to maturity, i.e., to manifestation of andesites and granodiorite intrusions. As a result of intense tectonic movements in these regions at the end of the Carboniferous, all the Palaeozoic formations were dislocated and the fold-overthrust structure of this zone was formed.

Volcanic eruptions of the island-arc type with mafic and felsic composition occurred in the western part of Kalimantan Island in the extreme south-eastern part of the Yunnan-Malaya system. Spilite-chert complexes accumulated in the east.

Information about the history of development in the Carboniferous period is still inadequate for the vast territory falling between the Kuen Lun zone and the Yunnan-Malaya system. Formations of Middle and Late Carboniferous are represented in this region by various shales, micritic limestones and products of submarine volcanism of mafic and intermediate composition. Deep troughs separated by island systems evidently prevailed here. Paralic coal-bearing Upper Carboniferous formations are known in the eastern part of Tibet.

E a r l y   P e r m i a n. In the eastern part of the Kuen Lun system troughs deepened again and comparatively deepwater terrigenous and sometimes clay-carbonate formations were deposited in them. Manifestations of island arc andesite and andesite-basalt volcanism continued over extensive areas. The southern part of the former Qilianshan system was involved in this subsidence, in which stable conditions of a platform regime had prevailed until then (from the Early Carboniferous). Conditions in the western part of the Kuen Lun system remained as before but volcanism almost died out and some eruptions were of felsic composition. Throughout almost all the territory of the Kuen Lun system, at the end of the Early Permian subsidence was replaced by uplift and Palaeozoic formations were affected by tectonic deformation of the Saalian epoch of orogeny. Intrusion of granodiorites commenced at this same time in the western part of the system.

Formation of carbonate and terrigenous deposits of the shelf type up to hundreds of metres in thickness continued in the territory of the Sinoburman massif. Conditions differed in the western part of the massif and thick terrigenous strata with bands of turbidites were formed. Their deposition occurred evidently in an intraplate deep trough separating the Sinoburman massif from the Indian Craton. Manifestation of felsic and intermediate volcanism of the island arc type commenced in the southern part of the massif (on Sumatra Island). A similar volcanism continued in the western part of Kalimantan Island. The eastern part of the Malacca peninsula with andesite-dacite island arc volcanism and a zone of turbidite sedimentation sloping westward is separated at present from the Sinoburman massif by a tectonic suture, along which it accreted to the massif at the end of the Permian. On the eastern

margin of the massif, tectonic deformations which resulted from its collision with the island arc of northern Thailand ceased but granite-granodiorite
intrusion continued.

Uplift intensified and the island zone enlarged on the Indosinian massif. Shallow-water carbonate-terrigenous formations accumulated in troughs
and continental-terrigenous (including coal-bearing) formations in the central part of the massif. Upward tectonic movements north of the Indosinian
massif and in the Shongda River zone led to formation of an extensive arch
uplift affecting much of northern Vietnam and adjoining regions of the People's Republic of China. Formation of this arch was accompanied by fold
deformation and faults with a north-western strike inheriting the much older
structural features and by intrusion of differentiated mafic plutons.

L a t e   P e r m i a n. Completion of the Hercynian stage of development of the Kuen Lun mobile system ended in uplift and extensive granitoid
intrusion. Numerous major plutons of normal and alkali granites arose in the
west, predominantly in the southern zone of the system. Granitic batholiths
were also formed in the central segment. Fold and fault deformation continued in the Late Permian. Upper Permian formations are not known in
the Kuen Lun system except in the eastern part, which represents an activated south-eastern section of the former Qilianshan system in which coarse
molasse up to 2 km in thickness was deposited in intermontane troughs.

In the Yunnan-Malaya system, the Sinoburman massif was covered, as
before, by a shallow sea in which carbonate and terrigenous formations of
small thickness were deposited except in small island regions. Subsidence
of the trough separating this massif from the Indian Craton continued and
terrigenous complexes with turbidites accumulated in it. A new phase of
magmatic manifestation commenced on the eastern margin of the massif.
Here, at the end of the Permian, eruptions of andesites, dacites, rhyolites
and their tuffs as well as intrusions of granite-granodiorites comagmatic with
them occurred. These volcanoplutonic complexes were formed under the
condition of a mature continental crust and are traceable in the south up
to the Malacca peninsula and south-eastern Sumatra Island. The eastern
zone of the Malacca peninsula was accreted to the Sinoburman massif and
experienced uplift and fold deformations.

Subsidence intensified on the Indosinian massif. Continental terrigenous and coal-bearing deposits were laid in a basin formed in its central
part. Subaerial felsic volcanism manifested here and granites intruded. Shelf
carbonate-terrigenous deposits accumulated in the marginal zones of the
massif and flyschoid strata (Lampang series) on its north-western slope.

Upheaval of the extensive arch in the Shongda zone continued in the
northern part of the Indosinian massif. But subsidence along faults of individual blocks and formation of deep grabens (rifts) commenced in its marginal
zones. The formation of these structures was accompanied by volcanic

activity whose centres lay along boundary faults. Alkaline olivine basalts and trachybasalts predominated among lavas. Their petrochemical features indicate the riftogenic nature of the grabens in the Shongda zone. Aside from basalts, littoral-marine tuffaceous-terrigenous strata were also formed in the grabens.

Permian formations within the territory between the Kuen Lun and Yunnan-Malaya systems are represented by submarine-volcanic and deepwater terrigenous formations. Volcanism was of mafic and intermediate composition. Extensive development of spreading and later, in the Triassic, of island arc systems is assumed in these regions.

E a r l y   T r i a s s i c. Conditions of an orogenic regime of development were maintained in the northern zones of the Kuen Lun system and uplift, evidently responsible for a rugged topography, prevailed. Products of its disintegration, i.e., coarse molasse, filled the Hexi (Guangxi) trough situated south of the Alashan block. Subsidence, however, renewed in the southern zones of the system, especially in the east, at the commencement of the Triassic and thick marine terrigenous formations of the flyschoid type were deposited on submarine slopes of the new deep troughs.

Uplift affected the eastern margin of the Sinoburman massif. It was accompanied by deformation of Late Palaeozoic volcanosedimentary complexes of an erstwhile island arc and its associated structures and the terminal phases of manifestation of Late Hercynian granitoid magmatism and subaerial eruptions of rhyolite-andesite lavas and tuffs. Marine and continental terrigenous (sometimes coal-bearing) formations accumulated in the troughs. Formation of shallow-water carbonate beds continued in the western part of the massif.

The Indosinian massif too experienced uplift while narrow shelf zones were preserved only along its fringes. In the central part of the massif volcanic activity was intensified, with rocks of felsic composition predominating in the composition of eruptions.

Subsidence of grabens (rifts) formed in the Late Permian intensified in the Shongda zone to the north and north-east of the Indosinian massif. Here a marine trough filled with products of andesite volcanism and clastic deposits was formed at the beginning of the Triassic.

Thus the easternmost part of the Mediterranean belt, Kuen Lun-Indochina region, was distinguished in this stage from the rest of the belt by a prolonged tendency towards subsidence with the formation of volcanic island arcs and deep troughs. Transition to an orogenic regime in the Kuen Lun system occurred only in the Middle Permian but southern zones of this structure were again involved in subsidence in the Early Triassic. Uplift commenced in ancient massifs of the Yunnan-Malaya system with its complex mosaic structure in the second half of the Permian also and intensified in the Early Triassic.

5.2.3 Ural-Okhotsk Belt

5.2.3.1 *Uralian system.* M i d d l e - L a t e C a r b o n i f e r o u s. After a brief Viséan-Namurian period of tectonic 'rest', the Uralian system entered the orogenic stage of its development. This commenced with a new phase of uplift and tectonic deformation at the very end of the Namurian to commencement of the Middle Carboniferous, i.e., with a phase corresponding to Sudetan orogeny of the Mediterranean belt. This phase was most distinctly manifest in the Eastern Urals zone where the formation of gneiss domes and granite batholiths commenced. However, in the axial parts of troughs inherited from the Early Carboniferous, i.e., Magnitogorsk and Eastern Urals, and also in some graben-like basins in the Eastern Urals zone, subsidence was maintained and Namurian limestones were conformably replaced by shallow-water Bashkirian limestones of small thickness.

Upheaval intensified in the Late Carboniferous. The entire eastern part of the Uralian system and also the Central Urals zone and eastern margin of the Zilair zone were uplifted above sea level and only narrow grabens near faults were filled with coarse continental molasse, partly red, in the south. Intrusion of granites continued during this period and formation of the fold-overthrust structures of western vergence of the Magnitogorsk synclinorium was completed.

However, the western part of the Uralian system, Western Urals zone, together with the western flank of the Zilair zone, continued to experience stable subsidence while maintaining marine conditions. But the accumulation of carbonate deposits was replaced in the Middle Carboniferous by deposition of terrigenous and terrigenous-carbonate flysch. These conditions were preserved even in the Late Carboniferous, albeit some changes did occur. The western Zilair zone experienced uplift at the end of the Carboniferous while the western part of the Western Urals zone adjoining the East European Craton, on the contrary, experienced sharp flexural subsidence with formation of an extended barrier reef above the flexure. Between this reef (in the west) and zone of flysch accumulation (in the east), a deep trough with deposition of thin carbonate-silica-clay sediments was isolated. Development of the Uralian foredeep commenced with this trough.

E a r l y P e r m i a n. Uplift of the interior zones of the Uralian system intensified at the commencement of the Permian, creating a montane topography and interrupting further accumulation of sediments, and began to extend increasingly into the Western Urals zone. Coarse clastic material in the Artinskian stage reached the inner flank of the Uralian foredeep and flysch deposits were replaced by marine grey molasse. In the Kungurian stage this trough compressed between the Palaeo-Urals montane edifice and the eastern part of the East European Platform that underwent emersion became a zone of accumulation of evaporites represented in the axial part of the trough by rock and potash salts. The thickness of the Lower

Permian in the southern and central parts of the trough exceeded 3 km. Within the northern link of the Uralian foredeep, i.e., in Vorkuta (Kosyu-Rogov) basin, humid climatic conditions prevented salt accumulation and caused replacement of salt-bearing molasse by coal-bearing ones, filling the Pechora (Vorkuta) basin.

Final formation of the nappe-fold structure of western zones of the Uralian system occurred in the Permian. In eastern zones of the system the main features of this structure had evolved much before and the main and final phase of intrusion of granite batholiths was manifest even in the Early Permian. This is supported in particular by the sharply expressed Early Permian peak of radiometric age determination. Unlike in the eastern zones, the phase at the boundary of the Artinskian and Kungurian stages evidently represented the first phase of Late Hercynian orogeny, including overthrust in the western Uralian system. The much stronger dislocation of Artinskian and much older formations compared to Kungurian on the eastern flank of the Uralian foredeep, confirms this conclusion. Even the above-noted peak of granite formation in the eastern Uralian system coincides in time with this phase, corresponding to the Saalian phase of the West European region.

L a t e   P e r m i a n. A new intense uplift of the Uralian system, after its slight weakening in the Kungurian epoch, took place. The Uralian foredeep had by this time almost wholly filled with evaporites and become a region of accumulation of red lagoon-continental molasse up to 2 km or more in thickness. As before, humid climate prevailed in the Pechora basin and thick terrigenous coal-bearing deposits were formed.

The concluding phase of the Late Hercynian orogeny manifested at the end of the Permian but its geological documentation is very poor. It may be assumed that formation of the fold-overthrust structure of the Western Urals zone was almost completed during this period. In the Transuralian and Eastern Urals zones the already consolidated Palaeozoic basement was divided in the Late Permian by subvertical faults along which effusions of basalts and rhyolites of the Tura series occurred. Volcanics were interstratified with continental molassoid sediments.

E a r l y   T r i a s s i c. Subsidence of the Uralian foredeep ceased in the western Uralian system and tectonic deformations affected its sedimentary filling. Only in the north, in Pechora basin, did formation of sedimentary strata still continue but littoral-marine coal-bearing complexes were replaced here by continental red molasse and mafic volcanism was manifest at the end of the Early Triassic (Vetluga time).

Development of faults continued in the Transuralian and Eastern Urals zones and made for subsidence of the system of grabens and basins while volcanic activity persisted in the Transuralian region. This tectonic regime may be regarded as taphrogenic, representing transition from an orogenic to a platform regime. Molasse and volcanosedimentary formations accumulated in separated basins undergoing subsidence at a different amplitude. In the Transuralian region these structures represented narrow (1 to 15 km), long (up to 200 km), comparatively deep (up to 200 m) grabens and more extensive (width up to 45 km) depressions. The latter were situated east of Kamyshlov uplift. Some of the basins became desalted epicontinental seas and periodically came into contact with open seas. Basins were fringed with sharply manifest mountain systems. During the Early Triassic fissure eruptions of lavas of mafic and, less frequently, felsic composition occurred in the Transuralian zone.

At the end of the Early Triassic, tectonic movements intensified somewhat in the Uralian system. A consequence of this phenomenon was the partial erosion of Lower Triassic formations, formation of gentle dislocations and renewal of volcanic activity.

Thus during the Middle Carboniferous to the Early Triassic, the Uralian system experienced the orogenic stage. In the western part of the system the Uralian foredeep completed the full cycle of its development, from commencement of subsidence to total compensation with molasse and deformation at the end of the Permian. In the eastern and central regions uplift and montane topography predominated and processes of granitisation and metamorphism developed for a long time. Formation of a Hercynian fold-overthrust structure occurred with gradual migration from eastern to western zones and actually concluded at the end of the Early Permian in the Saalian epoch of orogeny. Transition to a taphrogenic stage was marked throughout the area of the Urals in the Early Triassic.

5.2.3.2 *Southern Tien Shan system.* M i d d l e - L a t e   C a r b o n i - f e r o u s. The relatively quiescent development of the system was interrupted at the end of the Early Carboniferous by powerful tectonic movements including major horizontal displacements and also outburst of volcanic and intrusive activity. The system entered the orogenic stage in the Middle Carboniferous. The area of the Alay-Kokshaal trough shrank sharply and marine deposits in it were represented by carbonate, less frequently, siliceous and terrigenous rocks. Towards the end of the Middle Carboniferous, even the Zeravshan-Eastern Alay trough ceased to exist. It underwent fold-overthrust deformation and was transformed into a complex tectonic structure.

Orogenic processes were reactivated even in the region of Central Tien Shan immediately adjoining the Southern Tien Shan system from the north. Massive eruptions of andesite, trachyandesite and dacite composition occurred in the southern part of the Chatkal range. Intrusion of the major polygenic granitoid massifs of the Sandalash-Chatkal and Sonkul complexes is placed in the Middle Carboniferous. However, the main episode in the

Middle Carboniferous tectonic history of the Southern Tien Shan system was the charrette of its northern structural-formation complexes on the more southern ones. This process attained maximum intensity in the Moscovian stage. Nappe formation was a consequence of the collision of the Kazakhstan and Tarim-Alay continental blocks and squeezing of the mobile structures separating them. As a result, the general amplitude of reduction of the system was at least 150 to 300 km and the amplitude of individual nappes up to 100 km or more; the southern volcanosedimentary Middle Palaeozoic complexes were repeatedly overlapped by northern complexes. Thick olistostrome and flysch formations accumulated before the nappe front. These formations became younger towards the south from the Namurian to Moscovian.

In the Late Carboniferous collision of continental blocks, further overthrusting of volcanosedimentary complexes and piling up of their material continued. Much of Southern Tien Shan was already an elevated land and accumulation of molasse commenced in intermontane troughs. As a result of overthrust of the Kazakhstan continent from the north and underthrust of the Tarim-Alay microcontinent from the south, the volcanosedimentary series of troughs separating them were squeezed out onto the roof of the Tarim-Alay microcontinent. The complex nappe structure arising as a result of this process subsequently underwent intense folding and later disruption by subvertical, predominantly shear faults, upthrusts and sometimes overthrusts, with northern vergence prevailing. Simultaneously, volcanic activity manifested somewhat more northward, the Beltau-Kurama volcanic belt was formed and granite plutons intruded. These processes of folding, granitisation and metamorphism were protracted, their initial phases alone having proceeded up to the end of the Late Carboniferous.

E a r l y   P e r m i a n. The last brief marine transgressions in this epoch penetrated into the north-eastern part of the system where grey molassoid rock complexes were deposited. But even at the end of the Early Permian, these were replaced by typical coarse continental molasse. Formation of nappes had already ceased by the beginning of the Permian but a high activity of orogenic tectonic processes was preserved. Crumpling of Palaeozoic formations into folds continued and magmatic manifestations, i.e., volcanic activity and formation of granitoid intrusions including that of the huge Gissar granite batholith, intensified. A rugged topography was also preserved in the territory of the system and molasse accumulated in intermontane basins.

L a t e   P e r m i a n. The concluding deformations of the Southern Tien Shan system were confined to the end of the Permian. They affected the entire complex of ancient formations, including overthrusted sedimentary complexes and Permian molasse. Dislocations occurring in this epoch differed in the different zones of the system. Gentle, box-shaped forms were confined to the regions of Late Palaeozoic basins filled with molasse. More intense folding and steep faults were manifest in regions where such a cover of continental formations was lacking. In this phase the main activity of the Talas-Ferghana transcurrent fault was manifest and the major Ferghana horizontal flexure was formed on its south-western flank. Granitisation processes peaked in the Late Permian and their activity subsided by the end of this epoch.

Early Triassic. The orogenic stage of the Southern Tien Shan system was completed by commencement of the Triassic and was replaced by a platform regime. Cratonisation of the Earth's crust ended with intrusion of leucocratic granites into the Gissar-Alay zone and Alay complex of alkaline and nepheline syenites. This was followed by commencement of a prolonged process of peneplanation of this montane system with accumulation of its erosion products in the adjoining plains.

Thus the main events in the tectonic development of the Southern Tien Shan system during the Middle Carboniferous to the Early Triassic were overthrust in the Middle Carboniferous of its northern zones on the southern (Sudetan phase of orogeny) and concluding deformations affecting the system in the Late Permian. Moreover, magmatic and metamorphic processes were extensively manifest and rugged topography was formed, i.e., an orogenic regime prevailed. It was replaced by a platform regime in the Early Triassic.

5.2.3.3 *Kazakhstan region.* Middle-Late Carboniferous. A transitional regime from orogenic to platform prevailed on the territory and extensive superimposed separate interior basins were formed. These basins were: Teniz, Dzhezkazgan, Chu, Issyk-Kul, Aksuat and Syr-Darya. At the same time, the Karatau-Naryn and transverse Sarysu-Teniz zones were involved in uplift and experienced fairly intense deformation: fold-overthrust in the former zone and fold-block in the latter. The basins served as regions for accumulation of molassic, variegated terrigenous sediments distinguished by high facies variation. Banded and cross-bedded sandstones and siltstones of varying grain size predominating in the beds were formed under continental conditions—alluvial, lacustrine and sometimes eolian.

An orogenic regime manifested more distinctly in Northern and Central Tien Shan in the southern part of the region; a rugged topography was preserved here and the development of intermontane troughs continued. Molasse and sometimes marine carbonate-terrigenous formations accumulated in these troughs. The Beltau-Kurama volcanoplutonic belt with an andesite-dacite-rhyolite complex was formed along the southern boundary of Central Tien Shan. The composition of plutons ranged from granodiorites to alaskites.

The second volcanoplutonic belt—Balkhash-Ili—extended into the south-eastern part of the Kazakhstan region and its development was associated with tectonic processes occurring in the adjoining Junggar-Balkhash mobile system. Eruptions of felsic composition predominated in this belt and granites intruded.

E a r l y P e r m i a n. Palaeotectonic conditions within the northern and central parts of the Kazakhstan region changed little compared to the Late Carboniferous. Intensity of orogenic processes diminished and the general regime of development approached that of a platform. Subsidence of major basins—sea-lakes—continued from the Late Carboniferous; further, these were interlinked in the west and covered an extensive area. Clay-carbonate muds predominated in the deposits at commencement of the Permian while thick beds of salts and gypsum formed at the end of the Early Permian in addition to red and grey marls. The quiescent character of subsidence, absence of rugged topography and manifestation of magmatism and tectonic deformations suggest a gradual transition of this part of the Kazakhstan region to a platform regime.

Fairly high magmatic and tectonic activity was preserved in the south-eastern and southern parts of this region. Volcanism continued in the Balkhash-Ili belt and the role of basalts rose in the composition of its products. Intrusion of granitoid plutons also continued here. In Northern and Central Tien Shan filling of intermontane basins with molasse was completed and intrusive and eruptive magmatism manifested. Volcanic formation was trachyandesite-basalt in composition. Moreover, transcurrent displacements along major faults of north-western strike were sharply activated here. Among such strike-slip faults were Chinghiz with a displacement amplitude of up to 100 km and extent of over 700 km, Talas-Ferghana in the extreme south of the region and others.

L a t e P e r m i a n. The general uplift of the Kazakhstan territory intensified. The Teniz (in the north) and Chu (in the south) basins, representing isolated sea-lakes, were still preserved. Sandy-clay and less frequently calcareous deposits ranging from 200 to 500 m in thickness were deposited in the Teniz basin; gypsum-bearing rocks were additionally deposited in the Chu basin. At the end of the Permian subsidence in these basins was replaced by uplift and the entire complex of sedimentary formations of Late Palaeozoic basins experienced folding into gentle brachymorphic folds complicated by steep faults.

Orogenic development of the southern and south-eastern regions was completed. Topography flattened and manifestation of volcanic and intrusive activity had died out in the Peri-Balkhash and Northern and Central Tien Shan regions by the end of the Permian. Deformation of Upper Palaeozoic molasse into folds occurred here too and strike-slip and vertical displacements along faults continued.

E a r l y   T r i a s s i c. Permian deformations and magmatism marked the end of the protracted history of orogenic development of the Kazakhstan region. The whole of the area entered a platform regime at commencement of the Triassic, then later experienced only weak uplift and represented the Kazakh Shield of the young West Siberian-Turan Platform.

Thus this period in the Kazakhstan mobile zone was characterised by gradual weakening of tectonic and magmatic activity with total transition to a stable platform regime at commencement of the Mesozoic. Volcanic and intrusive manifestation in the Balkhash-Ili and Beltau-Kurama marginal volcanoplutonic belts was associated with processes of subduction of oceanic crust of adjoining mobile structures of Southern Tien Shan and Junggar-Balkhash. The same factors explain Late Permian tectonic deformations.

5.2.3.4 *Junggar-Balkhash system.* M i d d l e - L a t e   C a r b o n i - f e r o u s. Uplift covering the area of the system at the end of the Early Carboniferous marked transition to an orogenic regime and led to its separation into the residual Sayak-Junggar trough extending from the north-eastern Peri-Balkhash region into the northern part of Junggar Alatau, volcanic islands and land masses with a rugged topography. Accumulation of coarse terrigenous strata of the molasse type continued in the trough but marine conditions in the Late Carboniferous were replaced by continental ones. The thickness of formations in the northern part of the trough exceeded 2.5 km. The western and south-western parts of the trough revealed extensive volcanism. Its subaerial products were of complex composition and represented basalt-andesite-rhyolite formations. At the end of the Middle to commencement of the Late Carboniferous, one more minor phase of folding has been fixed within the system. This folding could be paralleled with the Asturian of West Europe. The intrusion of several granitoid plutons was associated with this phase of folding.

E a r l y   P e r m i a n. A new and concluding phase of orogeny affected the territory of the system at the beginning of the Permian. This phase, locally called Sayakian, roughly corresponds to the European Saalian, although the former could be somewhat older. The Sayak-Junggar trough experienced total inversion in the Sayakian phase. As a result of Sayakian tectonic deformation, the main structural elements of the Junggar-Balkhash system were conclusively formed. An intrusive diorite-granodiorite (Sayakian) complex intruded in this epoch but the area of volcanic activity diminished somewhat and the composition of eruptions became more uniform (rhyolitic).

L a t e   P e r m i a n. The belt of volcanic manifestation was displaced into more interior zones of the system and attained maximum width. The composition of eruptions became sharply contrasting, ranging from basalts to rhyolites. Intrusions continued with plutons of monzonites, alkali granites and syenites predominating among them. Leucocratic granites were very widespread within the volcanic belt.

At the end of the Permian, the south-eastern part of the Junggar zone (in the territory of the People's Republic of China) was affected by brief subsidence and transgression penetrated here from the east, from the side of the southern trough of the Mongolia-Okhotsk system. This subsidence gave rise to the Junggar basin in which marine terrigenous strata accumulated.

E a r l y T r i a s s i c. By the commencement of the Triassic, both volcanic and intrusive activity had ceased in the territory of the system. A fairly rugged topography was still preserved at this time but the general reduction in tectonic activity indicates the conclusion of the orogenic stage of development and transition to a platform regime. The Junggar basin experienced a partial regression and was transformed into an intermontane basin occupied by a lake. Consequently, in this part of the Ural-Okhotsk belt too the period from the Middle Carboniferous to the end of the Permian was one of manifestation of an orogenic regime replaced by a platform one at the beginning of the Mesozoic.

5.2.3.5 *Ob'-Zaisan and Gobi-Hinggan systems.* M i d d l e - L a t e C a r b o n i f e r o u s. Uplift encompassing the Ob'-Zaisan system at the end of the Early Carboniferous (Saurian phase) and accompanied by the intrusion of granitoïds of the Saur-Zmeinogorsk complex marked a change in general conditions of tectonic development and transition to an orogenic regime. Fold deformation giving rise to the Rudnyi Altay and Zharma-Saur zones manifested in the Saurian phase. These dislocations are quite variable in intensity and morphology of structures, ranging from linear and ejective to brachymorphic. In the Middle and Late Carboniferous intermontane troughs formed in the territory of the system and were filled with continental coal-bearing molasse and to a lesser extent with felsic volcanics. The thickness of molasse is 1.5 to 2 km. The main troughs were the Peri-Irtysh extending along the Irtysh zone of shearing and the Kuznetsk with its thick coal-bearing molasse. The Kuznetsk basin was isolated at the beginning of the Middle Carboniferous as a result of conclusive uplift of the Salair Range.

A residual marine trough was preserved even in the Middle Carboniferous south-west of the Irtysh zone (Kenderlyk trough). But, even at the end of the Middle Carboniferous, marine terrigenous facies were replaced in the Kenderlyk trough by continental molasse, including also the products of subaerial trachyte-andesite volcanism. This Middle Carboniferous molasse is overlain by an Upper Carboniferous one with some unconformity. The Upper Carboniferous molasse too was volcanic, coal-bearing and with oil shales in the upper horizons.

The Southern Gobi microcontinent collided with the Northern; this resulted in the closing of the Southern Mongolia ocean and formation of the Transaltay ophiolite suture in its place. Opening of a new ocean basin between this microcontinent (now marginal in respect to the northern continent) and the southern Sino-Korea continent) commenced at this time south of the Southern Gobi microcontinent.

In the Mongolia-Amur zone whose Late Hercynian phase of development had commenced in the Early Carboniferous, troughs were rapidly filled with sandy-clay deposits. At places they have a flyschoid character, often include bands of coarse clastic rocks and exceed 2.5 km in thickness. In the extreme west (Hantay trough) these formations accumulated under shallow littoral and continental conditions, suggesting the closing of the marine basin in this direction. Intensification of uplift joining the Kerulen-Argun and Bureya massifs by the Middle Carboniferous isolated the marine basins of the Mongolia-Amur zone from the West Pacific mobile belt. Evidently, it was in this period that the troughs of the Mongolia-Amur zone joined with the seas of the eastern part of the Gobi-Hinggan system.

E a r l y   P e r m i a n. In the Ob'-Zaisan system uplift intensified and the area of sedimentation shrank. Nearly freshwater lakes, which had lost contact with the open sea, continued to exist in the intermontane basins. The Kuznetsk basin was the largest and, as before, coal-bearing molasse was deposited in it. In the Kenderlyk trough, too, accumulation of continental coal-bearing molasse continued, reaching a thickness of 1.7 km. In the Early Permian granosyenites and granodiorites of the Tastau and Serzhikhin complexes intruded and, as a result of tectonic deformation, the structure of the system was increasingly complicated in this period. In particular, movements continued along the Irtysh zone of shear. Its activity lasted all through the Hercynian stage. On both sides of the main fault, adjoining rocks have been left to a width of up to 10 km and at places metamorphosed to crystalline schists of amphibolite facies.

The western part of the Gobi-Hinggan system, serving directly as the eastern continuation of the Ob'-Zaisan, was affected by uplift at commencement of the Permian. But a system of troughs and island areas continued to develop in the central and eastern segments. Further, a new phase of extension and deep subsidence manifested in the southern zone where subaqueous eruptions of basalts and andesites occurred and siliceous and clay deposits were formed. In the Hinggan zone andesite-rhyolite volcanism of the island-arc type was extensively manifest. The thickness of the Lower Permian went up to 3 km or more here.

Marine troughs in the western part of the Mongolia-Amur zone experienced inversion and the Carboniferous formations filling them were crumpled into moderately compressed folds. Subsidence was still preserved in the east and terrigenous rocks were deposited in a small Chiron trough and in a large isometric Middle Onon basin.

L a t e P e r m i a n. Subsidence of the Kuznetsk basin and the Kenderlyk trough of the Ob'-Zaisan system continued but weakened over time and was replaced by uplift at the end of the Permian. In these large intracontinental freshwater basins thick, comparatively fine clastic sediments accumulated. In the Kuznetsk basin these beds contain coal-bearing formations; the total thickness of the Upper Permian exceeds 3 km. At the end of the Permian, these formations experienced fold-fault deformation, which was strongest in the north-western and south-western margins of the basin as a result of overthrust by adjoining fold zones: Kolyvan-Tom' along the Tom' overthrust (nappe) and Salair along the Guriev overthrust. The Kuznetsk basin deposits along these upthrusts were crumpled into narrow asymmetrical folds. Further, deep within the basin, linear folding was replaced by ejective folding, later brachymorphic and even dome-shaped at the centre of the basin.

In tne rest of the territory of the Ob'-Zaisan system, uplift predominated as before but became more quiescent and the topography evidently smoothened. This is supported by the absence of coarse molasse in basins. An exception is the volcanic (tuffaceous) molasse which accumulated in the region of Saur range south of lake Zaisan. Subaerial eruptions of mafic and felsic composition occurred at places here. The Kalba complex of normal and subalkaline granitoids is comagmatic with the aforesaid volcanics. This complex includes, in particular, a huge Kalba-Narym batholith elongated along the Irtysh zone of shear. The youngest phase of the Kalba complex intruded at the end of the Permian to commencement of the Triassic and is represented by leucocratic and subalkaline granitoids. The period of formation of this complex was fairly protracted, from the end of the Carboniferous to commencement of the Triassic. These magmatic manifestations topped the orogenic stage of development of the Ob'-Zaisan system and a platform regime was later established here.

In the Gobi-Hinggan system filling of later marine troughs with deposits was completed and uplift forming a montane topography predominated everywhere at the commencement of the Late Permian. Disintegration products of the relief in the form of thick molasse completed the filling of troughs and basins. Manifestation of subaerial volcanism in the south-western part of the system had a bimodal basalt-rhyolite composition. In this same epoch intrusion of granitoid plutons occurred and most of the Upper Palaeozoic sedimentary and volcanic deposits were dislocated.

In the development of the Mongolia-Amur zone, even at the end of the Early Permian, transition to an orogenic stage took place in which tectonic conditions equated those of the Altay-Northern Mongolia system fringing it. In this system conditions of an orogenic regime prevailed throughout the Late Palaeozoic. Only in the Chiron trough was subsidence compensated as before, by deposition of marine terrigenous rocks. The manner in which this basin maintained contact with the open sea is not yet clear.

E a r l y T r i a s s i c. Over much of the Ob'-Zaisan system the orogenic regime was replaced by a platform one and its northern regions (Kuznetsk basin in particular) were affected by flood basalt volcanism whose main areas lay in Western and Eastern Siberia.

In the Gobi-Hinggan system formation of a montane topography continued and molasse accumulated in the intermontane troughs. Magmatic activity came to an end. Volcanism was characterised by eruption of felsic lavas and tuffs and intrusive magmatism by intrusion of alkali-granites and nepheline syenites.

Thus during the stage under consideration, an environment of orogenic regime with extensive manifestation of intrusive and eruptive magmatism predominated in the Ob'-Zaisan and Gobi-Hinggan system. The 'front' of orogenic processes constantly migrated from north-west to south-east and east while tectonic activity died out behind it, replacing the orogenic regime with a platform one.

5.2.3.6 *Altay-Northern Mongolia system.* At the end of the Early Carboniferous, tectonic and magmatic activity died out in the Altay-Western Sayan system; the orogenic regime was replaced by a platform one, and this region became attached to the Siberian Craton. As a result, only the Northern Mongolia part with an orogenic regime as before, remained of the Middle Palaeozoic Altay-Northern Mongolia mobile system by commencement of the Middle Carboniferous. This is supported mainly by high magmatic activity, especially in Mongolia-Altay, in the form of large intrusions of granites of the Gobi-Altay complex in the Middle-Late Carboniferous. In the Late Carboniferous subaerial eruptions of intermediate and mafic composition occurred north of Hangay-Hentey upland and granodiorites intruded. Moreover, in the Middle and Late Carboniferous a fairly rugged topography was preserved in the territory and terrigenous rocks, including bands of coarse clastic formations, accumulated in Hangay trough.

In the Permian the Hangay trough was filled with typical molasse. Development of a large and extended Mongolia-Transbaikalian volcanoplutonic belt occurred in this period. The composition of volcanic products in it changed over time. Bimodal basalt-rhyolite eruptions predominated in the Early Permian and volcanic products of high alkalinity in the Late Permian. Intrusive magmatism also experienced a similar evolution. Granite, adamellite-granite and granodiorite plutons intruded in the Early Permian and alkali-granite, syenite and other highly alkaline plutons in the Late Permian and Early Triassic.

Evidently Palaeozoic formations of the Hangay-Hentey upland were deformed in the Permian. As a result, complex fold and fault (to a lesser extent, overthrust) structures were formed. The intensity of these tectonic processes and magmatism diminished later, in the Early Triassic. Although some volcanic manifestation was still preserved, the general character of the tectonic regime of the Northern Mongolia region was already proximate to platformal.

5.2.4 West Pacific Belt

5.2.4.1 *Verkhoyansk-Kolyma region.* M i d d l e - L a t e   C a r b o n i-f e r o u s. Palaeotectonic conditions prevailing in the Early Carboniferous in the Verkhoyansk system in the western part of the region were preserved in this epoch too. Intense subsidence compensated by inflow of clastic material from the Siberian Craton continued in the axial zone of the system. This compensation was evidently incomplete and purely clay, deepwater sediments up to 2 km in thickness accumulated. Minor eruptions of basalts occurred in the Omulevka zone. The bulk of the mass and much of the coarse clastic portion of the sediments supplied from the craton were deposited in a broad shelf zone in the west immediately adjoining the craton. Here very thick (up to 4 km) uniform beds of sand and clay, sometimes containing bands of turbidites, were formed under shallow-water conditions. These formations gradually built up the prism of sedimentary complexes of the continental slope whose surface prograded eastward.

The Southern Taimyr zone represented the north-western continuation of the Verkhoyansk system. Intense subsidence commenced in it even in the Early Carboniferous and continued in the Middle and Late Carboniferous. Thick sandy-clay beds accumulated here too.

A new palaeotectonic environment arose in the area of the Kolyma-Omolon microcontinent in the Middle Carboniferous when its ancient blocks began to rise. The bulk of the Middle-Upper Carboniferous deposits were formed here under shallow-water or continental conditions. These formations in the Omolon massif were represented by shallow marine terrigenous strata with thick beds of carbonate rocks and continental sandstones with rare beds of basalts. In the Alazeya uplift tuffaceous-terrigenous formations accumulated. A distinctive feature of the Peri-Kolyma uplift was the mighty manifestation of basalt magmatism in the Late Carboniferous. Stable subsidence with the formation of clay and tuffaceous-silty rocks occurred in this region only in the Gizhiga trough.

E a r l y   P e r m i a n. The structural environment and tectonic development in the Verkhoyansk system remained as before. Sandy-clay deposits continued to accumulate in the central deepwater zone of the continental slope and rise. Their thickness exceeded 3 km in the Southern Taimyr and Southern Verkhoyansk zones and also in western Verkhoyansk.

Some changes occurred in the Kolyma-Omolon region. Volcanic activity ceased and subsidence of the Peri-Omolon (Sugoy) trough separating the

Omolon massif from the Peri-Kolyma uplift commenced. This trough and other basins were filled with terrigenous and carbonate deposits. Subsidence continued in the Gizhiga trough and fine terrigenous and sometimes tuffaceous sediments accumulated. A significant part of this region (western Omolon massif and others) experienced upheaval and represented large low islands. Marine as well as continental sand beds were formed in the north within the Yarakvaam uplift.

Late Permian. The nature of tectonic development of the Verkhoyansk system did not change and, as before, the prism of sedimentary rocks adjoining the Siberian continent grew in its western part. The Upper Permian here is 2 to 3 km thick and is represented by sandy-clay rocks. A similar composition and thickness of the Upper Permian are known in the Southern Taimyr zone but, at the end of the Permian, uplift commenced and the marine trough here shoaled while continental facies began appearing in the composition of deposits. Accumulation of clay rocks continued in the deepwater portion of the Verkhoyansk system.

Total absence of carbonate rocks is a characteristic of the Late Palaeozoic marine deposits of the Verkhoyansk system. The salinity of marine waters remaining normal, this feature could be explained by the low water temperature. This conclusion fully accords with palaeomagnetic data, which indicates a peripolar disposition of this region in the Permian period. The fine clay deposits of the Late Permian basin contain an admixture of coarse clastic material evidently brought in by floating ice. These glacial-marine sediments were developed over a very large area.

Subsidence intensified and transgression covered almost the whole area in the Kolyma-Omolon region. Sandy-clay and very coarse clastic (sometimes coal-bearing) deposits as well as carbonate deposits accumulated in shallow seas to a thickness not exceeding a few hundred metres. In the Sugoy trough, however, the thickness of the Upper Permian reaches 800 m and in the Gizhiga 2.5 km; in the latter the Upper Permian formations are represented almost exclusively by clay rocks.

Early Triassic. Uplift covering the Taimyr zone at the end of the Permian intensified in this epoch. Its northern part experienced fold deformations and mountain building. These regions, as also the territory of the Siberian Craton proper, were enveloped by manifestation of trappean magmatism. Subsidence and formation of a uniform terrigenous complex continued in the rest of the Verkhoyansk system.

Unlike this scenario, almost the entire territory of the Kolyma-Omolon microcontinent was involved in uplift. Its amplitude was not large and did not lead to formation of a rugged topography and major tectonic deformation. The tendency to quiescent subsidence was preserved only in the southern part of the Omolon massif and carbonate and terrigenous deposits of small thickness and platform type accumulated here. The rate of subsidence in the Sugoy and Gizhiga troughs was very high and Early Triassic deposits in them were predominantly clayey.

From the end of the Permian and in the Early Triassic, a major transformation occurred north of the Kolyma-Omolon region in the Chukchi system. A large deep basin was formed here and its subsidence was not fully compensated by sediments. The intrusion of numerous sills of diabase indicates the probable extension of processes making for the formation of the basin. The main episodes of development of the Chukchi system occurred in the much younger Mesozoic epoch.

Thus the Middle Carboniferous-Early Triassic development of the Verkhoyansk-Kolyma mobile region had nothing in common with development of the Mediterranean and Ural-Okhotsk belts with respect to the direction and chronology of the main tectonic processes. A very prolonged formation of thick terrigenous deposits proceeded in the Verkhoyansk system while quiescent uplift of the Kolyma-Omolon microcontinent was interrupted by brief episodes of subsidence, transgression and manifestation of mafic volcanism.

5.2.4.2 *Koryak system.* Upper Palaeozoic formations are known in the westernmost part of the Talovka-Pekulney zone of the system where they usually occur in the form of tectonic blocks, olistostromes and schuppens. The Middle and Upper Carboniferous are represented by greywackes and sometimes by reef limestones while Permian formations of terrigenous, siliceous and calcareous complexes are encountered very rarely. The continental slope probably extended along this zone in this epoch while a deepwater basin pertinent to the Palaeo-Pacific occurred in the east. This is confirmed by finds of cherts, spilites, fine clastic rocks and Permian limestone beds containing fusulinids in the eastern part of the system (Terekhova et al., 1979). Palaeontological data shows that in the Late Palaeozoic it represented a region containing a mixture of boreal and tethyan fauna.

5.2.4.3 *Nippon-Sakhalin system.* M i d d l e - L a t e C a r b o n i - f e r o u s. Accumulation of shallow-water terrigenous and carbonate sediments continued in the shelf zone in western Japan (Mazarovich and Richter, 1987). The total thickness of limestones (often oolitic) and clay deposits here reaches 300 to 400 m. Sedimentation was interrupted by uplift and deformation in the middle of the Middle and at the end of the Late Carboniferous epoch.

In Primorie limestones and terrigenous rocks with bands of cherts in the central regions of Sikhote-Alin' fall in the Middle-Upper Carboniferous. Recent investigations have shown that practically all of them occur in the form of olistoliths among Lower Mesozoic formations. Fragments of siliceous beds of Carboniferous age are encountered in the littoral zone in addition to limestones. A complex system of terrigenous and siliceous formations as well as mafic volcanics was formed in the Kur and Urmi River basins in the northern part of Primorie in the Middle-Late Carboniferous.

In the north-eastern part of Honshu Island where, too, the shelf zone has extended, reef limestones (up to 700 m) predominated among the Middle-Upper Carboniferous sediments. These shallow-marine carbonate-terrigenous formations are replaced south and east by volcanic-terrigenous-carbonate and later by terrigenous-volcanic-siliceous deepwater formations in which limestones form only thin lenses. Volcanic material is represented by mafic lavas and tuffs, which could be a second layer of oceanic crust.

E a r l y   P e r m i a n. Uplift of the Khanka massif in the western part of Primorie intensified at commencement of the Permian and sandy-clay formations with coal layers were formed within it under continental and littoral marine conditions. Subaerial and subaqueous eruptions of felsic and intermediate lavas and tuffs also occurred in this massif. The total thickness of the Lower Permian ranges from 1 to 4 km. In the south the Lower Permian in the Tumangan trough (north-eastern part of Korean peninsula) is represented by sandy-clay rocks of the flyschoid type. Terrigenous and siliceous rocks were deposited under marine conditions in the basins of the Kur and Urmi rivers in the Khabarovsk region. Further, substitution of terrigenous beds by silica-clay beds occurred in an easterly direction. Here too volcanism was manifest but under subaqueous conditions with basalts and andesites predominating in the composition of volcanic products.

In Sakhalin, blocks (olistoliths) of Permian rocks represented by limestones, plagiobasalts and banded cherts are bedded among Cretaceous siliceous strata. The faunal complex in Sakhalin limestones differs sharply from a tethyan fauna of Sikhote-Alin' and interior regions of the Japanese islands. The Permian fauna of Sakhalin represents a transition from a tethyan to boreal palaeobiogeographic province and suggests the primary remoteness of this region from the continental zones of Sikhote-Alin' and Japan.

Limestones and terrigenous strata in the north-western part of the shelf zone of Japan up to 600 to 800 m in thickness contain bands of coarse clastic continental rocks. These are replaced in the east and south-east predominantly by terrigenous and flyschoid formations, including debris cone facies of submarine canyons of the continental slope. Sedimentation breaks and minor deformation occurred in the middle of the Early Permian. A large amount (up to 2 to 2.5 km) of mafic volcanics and siliceous rocks are present in terrigenous strata in the Mino-Tamba, Ryōke and Chichibu zones.

L a t e   P e r m i a n. Uplift in Primorie affected the Khanka massif while intermediate and felsic volcanism was manifest and intrusions occurred in the southern part of the massif, indicating the existence of a marginal volcanoplutonic belt. Facies of montane and plain river-beds, peat bogs, floodplains, deltas and lagoons, shallow marine zones, reefs and subaqueous slopes and canyons have been reconstructed from the terrigenous strata accumulated in the east. Deepwater banded cherts, sandstones and mafic volcanics of oceanic island type with lenses of limestones were formed farther east (coastal zone). Folding occurred in the territory of western Primorie at the end of the Permian. Palaeozoic formations of the Tumangan trough in the north-eastern part of the Korean peninsula were crumpled into folds and complicated by overthrusts with formation of serpentinite melange and subsequently intruded by diorites and granodiorites.

The previous palaeotectonic conditions persisted in the territory of the Japanese islands. Shallow-water carbonate and terrigenous formations (Hida, Circum-Hida zones) and sometimes continental terrigenous facies accumulated in the north-western shelf zone. These were replaced towards the south and east by terrigenous complexes of mobile shelf and continental slope (including flysch) and later by volcanic-siliceous-terrigenous complexes of inner zones of a marine basin resembling a marginal sea (Sangun, Maizuru, Mino and Ashio zones). Numerous bands of conglomerates appearing at the end of the Permian in the composition of formations here point to the reactivation and differentiation of tectonic movements. The Kurosegawa-Ofunato island arc represented the outer boundary of this sea. Shallow-water terrigenous and tuffaceous rocks as well as lavas and tuffs of mafic and felsic composition (Ryōke and Chichibu zones) were deposited within this island arc zone. At the end of the Permian, gabbro and granitoids intruded into this same region. Finally, in the more eastern regions (Sambosan, northern Kitakami and eastern Chichibu) volcanosiliceous complexes were formed under deepwater, evidently oceanic conditions. The presence of a metaophiolite association, the nature of deposition, and absence of granites and gneisses in serpentinite melanges permits interpretation of these regions as a palaeo-oceanic realm.

Early Triassic. Tectonic movements at the end of the Permian and beginning of the Triassic intensified sharply in the Nippon-Sakhalin system.

Conditions close to an orogenic regime prevailed in western Primorie: uplift predominated and fold formation and intrusive activity continued. Land masses (Khanka and Bureya ancient blocks) were surrounded from the east by a marine basin and the composition of formations from west to east in the basin varied from coarse littoral facies to fine, deeper water facies.

Uplift commenced in the interior regions of the Japanese islands, which was caused by collision of the immense Palaeo-Asian mainland with the Honshu microcontinent (Faure and Charvet, 1987). At the site of the shelf and slope of the marginal sea, green schist metamorphism and schuppen-overthrust dislocations were manifest and the tectonic zones Hida and Circum-Hida arose. These zones were made up of blocks of various metamorphic and Palaeozoic rocks with serpentinite protrusions.

They were overthrust to the south and were accompanied by glaucophane schists of the Sangun zone in their footwall. Ungraded Otani conglomerates were deposited along the outer margin of these zones in the Early Triassic. Their formation is explained by a general upheaval of the northern regions of Honshu Island. In the west of the island in this period, zone of Maizuru serpentinite melange was formed. With this melange is associated metamorphism of Palaeozoic formations and diaphthoresis of much older metamorphic complexes and granites constituting the basement of the Honshu microcontinent. Tectonic deformation and manifestation of metamorphism in the Ryōke, Sambagava and Chichibu zones likewise fall in the Early Triassic.

In the south-eastern region of the Japanese islands subsidence and accumulation of marine terrigenous (from Chichibu zone) and volcanic-siliceous-terrigenous formations continued. Volcanic rocks are represented by basalt pillow lavas, spilites, volcanic breccia, hyaloclastics and mafic tuffs.

Thus tectonic development of the Nippon-Sakhalin system during the Middle-Late Carboniferous, Permian and Early Triassic was determined, as in the preceding epochs, by its marginal-continental position. The major event was extensive manifestation of tectonic deformation, magmatism and metamorphism at the end of the Permian to commencement of the Triassic. These processes led to an eastward shift of the zone of continent-ocean transition.

5.2.4.4 *Cathaysian system.* M i d d l e - L a t e C a r b o n i f e r o u s. As pointed out in the *Atlas of the Palaeogeography of China* (1985), a deep marine regime was established in the Early Carboniferous in the littoral strip in the south-eastern part of the country, partly encompassing the Zhe-Ming Caledonian orogenic uplift. This zone enlarged in the Middle-Late Carboniferous and was called the Fuding-Shantou trough. It is highly probable that this basin represents a continuation of the marginal Sangun-Yakuno sea of the Japanese islands.

There is proof of the existence in the Early Permian of a marine basin even on Taiwan Island in the form of blocks of limestones with fusulinid fauna. The presence of mafic and ultramafic schuppen in the host strata indicates that the axial part of the basin, again as in Japan, was underlain by an oceanic crust.

In the L a t e P e r m i a n land predominated on much of Taiwan Island except for the eastern coast. This was probably a consequence of uplift which was manifest even in Japan. It affected only the very narrow and interrupted strip along the continental coast, the main portion of which, as before, was occupied by the Zhe-Ming uplift. Somewhere south of Hainan Island, the Cathaysian geosyncline should have joined with the Viet-Lao branch of the Palaeo-Tethys.

5.2.4.5 *East Australian region.* M i d d l e - L a t e C a r b o n i -
f e r o u s. The Kanimblan orogeny manifest at the end of the Early
Carboniferous completed the orogenic stage of the Lachlan system and,
commencing with the Middle Carboniferous, almost all of its entire territory
became a unit of the Australian Craton. As a result, the high tectonic activity
at the end of the Palaeozoic was maintained only in the New England system
(Scheibner, 1987). Further, in the western part of the system, uplift and an
orogenic regime prevailed while development of marine basins continued in
its eastern part.

Intrusion of granitoid batholiths accompanied by metamorphism, tectonic
deformation and subaerial volcanism of felsic and intermediate composition
occurred in the northern part of New England and in Queensland. Marine
sedimentation was preserved only in the Yarrol trough but even here subsidence was unstable and replaced by uplift while marine and continental
facies alternated in the composition of formations. Further, volcanic eruptions
of mafic composition occurred in this trough in the Late Carboniferous.

Filling of the Tamworth trough with continental deposits was completed
in the south in the western regions of New South Wales. Tillites and other
glacial rocks as well as products of felsic volcanism played a major role
among these formations. This trough was bound in the east by Peel fault,
which developed in the Late Carboniferous as a deep fault that separated
the trough from uplift within which granitoid plutons intruded and deformation
manifested. In the extreme eastern part of New England a normal marine
environment continued to prevail throughout the Middle and Late Carboniferous and coarse clastic formations were replaced from west to east by
turbidites and clay sediments.

E a r l y P e r m i a n. By the beginning of the Permian, much of the
New England system was transformed into a region of uplift and erosion
with typical manifestation of an orogenic regime: granite formation, metamorphism, subaerial volcanism and tectonic deformation. Residual troughs were
still preserved however. The Sydney-Bowen zone of foredeeps arose before
the western front of the system and a transverse uplift divided it along the
boundary of Queensland and New South Wales. Formation of the northern
Bowen trough was accompanied by considerable reactivation of volcanism
whose centres were confined to a linear uplift in the form of a volcanic arc
in the boundary between this trough and the residual Yarrol trough. The volcanic basalt-andesite-rhyolite complex containing ignimbrites was overlain in
the Bowen trough by paralic and later continental coal-bearing molasse. The
thickness here reaches 3 to 4 km. In the Sydney trough conditions differed
and marine coarse terrigenous formations and subaerial felsic volcanics
accumulated in it. Here, too, the thickness is significant, reaching 3 km.

The Yarrol trough was filled with turbidites and marine sand-
conglomerate beds with participation of andesite volcanics in the Early

Permian. At the end of this epoch subsidence ceased here, the trough was closed and its sediments subjected to deformation. In the more eastern Gympie trough, a marine basin was preserved and comparatively shallow-water carbonate-terrigenous complexes accumulated.

In the eastern part of New South Wales formation of marine deposits continued, including turbidites and reef carbonates, and mafic and felsic volcanism manifest. However, at the end of the Early Permian, uplift intensified in these regions and the Hunter phase of orogeny commenced in Kungurian time. Major overthrust and strike-slip displacements occurred during this orogeny. The Peel fault, which had developed as a deep fault in the Late Carboniferous, was transformed into a major overthrust at commencement of the Permian and marked the obduction zone. Deformation in the fault zone was accompanied by metamorphism which attained amphibolite facies at places. Batholith bodies of granites arose simultaneously due to subduction of oceanic crust (east of New England) as also melting of the lower parts of the volcanosedimentary prism. Thrust deformation manifested within the central uplift of New England which by this time was divided into blocks.

In the northern part of New England granites intruded into the Hodgkinson zone in northern Queensland and in the adjoining reactivated parts of the Precambrian Georgetown block and Caledonian Lulworth-Ravenswood block. This process was preceded by felsic volcanism at the end of the Carboniferous to commencement of the Permian.

L a t e P e r m i a n. Filling of the Bowen and Sydney foredeep basins was completed with the deposition of continental coal-bearing molasse up to 2 to 3 km in thickness. At the end of the epoch these troughs experienced folding. Initially, the deformation of their sedimentary filling was synsedimentary. These dislocations characterised the concluding epoch of compressive deformation of the East Australian mobile region, i.e., the Bowen epoch. Zones adjoining from the east (Yarrol and Gundiwindi—Muki-Hunter overthrusts) were thrust on the Bowen and Sydney troughs. The Peel fault was transformed into a transcurrent fault with dextral displacement of about 150 km. In this epoch even marine Permian formations of the eastern zones of the New England system (Yarrol and other troughs) were deformed and granitisation and metamorphism were intensified. Regional metamorphism was manifest in greenschist, amphibolite and, at places, glaucophane facies. Volcanic eruptions of predominantly felsic composition were concentrated in the regions of manifestation of granitoid intrusive magmatism forming part of a marginal volcanoplutonic belt. Marine sedimentation was preserved only in the Gympie trough in the far eastern part of the system where Upper Permian formations were represented by terrigenous and carbonate rocks of the shelf type.

E a r l y T r i a s s i c. Disintegration of the fold system of New England with formation of a series of superposed intermontane basins, i.e., Esk graben, major (Clarence-Moreton and Maryborough) and minor (Tarong, Malgildie and Lorne) basins commenced in the Triassic. These basins unconformably overlie much older ones, including the Permian, and consist predominantly of molasse. Formation of the basins marking transition of the New England system from an orogenic to a taphrogenic regime, was accompanied by volcanic activity of basalt-andesite-rhyolite composition. The association of these volcanics with intrusion of the latest plutons of the eastern zones of the system is indisputable. This volcanoplutonic belt arose at the verge of the orogenic and taphrogenic stages of the region. It has been suggested that the basins were formed on a rift base, which complicated this belt in the Early Triassic. The maximum thickness of volcanics (up to 3 km) is noticed in Esk graben and the minimum in the Maryborough littoral basin in which marine formations still predominated in the Early Triassic. This indicates the proximity of an open sea basin from the east.

Thus from the Middle Carboniferous to the end of the Permian an orogenic regime was manifest in the East Australian mobile region, represented within the continent only by the New England system. The tendency towards extension of zones with a very active regime from the continent to the ocean, manifest in the much older Palaeozoic epochs, was preserved in this period also. Successive build-up of continental crust due to accumulation of sediments and products of volcanic activity, and their metamorphism and granitisation continued. Compressive phases caused by overthrust of oceanic crust from the east onto the continental crust coincided in time with the known phases of orogeny in Europe. As a result, the New England fold system may well be compared to the corresponding Late Hercynides, such as the Urals, where, too, a taphrogenic stage of development commenced in the Early Triassic.

5.2.4.6 *West Antarctic region.* In the northern part of the Antarctic peninsula and on Alexander Island, the lower part of the shale-greywacke Trinity series belongs to the Carboniferous and Permian. This series contains bands of turbidites and layers of intermediate and felsic volcanics. Evidently this complex was formed under conditions of a marginal sea. In the Antarctic peninsula siliceous-volcanic formations are widely developed in the Trinity series and in lithologic formations comparable to them. These formations consist of metamorphosed mafic lavas and tuffs, cherts and slates. This points to intensification of volcanic manifestation at the end of the Permian (judging from the admixture of volcanic fragments, it commenced early in the Lower Permian in the Ellsworth Mountains). Tectonic deformation and metamorphism encompassed this Late Palaeozoic Early mesozoic Trinity complex at pre-middle jurassic time (according to the latest data).

5.2.5 East Pacific Belt

5.2.5.1 *Cordilleran system.* M i d d l e - L a t e C a r b o n i f e r o u s. Orogenic processes of the Antler epoch attained culmination in the eastern zone of the Cordillera all along its extent from Alaska to California. Uplift affected almost the whole of Alaska and Palaeozoic formations of Brooks Range underwent intense folding but comparatively weak metamorphism. A major asymmetric anticlinorium with a gentle southern flank and upturned northern limb complicated by imbricated thrusts with an amplitude up to 20 km formed here. In the region of Richardson Mountains inversion and folding occurred and compensational basins of the marginal type were formed on both sides of the uplift, Old Crow in the west, on the margin of the Yukon massif, and Midchannel in the east, in the region of the Mackenzie River delta, on the margin of the North American Craton. Both basins were filled with molasse.

In the western part of British Columbia predominantly sand-conglomerate beds were formed in intermontane troughs and granitoids (aged 320–310 m.y.) intruded into the uplift. Filling of troughs in the northern Rocky Mountains with sediments was completed. Intense subsidence continued east of the zone of the Antler uplift, however. Lower Carboniferous flysch was replaced in these regions by arhythmic sandy-clay and carbonate sediments of significant thickness, exceeding 4 to 5 km in the Colorado basin. Within the Antler orogen, fold-overthrust deformations developed further. The amplitude of overthrusts of volcano-siliceous western complexes on the carbonate-terrigenous eastern shelf complexes reached 90 to 100 km (Roberts Mountain overthrust).

In the western zone of the Cordillera, formations of Middle-Late Carboniferous are represented by clay and siliceous beds of small thickness, subaqueous volcanics of mafic composition and sometimes micritic limestones, most probably formed under deepwater conditions. Island arcs of the Klamath Mountains and Alexander archipelago preserved their elevated relief but volcanism had ceased in them by the Early Permian. The nature of tectonic development of the Vancouver island arc in this period is not yet clearly understood.

E a r l y P e r m i a n. The palaeotectonic environment stabilised somewhat in the eastern zone of the system. The intensity of manifestation of orogenic processes lessened, accumulation of molasse ceased nearly everywhere and the area of marine basins enlarged. These basins shoaled and fine-grained terrigenous rocks and limestones were deposited in them. Their thickness usually did not exceed a few hundred metres but the rate of subsidence in central Utah in which carbonates predominated in the Lower Permian sequence remained high and attained 3 km in thickness. The supply of material into this basin was from the east (from the craton) as well as from the west (from the Antler orogen).

Thick (up to 4 km) volcanic and volcanosedimentary strata comprising spilites, andesite-basalts, andesites and their tuffs as well as greywackes and siliceous rocks accumulated in the western zone of the Cordillera. Submarine volcanic activity again intensified all along its extent from Alaska to California. Detailed palaeotectonic reconstruction of this zone presents great difficulty.

A continental crust already existed at the base of island arcs of the Alexander archipelago and Vancouver Island-western Cascade Mountains and Klamath Mountains at the beginning of the Permian. There were almost no lava eruptions within the island arc of Vancouver Island and reef limestones, terrigenous rocks and radiolarites accumulated. In the Alexander archipelago arc individual volcanic islands were formed, accompanied by minor eruptions of lavas of andesite composition. Turbidites were formed on its eastern slope. Volcanic activity intensified on the arc of the Klamath Mountains. Lavas of mafic and intermediate composition poured out time and again on the islands and submarine ridges. These lavas alternate with lava agglomerates, tuffs and tuffaceous sandstones. These volcanic products, in geochemical features, belong to the low-potash tholeiitic island arc series with a broad scale of differentiation (Lapierre et al., 1986). Reef limestones were deposited on the littoral zones of islands.

L a t e P e r m i a n. The second half of the Permian period in the eastern zone of the Cordilleran system is characterised by the preservation of an orogenic regime and new manifestations of uplift and folding representing the concluding phases of Hercynian orogeny. Intense uplift of the Richardson Mountains was accompanied by intrusion of minor granite plutons. Development of troughs was completed in the regions of Brooks Range and in north-eastern Alaska. Uplift intensified here too.

Residual troughs in the southern part of the system experienced regression. Their size was greatly reduced, and as before, carbonate and terrigenous rocks were deposited in them but the thickness of sediments no longer exceeded a few hundred metres. Unconformities at the base of the Upper Permian are noticed in central Nevada, between the Permian and Triassic in central Nevada and Oregon and also in the Cassiar Mountains (British Columbia). They indicate manifestation of tectonic deformation. In particular, new obduction with an eastward direction took place at the end of the Permian west of the Antler uplift and subaerial intermediate and felsic volcanism manifested in these regions. This epoch of deformation coincided in time with movements of the Alleghanian epoch of orogeny in the Appalachian system, marking the end of orogenic development of the latter. It was called the Sonoman in the Cordillera.

The southern part of the North American Craton, i.e., Mexican territory situated on the southern continuation of the orogenic zone of the Cordilleran system, experienced reactivation of tectonic activity. Here arose conditions of an epiplatformal orogenic regime with manifestation of granitisation, metamorphism and mighty uplift. As a result, the Cordilleran system adjoined the Ouachita-Mexican system and they, together with the Appalachian system, fringed the North American Craton in the form of a semi-circle of orogenic zones.

The magnitude of submarine volcanic eruptions decreased in the western zone of the Cordillera but the former conditions continued to prevail and volcanic and sedimentary strata were formed. The thickness of these strata in the northern part of the zone reached 4 km. Uplift of island arcs intensified. Much of the area of the Alexander archipelago arc and Vancouver Island-western Cascade Mountains represented an island mass in which folding and granitisation were manifest. The arc of the Klamath Mountains was elevated above sea level and volcanism (subaerial) continued within its boundaries. Further, its composition changed towards a predominance of andesites initially and rhyolites later.

E a r l y   T r i a s s i c. The activity of orogenic processes decreased in the eastern zone. Uplift predominated and small residual basins were preserved in the central Rockies. In the Cassiar Mountains (British Columbia), granite intruded (age 240–220 m.y.) and tectonic deformation (Cassiar orogeny) occurred, marking the end of the Hercynian era of orogeny of the Cordilleran system. In the southern part of the system (Nevada), Upper Palaeozoic formations experienced deformation, were displaced eastward and overthrust on the margin of the North American Craton (Babale, 1987). Deformation led to isoclinal folding and overthrust slices folded in turn into open gentle folds.

The elevated position of the massif of former island arcs with already formed continental crust, i.e., the Alexander archipelago and Vancouver Island-western Cascade Mountains, was maintained in the western zone of the Cordillera. Manifestation of subaerial felsic volcanism ceased in the similar massif of the Klamath Mountains. Between these massifs (microcontinents) and the eastern elevated orogenic zone of the Cordilleran system, the Lower Triassic is represented by fairly thick terrigenous and submarine-volcanic (intermediate and mafic in composition) strata.

Thus all through the stage under consideration, conditions of an orogenic regime were preserved in the eastern zone of the Cordillera. Uplift prevailed, tectonic deformation (most intensely at the end of the Carboniferous and in the Late Permian), granitisation and metamorphism were manifest and molasse were formed. The weak volcanic activity is worthy of note. In the western zone, contrarily, volcanism was very intense and here sedimentary formations of considerable thickness formed. The expanding tendency of island arcs of the western zone stabilised and, as a result, major elevated massifs with continental crust were formed.

5.2.5.2 *Andean system.* M i d d l e - L a t e   C a r b o n i f e r o u s. The Ecuador-Colombian segment experienced subsidence and extensive transgression. These processes began at the end of the Early Carboniferous as supported by the unconformable attitude of Lower Carboniferous sandstones and conglomerates on dislocated Devonian formations in several regions. Middle and Lower Carboniferous terrigenous and carbonate strata, on the contrary, conformably overlie Lower Carboniferous and represent formations of shallow shelf basins. Their thickness runs into hundreds of metres, sometimes up to 1.5 km. Only the Maracaibo massif in the northern part of the system preserved an elevated position. Littoral and continental sand-conglomerate strata were formed on its slopes.

In the Peru-Bolivian segment in which marine troughs already existed in the Early Carboniferous, subsidence intensified and the thickness of the Middle and Upper Carboniferous in the Abankaya region reached 2 km or more. Carbonate-clay lithofacies predominated here. Elsewhere in the region shallow-water limestones, their thickness hardly exceeding 500 m, played the main role in the composition of formations. On the Arequipa massif, as in the preceding epoch, uplift predominated and granitoid plutons intruded at the end of the Carboniferous.

Subsidence enlarged perceptibly in the Chile-Argentinian segment (Castanos and Rodrigo, 1980; Hervé et al., 1981). The band of troughs and basins extended from north of Chile to Patagonia. Commencement of subsidence and transgression occurred in some regions at the end of the Early Carboniferous. Lacustrine deposits accumulated in the basins of the eastern zone and, in the Late Carboniferous, fluvioglacial formations. In the troughs of the western zone formations of marine and continental terrigenous facies of the molassoid type alternated. In northern Chile a thick (over 1 km) flysch series formed in the Salar-de-Navidad region (Bahlburg et al., 1987). Accumulation of limestones with Fusulina predominated in the southern part of Patagonia, indicating quiescent shallow marine conditions.

Folding manifested in uplifts of the Chile-Argentinian Andes at the end of the Middle Carboniferous. Granitoid magmatism was also characteristic of these uplifts. Its magnitude was most significant within the Pampeanas massif and in Cordillera Coastal of Chile in which the Coastal batholith was formed. Judging from radiometric data and finds of granitoid fragments of this batholith in Upper Carboniferous conglomerates, its intrusion occurred in the Middle Carboniferous.

E a r l y   P e r m i a n. The magnitude of subsidence in the Ecuador-Colombian Andes remained as before and the areas of marine troughs and basins were almost as large as in the Middle and Late Carboniferous. The shallow-water carbonates and terrigenous formations deposited in them were similar to platform deposits in composition and small thickness. Subsidence gradually gave place to uplift at the end of the epoch and regression affected the entire area of the segment; this inversion of tectonic development was accompanied by fold and fault deformation of Late Palaeozoic sedimentary complexes.

Transgression penetrated at the beginning of the Permian into the south-eastern regions of the Peru-Bolivian Andes and carbonate and clay sediments with bands of gypsum accumulated here in a partially closed basin. The subsidence of this basin later slowed down, was compensated by sediments, and marine deposits in it were replaced by coarse clastic continental (molasse) formations, including fluvioglacial facies in the Artinskian and Kungurian stages. The material source was the elevated region of the Pampeanas massif in the south in which mountain glaciation developed. In the western regions of the system marine troughs were preserved within the former boundaries and deposition of fairly uniform shelf limestones (Copacabana formations) and clay rocks continued in them. However, uplift also manifested here at the end of the epoch, as a result of which sea regressed from almost the entire area of the Peru-Bolivian Andes.

In the northern part of the western zone of the Chile-Argentinian Andes, filling of marine troughs also gradually ceased and flysch formations were replaced by shallow-water limestone-sandstone strata and molasse. Stable formation of marine carbonate and terrigenous deposits continued only in southern Chile. But even here continental terrigenous facies were deposited in the marginal parts of the basin. The general intensity of tectonic movements in this segment of the Andean system was extremely high. Rugged topography with mountain glaciers developed in uplifts while the intermontane basins in the eastern zone of the system were filled with molasse and partly fluvioglacial formations exceeding 1 km in thickness. In the Early Permian magmatic activity became intense in the Chile-Argentinian Andes. At the end of the epoch, in the central part of the segment, centres of felsic volcanism arose. Granites intruded into the north and, judging from radiometric data, tectonic reactivation affected the extensive territory of the former Patagonia-Deseado platform block (Forsythe, 1982). Uplift intensified here and granitisation and high-temperature metamorphic processes developed.

Late Permian. Uplift predominated throughout this epoch and a rugged topography was formed in the territory of the Ecuador-Colombian Andes within the future Central and Eastern Cordillera. Red molasse strata (Palaeo-Jiron formation) accumulated in the large basins. The concluding Hercynian fold formation continued and reached maximum intensity and regional metamorphism was manifest. The intrusion of granitoids (age 250 m.y.) accompanied this epoch of folding and metamorphism in the western zones. Further, migration of uplift took place, folding and granite formation from south to north along the Cordillera right up to Goajira peninsula, in which mighty uplift and intrusion of granites commenced at the very end of the Permian.

An extended marine basin was still preserved at the commencement of the Late Permian in the northern Peru-Bolivian Andes and was rapidly filled with terrigenous and carbonate sediments, including gypsum bands. Uplift predominated later here, as in the rest of the area of the system. An intermontane trough extended in the eastern zone at the site of the Central Cordillera of Peru and Bolivian upland and red continental molasse containing thick members of volcanic rocks of intermediate and mafic composition (Mitu formation) were deposited in this trough. The total thickness of the Upper Permian exceeds 2 km. Formation of molasse and volcanism was preceded at the beginning of the Late Permian not only by a general uplift of the region, but also fold deformation, as supported by sharp unconformity in the attitude of the Mitu formation on underlying deposits. Further, intrusion of rather large granitoid plutons occurred in the Eastern and Central Cordillera of Peru (Carlier et al., 1982).

Further intensification of volcanic eruptions and formation of continental volcanic Calipui belt were important features in the tectonic development of the Chile-Argentinian Andes. Its thickest formations, felsic tuffs, ignimbrites and rhyolites, accumulated in the axial part of the system. Uplift at the end of the Early Permian was accompanied by intense tectonic deformation. As a result, felsic volcanics of the belt lie unconformably on the Late Carboniferous and Early Permian formations. In the eastern regions molasse accumulated in intermontane basins while processes of tectonic-magmatic reactivation, i.e., intrusion of granitoid plutons and manifestation of metamorphism, continued in the southern part of the segment. Granitoid magmatism accompanied volcanic activity in the northern regions of Chile and Argentina too.

E a r l y   T r i a s s i c. The intensity of tectonic processes diminished in the Ecuador-Colombian Andes, i.e., topography smoothened and fold formation and metamorphism were completed. In the north, within the Maracaibo massif, molasse still accumulated in a continental basin and minor eruptions of basaltic lavas occurred. However, a general transition to a platform tectonic regime has been remarked everywhere.

In the Peru-Bolivian Andes, on the contrary, high tectonic and magmatic activity was preserved. Coarse molasse, lavas and tuffs of varied composition were deposited over extensive areas with felsic and sometimes alkaline varieties predominating. Intrusions of granitoids and alkaline rocks continued. However, at the end of the epoch, manifestation of magmatism ceased and, evidently, the topography was smoothened since accumulation of molasse came to an end.

In the Southern Andean system formation of the continental volcanic Calipui belt continued, granites intruded on the former scale and metamorphism manifested. The thickness of volcanic and volcanoclastic Lower Triassic formations measured hundreds of metres. Rocks of felsic composition predominated among volcanics.

In general, tectonic conditions in the entire territory of the Andean region from the Middle Carboniferous to the Early Triassic corresponded to conditions of an orogenic regime. The intensity of its manifestation varied, however. In the Late Carboniferous and Early Permian very large areas of the region were affected by subsidence and transgression. However, the general predominance of uplift was maintained. Most powerful tectonic movements, maximum magmatism and most intense metamorphic processes took place in the Late Permian epoch.

5.3 PALAEOTECTONIC ANALYSIS

The structural plan of the Earth underwent very significant reorganisation in the stage under consideration. The Ural-Okhotsk belt throughout its extent, except for the easternmost Amur-Okhotsk segment, entered an orogenic stage and the hitherto isolated Laurussian (Euramerican), Siberian and Sino-Korea craton merged into a single continent, Laurasia. The Mediterranean belt with its American continuation also experienced a similar evolution; as a result, Laurasia merged further with Gondwana, forming the supercontinent Pangaea II. Only in the east with some southward displacement, was the Palaeo-Tethys replaced by the Meso-Tethys which largely opened towards the Proto-Pacific ocean. At the end of the Carboniferous (Fig. 5–6), break-up of Gondwana commenced in the form of continental rifting, in anticipation of its total disintegration. A similar process affected Laurasia too where two major rift systems arose, North Atlantic and West Siberian, which joined in the Arctic north of Novaya Zemlya. Destruction of continental crust in both systems proceeded at places up to the stage of spreading. In the West Siberian system this process ceased later while it manifested in the Atlantic as a forerunner of the formation of a new ocean. All of this, occurred in the Mesozoic, however.

5.3.1 Continental Cratons

5.3.1.1 *Gondwana.* At the commencement of this period Gondwana still represented a monolithic continental mass which experienced predominantly uplift. Only within South America, northern Africa and north-eastern Arabia, did weak subsidence (except in the Amazon basin) with accumulation of lagoon-marine sediments occur. The situation changed sharply at the end of the Carboniferous to commencement of the Permian when rift basins began to form in the southern half of Africa, India and eastern Antarctica (Fitzroy and Perth rifts in Australia were formed even before) and much of the supercontinent evidently continued to heave and was affected by glaciation. By the middle of the Early Permian (Fig. 5–7), glaciation disappeared and this resulted in a rise of sea level. However, Gondwana in general remained uplifted and only some peripheral regions of it in Arabia and India experienced transgression of a shallow sea which penetrated into the interior of the supercontinent in South America not long ago. However, in the Late Permian (Fig. 5–8) a marine strait formed along one of the rift zones, i.e., along the east coast of Africa and west coast of Madagascar and Australia, in anticipation of the future opening of the Indian Ocean and in particular its Somalia-Mozambique basin.

Almost all the margins of Gondwana, except for the Australian and possibly Antarctic, developed in this stage as passive ones. An outburst of tholeiite-basalt volcanism occurred on the Indian margin (Pir-Panjal Mountains), associated with the birth of the Neo-Tethys in the Early Permian.

5.3.1.2 *Laurasia.* While the eastern part of the Midcontinent and much of central Siberia represented in the Middle-Late Carboniferous regions of coal accumulation—paralic in America and limnic in Siberia, accumulation of marine carbonates predominated in western North America and eastern Siberia under conditions of slightly more intense but generally weak subsidence. The same contrast is noticed relative to the Sino-Korea and South China Cratons subsequently joined to Laurasia—paralic coal-bearing formations in the first craton and shelf carbonate and paralic coal-bearing in the south-eastern part of the second. Thus, either weak subsidence (predominantly) or evidently a correspondingly weak uplift were characteristic features for the entire Laurasian group of cratons in the Late Palaeozoic. A special feature in this background is the Late Permian tholeiite-basalt outpourings in the south-western margin of the South China Craton, associated with opening of the south-eastern segment of the Permian-Triassic Meso-Tethys.

Another important feature of the Late Palaeozoic tectonic development of Laurasia, this time similar to Gondwana, is the development of rift systems. In North America such a system had an east-south-east strike separating the south-western corner of the ancient craton and extending from the eastern Rockies (in Utah and Montana) through Colorado and New Mexico to the Wichita and Arbuckle mountains and Anadarko basin in Southern Oklahoma. This Southern Oklahoma rift has been traced from the Late Proterozoic to the Cambrian but shows almost no activity in the Middle Palaeozoic, with reactivation in the Pennsylvanian and Permian periods. This is manifest in the deep subsidence of the Paradox and Anadarko grabens with filling of the former with red-beds and evaporites and the latter with dark-coloured sandy-clay beds of enormous thickness and in a joint uplift of the San Juan, Wichita and Arbuckle mountains. This rift system in the east joins with the Ouachita-south-western continuation of the Appalachians and thus with the American continuation of the Palaeo-Tethys (Fig. 5–9).

The Southern Oklahoma rift was not unique in the North American Craton. In its extreme south-western corner, in triple junction with the further

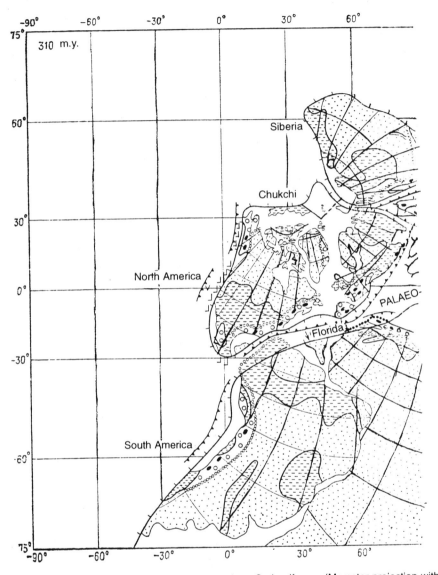

Fig. 5-6. Palaeotectonic reconstruction for the Late Carboniferous (Mercator projection with

This reconstruction shows the Florida block for the first time, which is missing in much older composition of the Indochina block but, in that event, the position of this block would have to islands.

south-western continuation of the Appalachian system, a complex Western Texas aulacogen consisting of two grabens-troughs separated by the horst

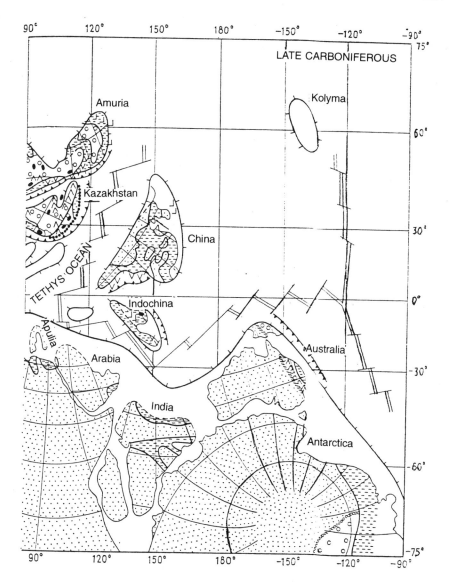

centre at 0° N lat. and 90° E long.). See Fig. 1–7 for legend.

projections. The western islands of the Indonesian archipelago require to be included in the
be changed since it adjoins Gondwana very closely and there is no space for inclusion of these

uplift of the Central Platform was formed in the Permian. In the Early and
Middle Permian relatively deepwater and thin black shale formations were

292

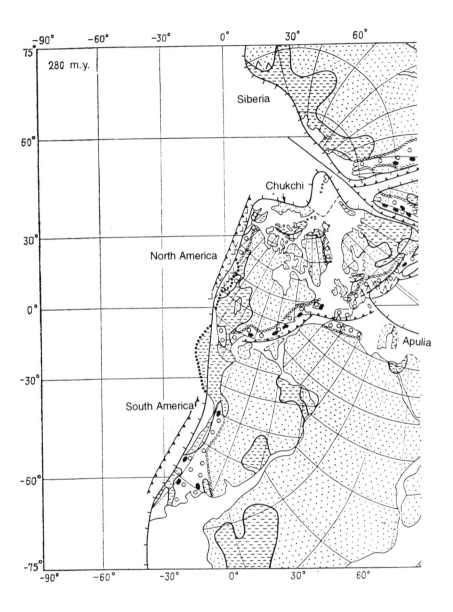

Fig. 5-7.

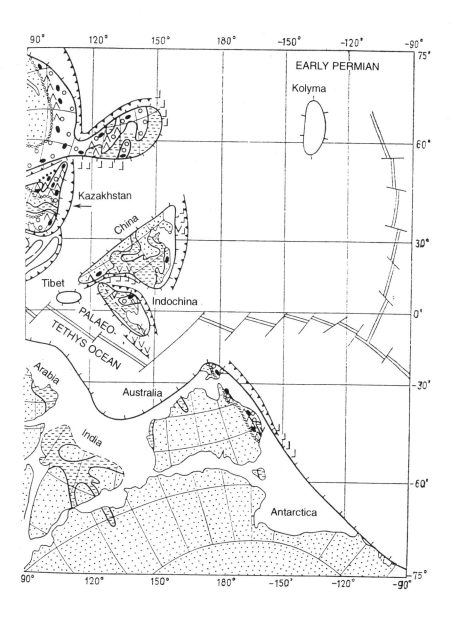

Fig. 5-7.

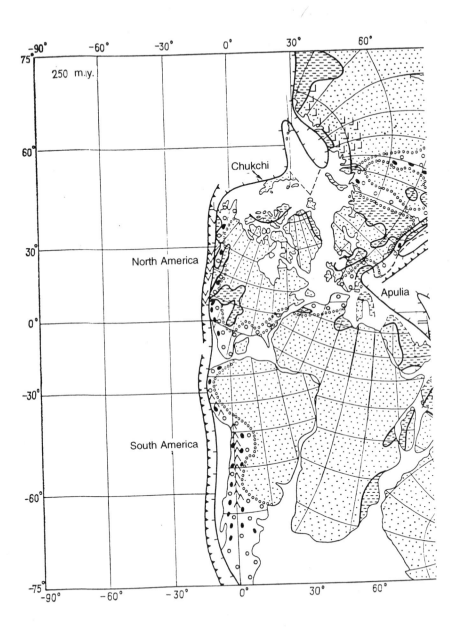

Fig. 5-8.

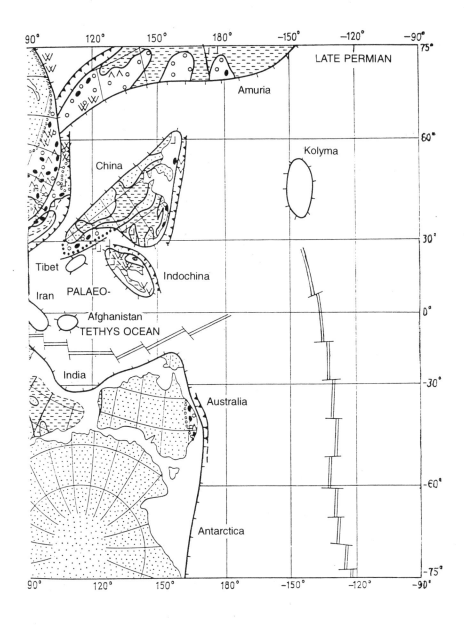

Fig. 5–8.

deposited in the troughs and reef limestones on their flanks and central uplift. In the Late Permian these troughs were rapidly filled with salt-bearing strata.

Another rift system (North Atlantic) arose in the Late Dinantian simultaneous with commencement of collision of the Central European microcontinents with Laurussia (Huszeldine, 1984). It extended between Greenland in the west and Scandinavia, the British Isles and Armorica in the east. Huszeldine associates the formation of dyke swarms in the Late Carboniferous (Stephanian) of northern Britain and southern Norway with development of this rift. It has been suggested that at the end of the Middle Carboniferous (Westphalian C/D), thinning of the continental crust between Greenland and Scandinavia proceeded until commencement of its break-up while this fault had extended by the Early Permian southward and was transformed into restricted spreading between Greenland, Britain and Armorica, and ended in the south on the Biscay transform. The proximity of disposition of this assumed intercontinental rift on the one hand to the ancient Iapetus suture, and on the other to the future Central Atlantic rift, must be emphasised.

Within the East European Craton, the Oslo graben represents a rift type structure of the same age known for its alkaline magmatism. This graben continues into Skagerrak strait and, through the adjoining part of the North Sea, probably joined with the above-mentioned North Atlantic rift (triple junction). In the Late Permian the North Atlantic rift, probably also extending into

---

Fig. 5-7. Palaeotectonic reconstruction for the Early Permian (Mercator projection with centre at 0° N lat. and 90° E long.). See Fig. 1-7 for legend.

In our view, the joining of North and South America in this scheme of L.P. Zonenshain and coauthors is not satisfactory. Here, space is clearly inadequate for all the known Early Permian structures with a continental crust. According to palaeogeographic data, Eurasia and Laurussia already represented one single continent of Northern Laurasia by the beginning of the Permian and there were no ocean zones between the East European Craton, Kazakhstan and Siberia. In this scheme too, as in all the preceding ones of this series, there is no Balkan region nor Moesian plate in Southern Europe and the Peri-Caspian basin has been separated from the continent irrationally. The absence of continuations between sedimentary basins of Marañao and central Argentina into African territory is striking.

Fig. 5-8. Palaeotectonic reconstruction for the Late Permian (Mercator projection with centre at 0° N lat. and 90° E long.). See Fig. 1-7 for legend.

Our comments on this reconstruction by L.P. Zonenshain and coauthors are a repetition of those given for similar preceding schemes: absence of Balkan region and Indonesian islands and position of Kolyma. Moreover, the Caucasian zone, which had experienced granitisation by this time, and the Turan plate are not taken into consideration here. By the end of the Permian, China should have adjoined Amuria. Marine sedimentary basins of Anti-Atlas do not find direct continuation into the Appalachians. Also not reflected are the territories of the Japanese islands and New Zealand in which complex geological processes took place in the Permian and blocks with a continental crust existed beyond doubt.

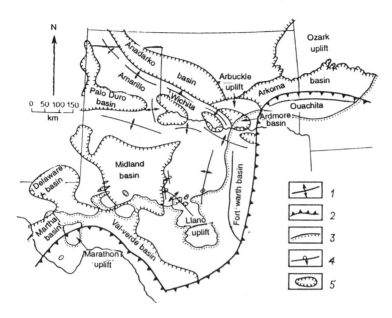

Fig. 5–9. Generalised tectonic map of the continental margin of North America in the Wichita-Marathon and adjoining regions (after Frasier and Schwimmer, 1987).

1—axes of uplifts; 2—overthrust front of Wichita-Marathon system; 3—contour of zone of Mesozoic-Cenozoic subsidence; 4—axes of recumbent folds; 5—contours of basins.

the Barents Sea, served as a channel for penetration of waters of the boreal basin into the North Sea and the Polish-German basin. According to Ustritsky (1990), the lapetus residual basin was preserved in the Barents Sea in the second half of the Palaeozoic.

The Donets basin is another Late Palaeozoic rift of the same craton. It is situated in the continuation of the Pripyat-Dnepr-Donets aulacogen ending the rift stage of its development in the Visean epoch. Rift formation in the Donets basin continued in the Middle-Late Carboniferous and beginning of the Permian accompanied by accumulation of an extremely thick paralic coal-bearing formation, further followed by formation of copper-bearing sandstones and evaporites. In the Saalian phase the Donets basin experienced inversion and the Donets foredeep was formed north of it.

Late Palaeozoic rift structures are also known in the Siberian, Sino-Korea (Pyongnam and Ogcheon aulacogens) and South China (in its southern part) cratons.

The predominance of extension and rifting processes, unlike in intercontinental mobile belts (Mediterranean belt and Ural-Okhotsk) for which it was a period of collision and intense compression—was a characteristic feature of Late Palaeozoic evolution in general for all continental blocks from which the Late Palaeozoic-Early Mesozoic Pangaea was formed.

5.3.2 Mobile Belts

5.3.2.1 *Mediterranean belt*. W e s t E u r o p e a n r e g i o n. This region had entered a collision, orogenic stage already in the second half of the Early Carboniferous. This process initially led to drying out of the Saxothuringian basin and its Armorican continuation and later, commencing from the Namurian, of Rhenohercynian, including Cornwall. It extended also into the Iberian part of the orogen. At this time its complex nappe-fold structure was finally formed. Until quite recently this structure was regarded as rather simple. In the Middle Carboniferous, ahead of the growing orogen, a foredeep of the 'European Coal Channel' was formed. It extended from southern Wales through Kent, northern France and Belgium into the Ruhr basin of Germany. The Upper Silesian basin of Czechia and Poland represents its eastern continuation or homologue. Paralic coal-bearing formations of Middle Carboniferous-Westphalian and variegated Upper Carboniferous-Stephanian, accumulated in this trough; these are separated by unconformity of the so-called Asturian phase which practically ends here the Hercynian orogeny but lithologically the Stephanian still pertains to molasse.

Apart from the system of foredeeps, a large number of intermontane basins were formed within the orogenic belt itself, mainly oriented along the strike of the belt. Some of the largest are the Saar-Hesse and Asturia basins. There are also narrow basins with a transverse or oblique-transverse direction, however. These are the coal trough of the French massif Central and an analogous structure of the Bohemian massif. These differences in the strike of post-Hercynian molasse troughs reflect a complex pattern of numerous fault dislocations breaking up the Hercynian fold system of Central Europe into a complex mosaic of blocks (Lorenz and Nicholls, 1984). Further, shear displacements, especially the dextral shear displacement along the Teisseyre-Tornquist line separating Palaeozoic and Precambrian Europe, occurred in the Late Palaeozoic along many of these faults. Strike-slip faults parallel to this major fault ('Varta line' and 'Odra line') play the main role in the set of faults of the eastern part of the Central European Hercynides and their foreland. In so far as faults with meridional or north-north-east strike are concerned, they correspond to the main direction of future Mesozoic rifts of the West European region.

Formation of all these structures was accompanied by intense volcanism, especially at the beginning of the Early Permian (Autunian). Intense granitoid plutonism, mainly monzodiorite and granodiorite types, preceded this volcanism in the Middle-Late Carboniferous (Matte, 1986). Early Permian orogenic volcanism has preserved its calc-alkaline character while bimodal volcanics of high alkalinity developed in the foreland of the Hercynides. Alkalinity increased with increasing distance from the Hercynian front. The area of this volcanism reached northern Scotland and southern Norway (Oslo graben), encompassing the North Sea, while the Hercynian orogenic magmatism extended even into the Hercynides, which later became a part of the Alpine belt.

The western, more accurately south-western continuity of the West European Hercynian orogen lay in the Southern Appalachians of North America and the Mauritanides of north-western Africa. In the Northern Appalachians, Hercynian deformation, magmatism and metamorphism manifested only in a narrow littoral belt of the Maritime provinces of Canada and New England in the USA. In the interior zones of Northern as well as Southern Appalachians (Piedmont), this Alleghanian diastrophism occurred in several phases and was manifest predominantly in granite plutonism, regional metamorphism reaching amphibolite facies and in tectonic nappes folded into anti- and synforms. These tectonic nappes were formed in much earlier (Taconian and Acadian) epochs of orogeny (Fig. 5-10). The main period of granite formation and metamorphism lay in the interval 315–285 m.y., i.e., from commencement of the Middle Carboniferous to commencement of the Early Permian, and thus corresponds in general to the main Sudetan-Asturian period of Hercynian diastrophism of West Europe. But deformation and metamorphism continued in a weakened form until the end of the Permian, i.e., up to 250–240 m.y. (Dallmeyer et al., 1986; Farrar, 1985; Ross, 1986; and Secor et al., 1986).

In the outer zone of the Southern Appalachians all Palaeozoic formations, up to the top of the Carboniferous and the basal part of the Lower Permian in the Appalachian foredeep (Dunkard formation), lie conformably; their deformation had already occurred in an epoch corresponding to the Saalian phase of Europe.

In the Ouachita zone, at the level of the upper Mississippian, i.e., roughly at the end of the Viséan epoch, a sharp change in the nature of sedimentation took place, with thick turbidites replacing deepwater and thin siliceous deposits in the flysch at the end of the Mississippian and Early Pennsylvanian. The sources of clastic material of these turbidites are located in the Appalachians as well as in the volcanic arc within the Ouachita zone. Flysch was replaced by molasse in the Late Carboniferous; a thick olistostrome is confined roughly to this period at places. Simultaneous with the deposition of molasse, tectonic nappes were displaced northward in Ouachita and to the north-west in the region of Marathon (Fig. 5-11) and Arkoma, Val Verde and Martha foredeeps were formed ahead of the front of tectonic nappes in the Late Carboniferous. Permian-Pennsylvanian formations accumulated in these foredeeps. The inner flanks of these troughs were ultimately affected by fold-overthrust deformations. At the same time, in the southern part of the

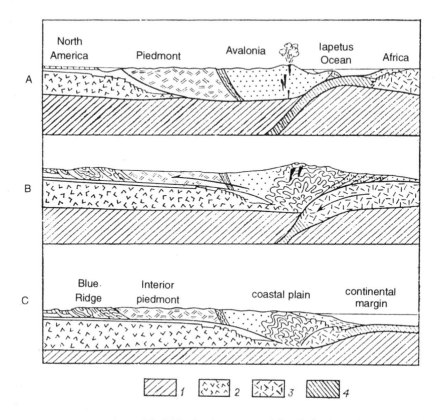

Fig. 5-10. Plate-tectonic model of Alleghanian orogeny (after F. Cook et al., 1983, modified). A—tectonic units of continental crust of eastern North America after Acadian but before Alleghanian orogeny; B—collision of North America and Gondwana lead to the development of intensely dislocated and metamorphosed orogen and to thrusting of crystalline naffe back on Early and Middle Palaeozoic foreland basins which simultaneously underwent a 'thin skin' deformation; C—configuration of tectonic units of present-day eastern North America. 1—Upper mantle; 2 and 3—continental crust (2—North America, 3—Africa); 4—oceanic crust.

Ouachita zone deformation had already ended by the Late Carboniferous. All these are regarded as a consequence of collision of the north-western, South American salient of Gondwana with Euramerica but various alternatives of this process have been proposed.

The nature of the junction of the Ouachita zone with the Cordilleran system in Mexico and Central America is a controversial subject. According to one point of view (Shubert and Cebull, 1987), the Marathon salient was bound by two transform faults, south of which lay the continuation of the ancient North American Craton reaching southern Mexico. According

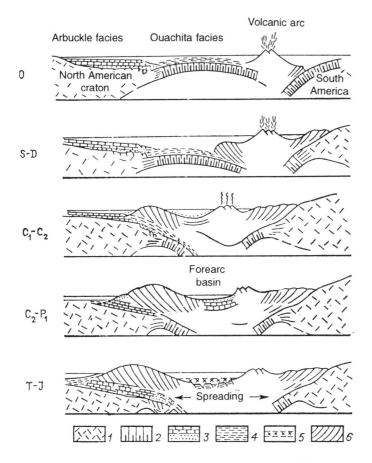

Fig. 5-11. Tectonic model of Ouachita orogeny (after J. Wickham et al. In: Frasier and Schwimmer, 1987).

1—continental crust; 2—ocean crust; 3—shelf facies; 4—deepwater facies; 5—evaporites; 6—accretion wedges.

to another view (Stewart, 1988), the North American Craton did not extend south of 28° N lat. while the continuation of Ouachita fold zone extends west of Rio Grande in a latitudinal direction, towards the Californian peninsula, in fact merging with the southern continuation of the Cordillera. Correspondingly, the outcrops of the Precambrian crystalline basement in central and southern Mexico are regarded as an independent microcontinent. The existence of a sinistral megashear, Mohave-Sonora, with a north-west strike plays a definite role in these structural reconstructions.

New data for southern Mexico concerning the Acatlan complex adjoining from the western Oaxaca block with its Grenville granulite basement has led to yet another, newer interpretation of these events (Yanez et al., 1991). Deformation and granulite magmatism of the Early Devonian period and manifestation of subsequent Late Palaeozoic diastrophism established in the Acatlan complex showed a distinct similarity with the Appalachians and confirmed its association with them. In so far as the Oaxaca massif is concerned, it is regarded as a continuation not so much of the Grenville belt of North America, as massifs of the same Grenville age in the northern extremity of the Andes (Sierra Nevada de Santa Marta and Garzón). Accordingly, the Acatlan complex should also have been in contact with palaeozones of the Northern Andes. It has been assumed that its Acadian deformation is a consequence of collision of the South American salient of Gondwana with North America, following which Gondwana departed southward together with southern Mexico and later a new collision of South America with North America occurred in the Late Palaeozoic. This interpretation is evidently tentative but even so the Oaxaca massif together with the Colombian Grenville massifs should be regarded as a microcontinent in a Palaeozoic mobile belt.

In any case, the Hercynian fold system continued south of 28° N lat. along Mexico and later traced through Central America (Guatemala, Honduras and Belize) and, after an interruption associated with wedging of the younger Caribbean basin, into the Northern Andes (Venezuela, Colombia and Ecuador). Association with the main part of the belt is confirmed by the similarity of Late Palaeozoic fauna of the Northern Andes and the Appalachians. At the same time, the south-western tip of the northern Andes became a part of the East Pacific belt. Scarce data (see Section 5.2) shows that this western segment of the Palaeo-Tethys developed similar to Ouachita and, like the latter, experienced conclusive deformations in the Middle (Early?) Permian, leading to nearly total regression of sea and establishment of a continental regime.

Caucasus-Pamirs and Kuen Lun-Indosinian regions. Events in the Balkans, Western Pontides and northern Caucasus proceeded in the Late Palaeozoic generally similar to those of West Europe. Commencing from the Late Viséan, deformation and uplift sharply intensified here too and intrusion of granites commenced except in the Outer Dinarides and southern slope of Caucasus Major. Towards the latter regions, these phenomena attenuate and marine formations of the Upper Palaeozoic, additionally conformably replaced by the Triassic, are developed. Similar changes are noticed from north to south in the northern Pamirs and western Kuen Lun, in the southern zones of which marine deposits of the Upper Palaeozoic are also developed and where significant deformations and intrusion of granites proceeded even at the end of the Triassic. From Kuen Lun, the deepwater basin continued south-east through Yunnan, Laos and Thailand into Malaya. Directly east of Kuen Lun, in Qinling, deepwater conditions in the Middle Carboniferous were replaced by shallow marine and continental, while in the Late Permian uplift already predominated here and extended even into Kuen Lun (*Atlas of the Palaeogeography of China*, 1985).

If Hercynian diastrophism and its associated phenomena in the latitudinal belt from the Balkans to Kuen Lun can be explained by subduction of oceanic crust from the south, the nature of metamorphism revealed in the Palaeozoic strata in southern Anatolia and the central part of Iran remains imprecise. Possibly, the reason for this is the existence of a more southern zone of subduction since data has been reported on the presence of ophiolites in the zone of the south-western boundary of central Iran. If this is so, isolation of the Anatolia-Iran microcontinent should be placed in the Late Palaeozoic.

A similar question also arises relative to Hercynian metamorphism and granite formation in Transcaucasus. According to Adamia (1984) and others, this subduction proceeded from the Sevan-Zangezur oceanic basin whose existence is assumed even in the Palaeozoic, a fact recently confirmed by the discovery of Upper Palaeozoic Conodonts in the Sevan zone. Similar suggestions have been made relative to the western continuation of the Sevan-Zangezur zone, i.e., Ankara-Izmir and Vardar. Sengör and coauthors (1985) and Gamkrelidze (1984) express a somewhat different view. They assume the existence of an oceanic basin in the Permian and Triassic slightly north of Izmir-Zangezur with continuation eastward into northern Iran, into the region of Resht and Meshhed, where ophiolites or, in any case, rocks which could have affinity to ophiolite association are known.

5.3.2.2 *Ural-Okhotsk belt*. For almost the whole of this belt, except the Far Eastern Amur-Okhotsk segment, the Late Palaeozoic represented the concluding stage of active development, i.e., the stage of collision of continental masses restraining the belt and disposed within it.

The internal eastern zones of the Urals experienced deformation, uplift and intrusion of granites in the Late Palaeozoic. These processes continued up to the Early Permian inclusive. In the extreme east the Valeryanov volcanoplutonic belt continued to exist at the beginning of the Middle Carboniferous. This volcanoplutonic belt arose in the Middle Viséan stage and is associated with subduction of the Uralian oceanic crust under the Kazakh microcontinent according to most researchers (in: *History of Development of Urals Palaeo-Ocean*, 1984), although there is also a view regarding its rift origin (Ivanov et al., 1986). A deepwater basin began to form in the Middle Carboniferous on the contemporary western slope at the border with the craton. Flysch accumulated on its Uralian slope and a barrier reef existed on its craton side. It began to be filled with massive salt-bearing strata in the Kungurian; these strata were replaced by coal-bearing ones north of the

Timan-Uralian junction. Much younger formations were already represented by continental molasse, which continued to accumulate on the northern and southern extremities of the foredeep even in the Early Triassic. Deformation of the outer zone of the Urals and foredeep, including formation of gravitational nappes (Zhivkovich and Chekhovich, 1985), occurred in the Early Permian and in the extreme north and extreme south already by the Middle Triassic.

In the Southern Tien Shan area a sharp increase in tectonic movements occurred from the middle of the Middle Carboniferous (Moscovian stage). A thick flysch-olistostrome series including major olistoplaques and gravitational nappes was formed, encompassing also the Upper Carboniferous. These nappes are made up of metamorphics, ophiolites, Silurian cherts and Silurian-Middle Carboniferous limestones. This entire material was displaced from north to south. In the Permian flysch was replaced by coarse clastic molasse; the sea receded south within the Pamirs. Plutons of granitoids were formed. This period of tectonic activity and orogeny had already ended in the Triassic. It is quite evident that it was associated with subduction from the south and later with collision of the Tarim microcontinent with Kazakhstan. Along the other side of the latter (south-east in contemporary co-ordinates and north-east in palaeoco-ordinates) in the Late Palaeozoic, commencing from the Late Visean (the so-called Saurian phase), the Junggar-Balkhash basin (marginal or interior sea) closed and a marginal volcanoplutonic belt formed simultaneously along the periphery of the Kazakhstan microcontinent. This belt was displaced towards the east and south compared to the similar Devonian belt due to accretion of the Kazakhstan microcontinent in the Late Ordovician-Middle Devonian.

In this continent secondary orogenic basins, Teniz and Dzhezkazgan, were formed in the Late Palaeozoic. These basins were initially filled with coal-bearing (Carboniferous) and later red coarse clastic (Permian) continental molasse. Very similar basins but of smaller dimension and with manifestation of volcanism developed in the south in the Northern Tien Shan part of the microcontinent.

Between the Kazakhstan microcontinent and Altay-Sayan part of the Siberian continent, the Ob'-Zaisan marine basin still existed in the Middle-Late Carboniferous and extended northward into the future West Siberian Platform. The oceanic crust of this basin experienced absorption under both the adjoining continental massifs with the formation of marginal volcanoplutonic belts. The basin had totally dried up by the beginning of the Permian as a result of collision between three continental blocks: Euramerica, Kazakhstan and Siberia. A single Laurasian supercontinent was formed and the Sino-Korea continent was attached to it even in the Permian period. At the end of the Permian to commencement of the Triassic, however, this continental mass was subjected to splitting as a result of formation of the 'Ob' palaeo-ocean' (Aplonov, 1987), an intercontinental rift of the Red Sea type opening up in a northerly direction towards the Kara Sea basin and closing in the south in Omsk region. This attempt of Laurasia at splitting is analogous to a similar attempt of Laurussia in the Carboniferous, expressed in the formation of the North Atlantic rift.

Elsewhere in Western Siberia independent basins with in accumulation of limnic coal-bearing formations existed in the Permian, the largest of them being the Kuznetsk basin.

The south-eastern continuation of the Ob'-Zaisan residual basin of the Palaeo-Asian ocean is represented by the Gobi-Hinggan basin opening into the Palaeo-Pacific through Girin region in southern Dunbei and southern Primorie. This basin continued to exist in the Early Permian, its northeastern branch extending north-east to the junction of the Palaeo-Pacific with the Amur-Okhotsk bay (*Atlas of the Palaeogeography of China*, 1985; Wang Quan and Liu Xueya, 1986). The collision of the Sino-Korea continent with the Siberian continent occurred at the end of the epoch, however; marine sedimentation was replaced by continental and differences vanished between Angara and Cathaysian flora. Felsic calc-alkaline volcanism and granite plutonism were manifest south of the suture between the continents in the Permian of northern China. Permian magmatism was developed even more extensively in Mongolia and western Transbaikalia. A volcanoplutonic belt formed north of the main Gobi-Hinggan basin of the Palaeo-Asian ocean in Mongolia (Fig. 5-12). Hangay-Hentey marginal belt (interior basin) continued to exist at the rear of the above volcanoplutonic belt in the Middle-Late Carboniferous; only its eastern Transbaikalia part survived in the Permian but even that was soon lost in folding. Another volcanoplutonic belt, Orkhon-Selenga (Fig. 5-13), extended in the north-west but it bears features of the riftogenic type (Gordienko, 1987).

Hercynian folding in Transbaikalia attenuated in an easterly direction and a continuous and conformable transition from the Palaeozoic to the Mesozoic in deep marine facies is noticed in the Amur-Okhotsk fold zone. This segment joined with the Sikhote-Alin' segment of the West Pacific belt and developed from the Late Permian as a bay of the latter. In the south the crust of the Amur-Okhotsk basin was absorbed under the Hinggan-Bureya microcontinent with manifestation of corresponding magmatism.

5.3.2.3 *West Pacific belt.* The area of the north-western part of this belt enlarged perceptibly in the Late Viséan as a result of intense subsidence of the Verkhoyansk-Kolyma system. From this time a thick prism of terrigenous sediments of the Verkhoyansk complex began to accumulate over the former extensive carbonate shelf of the Siberian continent. This continental margin passed to the north-west into a basin with oceanic crust and continued through the central part of Taimyr peninsula into the Kara Sea. In the

306

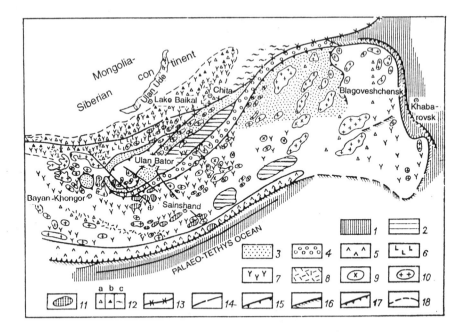

Fig. 5-12. Palaeotectonic-magmatic scheme of Sayan-Baikal-Mongolian region in the Late Palaeozoic (after Gordienko, 1987).

1—predominantly oceanic complexes including ophiolite complexes; 2—deepwater flysch silt-clay sediments including turbidites; 3—shallow-water terrigenous and terrigenous-carbonate sediments; 4—coarse clastic continental deposits, at places littoral marine; 5—littoral-continental volcanogenic strata of calc-alkaline composition with greywackes; 6—continental volcanic strata, predominantly of alkali-basalt composition; 7—continental volcanic strata of calc-alkaline composition with intermediate and felsic volcanics; 8—continental-volcanic beds with a bimodal composition; 9—calc-alkaline granites and plagiogranites; 10—calc-alkaline granites and granodiorites; 11—calc-alkaline gabbro, diorites and monzonites; 12—subalkaline or agpaitic (a), alkaline (b) and ultrametamorphic (c) granitoids; 13—zone of Mongolia-Okhotsk fault; 14—other faults; 15—boundaries of epicontinental shelf seas; 16—suggested boundary between continent and ocean; 17—fossil seismofocal Zavaritsky-Benioff zone; 18—boundary of Selenga-Vitim volcanoplutonic belt on active continental margin.

south-east (all in present-day co-ordinates) this basin and the Kolyma-Omolon microcontinent were separated from the open ocean (Palaeo-Pacific) by the Udino-Murgal volcanic arc. Upper Palaeozoic carbonate and siliceous formations with tethyan fauna and also ophiolites well known in the east among rocks of tectonic nappes of Koryak upland, according to present-day data represent exotic blocks shifted from the periequatorial region of the Palaeo-Pacific at the end of the Mesozoic to commencement of the Palaeogene. The same can be said of similar formations in eastern Sakhalin and central and eastern Hokkaido.

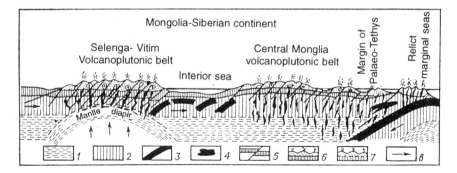

Fig. 5-13. Schematic reconstruction of Sayan-Baikal-Mongolian region in the period of commencement of formation of the Late Palaeozoic volcanoplutonic belts on the active margin of the Mongolia-Siberian continent (after Gordienko, 1987).

1—asthenosphere; 2—lithosphere; 3—oceanic plates of Palaeo-Tethys (Palaeo-Asian ocean; V.E. Kh. and K.B.S.); 4—blocked oceanic plates of marginal and interior seas; 5—blocked continental plates; 6, 7—fragments of volcanoplutonic belts formed (6—on continental crust, 7—on transitional crust); 8—direction of stresses in plates and fluid movements in mantle diapirs.

The Late Palaeozoic history of Sikhote-Alin' as a result of reassessment of the age of Upper Palaeozoic rocks (see Section 5.2.4.3) is not yet clear. However, available data nevertheless suggests the manifestation here of Late Hercynian deformations and orogenic magmatism. This is wholly evident for southern Primorie and the adjoining Tumangan zone on the Korean peninsula. The fact that the adjoining Hinggan-Bureya and Khanka massifs were affected by magmatic processes and that molasse troughs arose along them in the Late Permian suggest the probable relation of all these processes with the activity of a seismofocal zone gently inclined from the eastern Sikhote-Alin' synclinorium towards the west under the continental framework.

More definite data on the Late Hercynian tectogeny is available for the Japanese islands forming at this time a direct continuation of Sikhote-Alin'. The first significant deformation and uplift occurred in the middle of the Viséan epoch in the north-eastern part of Honshu Island and northern zones of south-western Japan (Abe phase); they repeated before the Namurian and Moscovian epochs, at the verge between the Carboniferous and Permian, in the middle of the Early Permian and at the verge between the Permian and Triassic, leading to considerable shoaling of the sea and manifestation of coarse clastic sediments, i.e., evidently to accretion of continental crust (Fig. 5-14). This tectonic activity was probably associated with the seismofocal zone dipping to the east. At the same time, in the outermost zones of south-western Japan, right up to the middle of the Permian, marginal sea conditions (Yakuno ophiolites C_3-P_1) prevailed. This sea was set off from the ocean by the Honshu microcontinent whose basement was made up of ancient granites and cover of Permian clastic sediments of the Maizuru zone (Faure and Charvet, 1987).

Collision of the Honshu microcontinent with the margin of the Sino-Korean mainland represented by the Hida zone commenced in the Late Permian. As a result of this the Hida zone was overthrust on formations of the marginal sea by or at commencement of the Triassic. The marginal marine formations were partially subjected to high-pressure metamorphism that formed the Sangun zone and partly obducted on the Honshu microcontinent with the formation of olistostromes in the cover of the latter.

The southern continuation of this Nippon-Sakhalin geosyncline encompassed the narrow strip south of the Korean peninsula (?), south-eastern

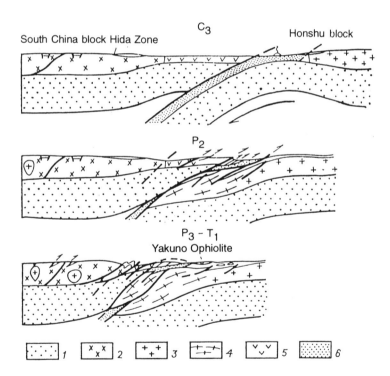

Fig. 5-14. Geodynamic evolution of the Late Palaeozoic (Early Permian) orogeny in Japan (after Faure and Charvet, 1987).

1—upper mantle; 2 to 4—continental crust (2—of South China block, 3—Honshu block, 4—transiting into plastic state); 5 and 6—oceanic crust of (5—South China plate, 6—Honshu microplate).

coast of continental China, Taiwan Strait and Taiwan Island and also the Ryu-Kyu archipelago, being separated from the epicontinental sea of the South China Craton by the Zhe-Ming marginal uplift (*Atlas of the Palaeogeography of China*, 1985). Farther south, already within the present-day shelf of the South China Sea, this geosynclinal system was evidently joined with the south-eastern continuation of the Viet-Lao branch of the Palaeo- and Meso-Tethys. In the latter only the narrow Shongda zone experienced active development in the Late Palaeozoic. Gatinsky (1986) regarded the Shongda zone as an intercontinental rift, a product of partial regeneration of the mobility of this system consolidated at the end of the Devonian. More southward, the main branch of the Palaeo- (Meso-)Tethys opened into the Palaeo-Pacific ocean, having traversed through Thailand, Cambodia and Malaysia.

The history of this branch reveals some similarity with the history of the Viet-Lao branch of the Palaeo- and Meso-Tethys. The point is that it, too, initially underwent Hercynian orogeny in the Middle-Late Carboniferous, followed by some stabilisation of tectonic regime in the Permian and riftogenic regeneration of mobility at commencement of the Triassic (Helmcke, 1985). The final cratonisation of both branches, however, occurred already in the Late Triassic to commencement of the Jurassic in the Indosinian (Early Cimmerian) epoch of orogeny.

Farther south, the New England geosynclinal system of Eastern Australia, encompassing also the central part of New Guinea, New Caledonia and New Zealand, belongs to the West Pacific belt. During the Late Palaeozoic there was a gradual build-up of uplift here and its spread towards the east. As a result, in the Late Permian a marine regime was preserved only in the far eastern salient of the Australian continent, in New Guinea, while deepwater conditions prevailed in New Caledonia and New Zealand. The eastern part of the New England system was involved in orogeny while the Sydney-Bowen foredeep was formed in front of it from the beginning of the Carboniferous at the boundary with the Lachlan system that was cratonised in the Devonian. The marginal eastern part of the Lachlan system adjoining the trough experienced uplift and served as a zone of subaerial volcanism (Yarrol belt). This volcanoplutonic belt extends through Cape York peninsula into New Guinea and for 1000 km farther westward deep into Indonesia while its south-eastern prolongation is recognised in the south-west of New Caledonia and west of New Zealand. The disposition of the more north-eastern and eastern deepwater zone on oceanic crust is confirmed by development of ophiolites on the South Island of New Zealand obducted in the Mesozoic on the margin of the eastern continent. The further continuation of this active margin of Gondwana is found in western Antarctica where traces of volcanic activity of the island-arc type are known at least in the Permian.

5.3.2.4. *East Pacific belt.* The northern part of the belt pertaining to the North American Cordillera entered the paroxysmal phase of the Antler epoch of orogeny in the Middle-Late Carboniferous with the formation of a belt of fold-overthrust dislocations and overthrust of formations of continental slope and rise on the shelf of the margin of the North American continent. At the verge between the Middle and Late Carboniferous, the Antler orogen was built up by new thrusts from the west (Fig. 5–15). This new compressive impulse was called the Humboldt orogeny. In the Early Permian uplift and deformation weakened and there was some diminution in contrast of movements, which was replaced by a new growth of tectonic activity in the Late Permian to commencement of the Triassic in the Sonoman epoch of orogeny. As a result, the belt of Antler dislocations grew from the west with the addition of a belt of Sonoman deformations; the Golconda overthrust served as its front. Oceanic and island arc formations of the western 'eugeosynclinal' zone of the Cordillera were obducted along the Golconda overthrust on the continental margin represented by the Antler orogen. The westernmost zone in the Late Palaeozoic, before the Sonoman events, continued to develop in general according to the same plan as in the Middle Palaeozoic. Here the existence of a volcanic arc which included in the south the eastern part of the Klamath Mountains and the northern part of Sierra Nevada is observed. Further westward, there is an accretionary complex with blocks of Permian limestones containing tethyan fauna of fusulinids and corals which differ from the island arc fauna. Even more westward and southward (?), at some unknown distance from the volcanic arc, there should have been another island arc (or arcs? or intraoceanic uplifts?), at present represented by south-eastern Alaskan terranes, the western coast of British Columbia and the Alexander and Vancouver archipelagos opposite them. Here, too, volcanic activity continued, active or quiescent at different places at different times.

Interesting events occurred in the southern part of the system in the Early Triassic. It has been assumed that a major zone of transcurrent faults with a north-west strike was formed in this period and displaced the southern continuation of the Antler and Sonoman orogens. This zone, known as the Mohave-Sonora megashear fault, traverses the boundary of Nevada and California and extends south-east into Mexico.

Like the North American Cordillera, the South American Cordillera (Andes) developed very actively in the Late Palaeozoic. The first major epoch of Hercynian deformation occurred at the end of the Devonian; in the Central and Southern Andes it generated an axial uplift in the intracratonic geosynclines of these segments (see Section 4.3.2.5) and led to the regression of sea at the beginning of the Early Carboniferous. In the Middle and Late Carboniferous subsidence again intensified as a result of sinking along faults. This has been observed even in the Northern

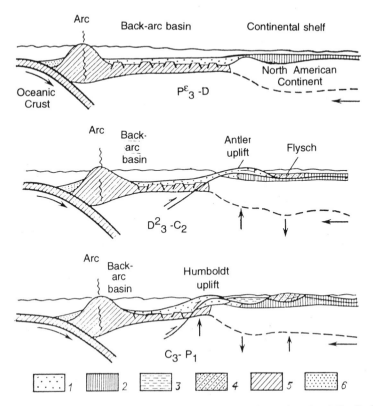

Fig. 5-15. Tectonic model of the Humboldt orogeny with back-arc thrusting (after B. Skipp. In: Frasier and Schwimmer, 1987).

1—deposits of back-arc basin; 2—shelf carbonates; 3—sandstones; 4—flysch; 5—oceanic crust; 6—deposits of forearc basin.

Andes where only the Maracaibo massif escaped subsidence. The Early Permian was everywhere characterised by a reduction of tectonic activity, as supported by the extensive development of neritic limestones from Venezuela to Patagonia throughout the Andes. On the whole, the Late Palaeozoic development was distinguished from the Early Palaeozoic by high tectonic differentiation of the Andean system and the much smaller role of subsidence. In the second half of the Early Permian, this development was interrupted by a new phase of compressive deformation. This deformation was more extensively spread than the Early Hercynian deformation (at the end of the Devonian) but was less intense and accompanied neither by metamorphism nor even flow cleavage while granite plutonism affected only the Southern Andes and the Deseado massif in Patagonia. Nevertheless it did cause a general uplift, which was followed by a mighty outburst of orogenic magmatism in the Late Permian and Early Triassic. In the Central and Southern Andes, subaerial volcanics, i.e., basalts, andesites, rhyolites and ignimbrites, are interstratified with red, mainly continental, molasse filling the extensive Peru-Bolivian intermontane trough. The characteristic features of distribution and composition of molasse point to block movements of the substratum in the period of their accumulation. The volcanics are of calc-alkaline and alkaline types. The major plutons of the Late Permian granitoids, among which normal granites predominate, bear similar characteristics. Compared to the Early Hercynian granitoids, the Late Hercynian ones are distinguished by higher K_2O content.

The origin of the Permo-Triassic magmatites of the Central and Southern Andes has been variously interpreted (Dalmayrac et al., 1980a and b). Some researchers associate it with subduction of the Pacific Ocean crust. This view is confirmed by the development of glaucophane schists of the same age on the coast of Chile between Concepçion and Valdivia. Others regard this magmatism as riftogenic, emphasising its essentially bimodal character, rarity of andesites, high alkalinity and remoteness to possible subduction zones. A compromise point of view has also been expressed: rifting at the rear of an island arc. This appears to be the most probable explanation.

In the Northern Andes the Late Permian epoch was also characterised by intense folding, granitoid plutonism and regional metamorphism within uplifts and accumulation of red molasse in intermontane basins. But volcanism is not manifest here and, unlike in the Central and Southern Andes, tectonomagmatic activity in this segment of the Andes distinctly attenuated in the Early Triassic.

6

General Regularities of Development of the Earth's Crust in the Palaeozoic

The Palaeozoic era represents a major cycle, a megacycle, of development of the Earth's crust, which, however, also includes a certain time period of the Late Proterozoic, commencing roughly from 1000–850 m.y. ago. In the present volume this cycle covers the duration from the existence of one Pangaea, Pangaea I, to the formation of another, Wegener's Pangaea II; the intervening period corresponds to the disintegration of the first Pangaea and the amalgamation of its fragments accreted by a new continental crust into a new Pangaea.

This megacycle consists of three cycles of a lower rank, long known as the Baikalian, Caledonian and Hercynian cycles. As will be demonstrated at the end of the chapter, these cycles are not identical in importance: the Baikalian cycle represents the preparatory, the Caledonian the main and the Hercynian the concluding cycle. The Baikalian cycle belongs predominantly to the Proterozoic, especially if the Vendian period is also placed in the latter aeon. This subject has already been discussed in volume I, *Historical Geotectonics. Precambrian* (Khain and Bozhko, 1988). In this chapter we revert time and again to the events of this cycle since the sources of processes which developed not only in the mobile belts, but also on cratons, pertain to the Late Riphean and Vendian. However, scant reliable information for mobile belts has led to different interpretations of the nature of tectonic development at the end of the Proterozoic, a circumstance which is reflected in the literature.

6.1 DEVELOPMENT OF CONTINENTS (CONTINENTAL CRATONS) AND THEIR MARGINS (MIOGEOSYNCLINES)

On the geochronological scale adopted by us, the existence of supercontinent Pangaea I may be placed towards the end of the Early Proterozoic, Early and (with a question mark) Middle Riphean, i.e., towards the Middle Proterozoic, in the terminology of most western researchers. Even in the first half of the Late Riphean, roughly between 1000 and 850 m.y., if not somewhat earlier, the entity of this supercontinent was disturbed as a result of the emergence of several mobile belts with a newly formed (oceanic) crust. Further, the mobile belts which separated the continents of the future Laurasian group from each other and this group from Gondwana developed in the Early Palaeozoic into true oceans—Iapetus, Ural-Okhotsk and Mediterranean—with a protracted subsequent history. At the same time, most mobile belts, rather systems, arising within the future Gondwana were characterised by limited width (less than 500 km according to palaeomagnetic data) and a relatively brief period of development falling within the frame of the second half of the Riphean and the Vendian. On a Phanerozoic scale this was hardly a small period and wholly comparable with the duration of the Caledonian and Hercynian cycles.

Accordingly, the development of these two groups of continents proceeded from the very beginning of the Palaeozoic, maybe even earlier, by significantly different ways, albeit there were no such differences in the Early Precambrian and Early to Middle Riphean. Let us first consider the Gondwana group.

In the middle and second half of the Late Riphean (Karatavian, according to M.E. Raaben, 1975), a comparatively large number of relatively small continents existed in the area of future Gondwana. These were separated by rather narrow linear deepwater basins, i.e., geosynclinal systems, with oceanic or transitional crust. In the first volume of this series, *Historical Geotectonics*, all of them were called intracratonic geosynclines and it was emphasised that they evolved on continental crust. In the axial zones of most of these mobile systems, however, ocean-type crust was undoubtedly newly formed, the presence of ophiolites providing positive support to this feature. This is true of Ribeira-Damara, presently separated by the South Atlantic, Mauritanides, Western Congolides, Libya-Nigerian and Mozambique belts. It is interesting that the break-up which led to the disintegration of Gondwana in the Mesozoic in most cases occurred along these mobile systems.

At the end of the Riphean to the Early Vendian (a variation of this age frame is possible), collision of the Gondwana micro (meso?) continents[1] occurred while the intermediate mobile systems experienced intense compression, fold-overthrust deformation, granitisation and metamorphism and were transformed into folded mountain structures. At many places this process also involved the remobilised ancient sialic substratum which was preserved in such systems, as exemplified by the Mozambique and Libya-Nigerian belts. The newly formed fold systems joined together the ancient continental cores and led to formation of the Gondwana supercontinent. These systems themselves, however, preserved high mobility for a very long period: molasse accumulated in their intermontane and marginal troughs in

---

[1] In western literature these are often called cratons (for example, West African, San Francisco etc.) and in Russian literature protoplatforms.

the Late Vendian and Early (and Middle?) Cambrian while late orogenic, post-collisional granitoids, often highly alkaline, intruded into uplifts. Peneplanation of mountain structures occurred only in the Late Cambrian to the Ordovician when a genuine platformal (orthoplatformal) regime was established everywhere. Further, such a reduction of tectonic activity manifested first in the periphery of Gondwana and only later affected its more interior regions. This may be seen from the distribution of marine sediments of platform cover, earliest (in the Cambrian) in the northern (in present-day coordinates) margins of Africa and India, in Arabia and Australia. In the interior regions of Gondwana, a platform cover began to form only at the very end of the Ordovician and in the Silurian (South America, Africa and Antarctica).

The area of Gondwana had enlarged already by commencement of the Palaeozoic due to inclusion of the Mediterranean belt, Proto-Tethys, whose active development ceased during the Baikalian cycle. The width of this Peri-Gondwana Epibaikalian Platform was particularly large in the region of the Near and Middle East (Turkey, Iran and Afghanistan) and encompassed north-western and northern parts of India, including the future Himalaya. The Sinoburman and possibly South China (Yangtze) Platform could be regarded as fragments of the above Epibaikalian Platform. The platform cover commences everywhere here with the Upper Vendian-Cambrian (upper portion of the Upper Riphean in the Yangtze).

Not only the eastern part was affected by the process of separation of microcontinental blocks from the northern periphery of Gondwana in the Early Palaeozoic. In the west continental rifting and subsequent spreading of the Palaeo-Tethys led to separation of the Armorican, English Midland and probably also the Bohemian and Rhodopian massifs.

The mature stage of development of Gondwana corresponds to the Middle Palaeozoic. There was maximum spread of marine conditions and platform cover at this time (Silurian-Early Carboniferous), indicating to minimal height above sea level which, in turn, suggested maximum cooling of the tectosphere. There were signs of elevated endogenic activity, i.e., manifestations of felsic and alkaline magmatism only in some interior regions, mainly in southern and eastern Sahara and in the interval Silurian to Early Devonian (echo of an earlier magmatic activity?). Formation of aulacogens took place also in the Devonian: Saoura-Ougarta in Africa, Fitzroy and, even earlier, Carnarvon and Perth in Australia. Much of the periphery of Gondwana adjoining the Mediterranean belt, Palaeo-Tethys, from Venezuela to India, long remained a passive margin, unlike the periphery of the Pacific Ocean (South America and Australia). At the end of the Devonian, however, the position began changing and the Mauritanides and Atlasides (Maghrebides) arose and experienced overthrust on the north-western African margin of Gondwana.

Gondwana entered its last stage of existence at the end of the Early Carboniferous. On the one hand, this was the stage of merging of Gondwana with Laurasia to form a single supercontinent, Pangaea II. This occurred as a result of collision of the north-western part of Gondwana (South America and Africa) with the western part of Laurasia (North America and Iberia). On the other hand, this stage immediately preceded the disintegration of Gondwana and covered processes and events preparatory to this disintegration. These events were manifest in the formation of rift systems, most extensively represented in the southern half of Africa, India and partly in Australia and Antarctica. Development of these rifts occurred in the background of a general uplift of the surface of these continents, affected additionally by continental glaciation commencing in the Namurian (in Australia; Veevers, 1986). Under the influence of tangential compression by the Mauritanides and Atlasides, the northern half of Africa falling outside the rift zone experienced deformation of a larger radius, leading to formation of present-day anteclises, arches, syneclises and basins. The major portion of these deformations occurred in the Permian (Fabre, 1988) and was imprinted on the pre-Mesozoic structure (Fig. 6-1).

Let us now turn to the Laurasian group of continents. Their separation took place as early as the Riphean but at the end of the Riphean to Early Vendian, their rapid amalgamation into a new single supercontinent possibly occurred as a result of Baikalian orogeny only if the central parts of some geosynclinal (palaeo-oceanic) basins did not continue to develop along lines inherited in the Early Palaeozoic, as assumed, for example, for the Tethys (transition from Proto- to Palaeo-Tethys; Belov, 1981). At the same time, the interior regions of all Laurasian continental platforms experienced the 'aulacogen' rifting stage of development in the Riphean and Early Vendian with the formation of numerous rift grabens and manifestation of trappean magmatism. The internal structure of the crystalline basement had formed almost completely already by the Early Riphean. The Grenville belt of North America and its continuation in south-western Scandinavia in Europe constitutes an exception but even this belt concluded its active development 1000–900 m.y. ago, i.e. very much earlier than the mobile systems constituting Gondwana. Transition to the platform stage was also completed in the Laurasian group of platforms much earlier than in the Gondwana group: in East Europe, Siberia, China and the Korean peninsula already by the Middle Vendian and in North America by the Late Cambrian. Correspondingly, the first maximum flooding of Laurasian platforms set in earlier, i.e., already in the Ordovician. In the Late Vendian, Riphean rifts were transformed into troughs and in the Early Palaeozoic into syneclises. Cooling and subsidence of platforms in the Early Devonian were interrupted by tectono-magmatic reactivation widely manifest in East Europe and Siberia. It was marked by regeneration of parts of the Riphean and formation of new rifts

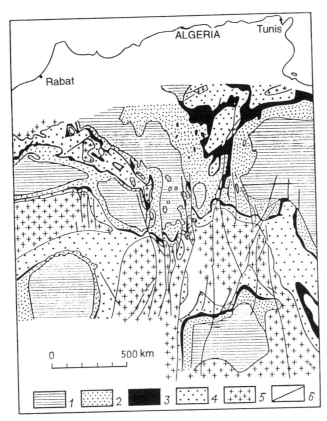

Fig. 6-1. Pre-Mesozoic palaeogeological scheme of central and western Sahara (after Fabre, 1988).

In the north-eastern and northern Tuareg Shield, Triassic overlaid Palaeozoic, in particular Permian deformations, which later became intensified in the Austrian phase. In the north-western part of the shield and in western Sahara, the first formations transgressing onto the Palaeozoic surface are usually of Cretaceous age and hence formations underlying them experienced Palaeozoic and Austrian deformations.

1—Carboniferous and Permian; 2—Devonian; 3—Silurian; 4—Cambrian and Ordovician; 5—Precambrian; 6—major faults.

and a new outburst of trappean magmatism. By this time, significant portions of the margins of the Laurasian platforms had transformed from the passive to the collision type and North America and East Europe (Baltica) merged into a single large continent, i.e., Euramerica or Laurussia. The Greenland Caledonides and Northern Appalachians were overthrust on the north-eastern margin of North America (Laurentia) and Scandinavian Caledonides on the north-western margin of East Europe (Baltica). The interior zones of the Uralian system experienced the first phase of fold-overthrust deformation. Even earlier, at the end of the Vendian to commencement of the Cambrian, the Baikalides of Rybachyi peninsula-Kanin-Timan were overthrust on the north-eastern margin of the same craton. It was at this same time that the Baikalian frame of northern, western and southern peripheries of the Siberian continent arose.

The main stage of subsidence of the East European and partly Siberian and North American platforms and simultaneously the stage of formation of their most important structures, i.e., syneclises and anteclises, set in following Devonian reactivation (Early Palaeozoic stage is no less important in Siberia and North America). Syneclises (basins) arose initially above aulacogens conforming to Shatsky's law, i.e., under the influence of deep internal processes occurring within the platform itself. At the end of the Palaeozoic, however, these syneclises experienced transformation under the influence of forces from the adjoining fold-overthrust structures. These forces from the Appalachians amalgamated Michigan and Illinois syneclises and also the Appalachian foredeep into a single large region of subsidence, i.e., the Interior basin of the Midcontinent, USA. A similar phenomenon occurred in the eastern part of the East European Platform where an extensive basin including the Moscow syneclise and older Volga-Uralian anteclise was formed under the influence of overthrusts by the Uralian edifice.

Since all the northern platforms are surrounded practically from all sides by fold-nappe structures overthrusting on them and with an amplitude running into hundreds of kilometres at places, these platforms appear as though divided into zones of influence of adjoining structures. This would explain the characteristic feature of their structure whereby at least three of these platforms possess an axial zone of uplifts forming their backbone. These axes are Boothia-Nemaha on the North American Platform, Baltic-Polessian axis on the East European Platform and Anabar-Irkutsk on the Siberian Platform. Unlike these platforms, such an axis on the Sino-Korea Platform has a latitudinal strike and extends along the northern margin of this platform, i.e., the axis of Inner Mongolia. As the fold-overthrust structures of the platform frame vary in age, the 'zones of influence' were formed in different periods, the structure of the platform periodically experiencing reorganisation; the contemporary structure thus represents a somewhat integrated one, reflecting the ultimate result of these reorganisations. The Transcontinental arch with a north-east strike played the leading role in the structure of the North American Platform right up to the Middle Carboniferous. This arch connects the south-western saliant of the Canadian Shield with the eastern Rockies and serves as a divide of the zone of influence of Cordilleran and Appalachian geosynclines, more correctly of Palaeo-Pacific and Palaeo-Atlantic oceans. It was only later, at the end of the Palaeozoic and Mesozoic, that the meridional axis of Boothia-Nemaha came to the fore. On the Russian Platform the Moscow syneclise initially (before Devonian) opened in the west into the Baltic syneclise and thus the Baltic-Polessian axis still played no significant role. On the Siberian Platform, practically throughout the Palaeozoic, the Central Siberian (Nepa-Botuoba) arch played a major role and the Anabar-Irkutsk axis acquired considerable significance only in the Mesozoic with development of the Verkhoyansk-Kolyma fold-overthrust system and West Siberian megabasin.

On the whole, the history of the platforms confirms the rule (law) of Karpinsky according to which, in terms of contemporary interpretation, that part of a platform tending directly towards the most actively developing mobile belt of the frame in a given stage experiences maximum subsidence (the above-mentioned zones of influence) at each stage of its development.

After the merger of Laurasia with Gondwana into the supercontinent Pangaea II in the Late Palaeozoic, the Laurasian continents were affected by the same process of destruction and rifting as the Gondwana continents. This feature was particularly well manifest in North America where the large Wichita-Ancient Rockies rift system was formed. The Oslo graben in Europe and Siberian Kutungdy aulacogen and Viluy aulacogen that continued to develop serve as other examples of such structures.

6.2. DEVELOPMENT OF OCEANS AND MOBILE (GEOSYNCLINAL) BELTS

Division of the Earth's crust into two main segments, continental and oceanic, i.e., Pangaea and Panthalassa, was distinctly marked at the beginning of the Late Proterozoic. The future Pacific Ocean, Palaeo-Pacific, represents the nucleus of the oceanic segment. The question of age and origin of the Pacific Ocean, already dealt with in the preceding volume (Khain and Bozhko, 1988), remains an open issue as before and a fairly old age can be assigned to the Pacific Ocean basin only from indirect data. The first relatively direct proof of its existence in the Riphean is provided by Lower (?) Riphean ophiolites of south-eastern China and Upper Riphean ophiolites of Tasmania. Data on the formation of a riftogenic passive margin of the Proto-Pacific in the Cordillera (Devlin and Bond, 1988) and Australia (Lindsay et al., 1987) is even more definitive. Two phases of marginal-continental rifting have been established in both these regions: the first in the Late Riphean at 900 (Australia) to 770 (Cordillera) m.y. and the second at the end of the Vendian to commencement of the Cambrian, about 600 (Australia) to 575 ± 25 (Cordillera) m.y.

The similarity of sequences of the Upper Proterozoic and Lower Cambrian of western North America on the one hand, and eastern Australia and Antarctica on the other, led Moores (1991) and Dalziel (1991) to the conclusion that they may have formed opposite flanks of the same continental rift.

Consequently, these continents were initially parts of a single supercontinent and only a further opening of this rift led to formation of the prototype of the contemporary Pacific Ocean. The hypothesis of Moores-Dalziel was at once developed by P. Hoffman (1991), who proposed palinspastic reconstructions. According to him, Eastern Gondwana initially adjoined North and South America from the west in an inverted position; then turning anticlockwise, it assumed the present position only in the Cambrian, relative to Western Gondwana having collided with it to form the Mozambique belt.

In our opinion, one has to approach this bold hypothesis with great care. The similarity of sequences per se is not sufficient proof of the recent joining of continents; much after the opening of the Atlantic, its margins developed quite conformably. Boucot (1992), a leading specialist in the biogeography of the Early Palaeozoic, in his critique of the Moores-Dalziel views, pointed out that the deposits under comparison were formed under different climatic conditions and contain extremely distinct fauna.

The existence of the Pacific Ocean and probably its mobile frame has already been established very reliably from the Late Riphean to the Vendian. Cambrian ophiolites of eastern Australia and New Zealand, Ordovician ophiolites of Koryakia, California (Klamath) and Southern Andes and Silurian ophiolites of Japan (northern Honshu Island) provide us with further confirmation of this event for the Early Palaeozoic. The following feature, first reported by Bozhko (1984), is characteristic of the early stages of development of the Peri-Pacific mobile systems: their development was accompanied by the separation of some blocks in the form of microcontinents from bordering continents. As a result, corresponding fold systems formed in the process of collision of these microcontinents with bordering continents. Such a picture was recently deciphered for the Japanese islands (Faure and Charvet, 1987), Tasman belt of Australia (Scheibner, 1987) and central and southern Andes (Dalla Salda, 1982). The Kolyma-Omolon, Sea of Okhotsk and central Kamchatka massifs, as well as the massifs of Honshu, Kurosegawa and Kurosio in Japan and Salinia in California, represent such microcontinents colliding with continents even in the Mesozoic or Cenozoic. A fairly logical assumption is that these microcontinents represent fragments of not the continents presently fringing the Pacific Ocean, but of some hypothetical continent—Pacifica—covering the central part of the present-day Pacific Ocean (Nur and Ben Avraham, 1982). Without totally rejecting such a possibility, we nevertheless regard as highly probable that the microcontinents primarily belonged to margins of continents existing at present on the basis of two premises. First, the process of separation of microcontinents is reflected in some well-studied more recent examples—separation of the Japanese islands during formation of the Sea of Japan, or Lord Howe and Norfolk microcontinents during formation of the Tasman and Coral seas and New Caledonian basin, or the events in southern Patagonia in the

Late Jurassic with formation of the 'Rocas Verdes' basin. Secondly, some microcontinents are set off from the 'mainland' continent by a basin with not oceanic but a transitional type crust, for example the Arequipa massif in the Peruvian Andes.

Volcanic arcs, ensialic (building up on microcontinents) and ensimatic (formed in marginal seas or in the peripheral zones of the open sea) besides microcontinents, played a major role in the Palaeozoic development of the Peri-Pacific mobile systems. Repeated collision of these arcs with the continental (or microcontinental) margin occurred time and again, leading ultimately to the build-up of continental crust. This process is most distinct in the Tasman belt of Australia where, by the end of the Palaeozoic, it led to progradation of the eastern margin of the continent to the present-day position, excluding only the ensimatic Gympie zone in the south-eastern part of Queensland. This accretion of continental crust occurred in two main stages, tentatively parallel to the Caledonian and Hercynian cycles, and consists of a series of independent impulses, i.e., phases of orogeny, known in Australia under local names (Delamerian, Bowning etc.). Significant accretion of continental crust occurred in Cathaysia (south-eastern China) where it belonged to the Caledonian cycle and in Antarctica where it concluded somewhat later, already in the Mesozoic. Accretion of continental crust was noticed on a smaller scale in the Cordillera as a result initially of Antler (C/P) and later Sonoman (P/T) orogenies. The complex Palaeozoic history of the Andes (Figs. 6–2 and 6–3) leads more to the regeneration and some build-up of much older continental crust than to its actually new formation. A very similar situation prevailed in Japan. Some accretion at the end of the Palaeozoic occurred in southern Primorie and north-eastern Korea.

Destruction of Pangaea I at the end of the Proterozoic to commencement of the Palaeozoic was not restricted to its Pacific periphery, but also affected its interior regions, leading ultimately to the formation of new intercontinental oceans and associated mobile belts—Iapetus (North Atlantic belt), Palaeo-Asian (Ural-Okhotsk belt) and Proto- and Palaeo-Tethys (Mediterranean belt).

Development of all these oceans and mobile belts commenced in the Late Riphean while the much older Riphean history remains either factually not known (Mediterranean belt) or points mainly to continental rifting (Uralian portion of the Ural-Okhotsk belt). The history of the North Atlantic belt differs somewhat in this respect. To some extent, its predecessor is the Grenville-Dalslandian belt on which the North Atlantic belt was superimposed, maintaining its general strike. The Grenville-Dalslandian belt itself developed during the Early and Middle Riphean.

In so far as the North Atlantic belt itself or the Iapetus ocean is concerned, data for Newfoundland and the Appalachians given in

322

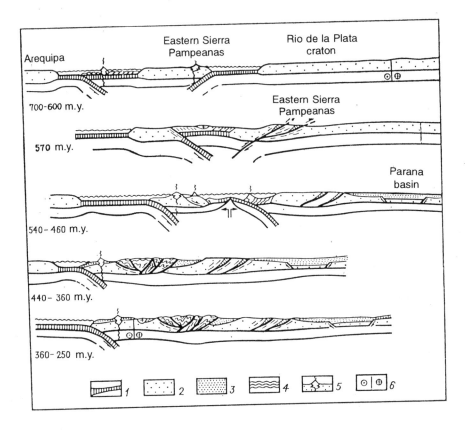

Fig. 6-2. Schematic profiles illustrating evolution of the Southern Andes during the Late Proterozoic and Palaeozoic on traverse 23 to 25° S lat. (after Ramos, 1988).

1—oceanic crust; 2—continental crust; 3—sediments; 4—deformed sediments; 5—magmatic arcs; 6—boundaries of terranes.

Section 1.3.2.1 shows that it originated as a continental rift in the Late Riphean to Vendian and that transition from rifting to spreading coincided almost precisely with the verge between the Vendian and Cambrian. This data pertains to the north-western margin of the belt, however. Relations on the opposite margin, primarily on the Avalon peninsula of Newfoundland and Anglesey Island opposite the Welsh coast, indicate that formations of the Early Palaeozoic Iapetus here were laid on much older but also geosynclinal formations of Late Riphean-Vendian age. Their similarity is suggested by the character of sedimentary (flysch and olistostromes) and volcanic formations (tholeiite-basalts), their dislocation, metamorphism and presence of granitoid plutons (O'Brien et al., 1983); among the latter, in particular, is the Holyrood

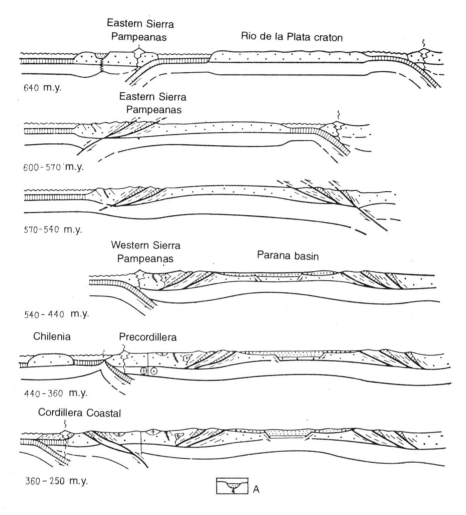

Fig. 6-3. Schematic profiles illustrating evolution of the Southern Andes during the Late Proterozoic and Palaeozoic on traverse 32° S lat. (after Ramos, 1988).

See Fig. 6-2 for legend. A—back-arc basin.

granite pluton intruding into the Avalon peninsula in the verge between the Vendian and Cambrian.

This data justifies the assumption that the Early Palaeozoic Iapetus should have been preceded by the Late Proterozoic Proto-Iapetus; Cogné and Wright (1980) proposed the name Celtic palaeo-ocean for its European part; it was bound from the south (south-east) by the Midland-Armorican salient of Gondwana, a future microcontinent. Proto-Iapetus was probably not so extensive as Iapetus and its development less active. Iapetus was displaced relative to Proto-Iapetus in a north-western direction. Evidently closure of the Proto-Iapetus should have occurred simultaneous with opening of the Iapetus; possibly, a block of continental crust, a microcontinent, lay between them. The existence of this microcontinent is assumed under the Midland Valley graben of Scotland based on the finds of granulitic xenoliths of Grenvillian (?) age in the Lower Carboniferous basaltic dykes.

The earliest stage of development of Iapetus proper was the stage of opening of the ocean basin. Its commencement is placed at the beginning of the Cambrian and, as pointed out above, documented by sequences of western and partly of eastern continental margins in North America, while its end coincided with the first phase of compression in the Arenigian-Llanvirnian, with relatively restricted manifestation in the Appalachians where it was known as the Oliverian or Penobscotian and was much stronger in the Grampian uplands of Scotland (Grampian phase) and in northern Norway (Finnmarken phase). The formation of oceanic crust represented by ophiolites outcropping in the strip from the Southern Appalachians to northern Norway but most fully in Newfoundland corresponds to this period in the history of the Iapetus (570–480 m.y. ago). However, radiometric age determinations do not provide values older than the Late Cambrian (505 m.y.) and correspond mainly to the Early Ordovician. In any case, it was at commencement of the Ordovician that Iapetus could attain its maximum width of 1000 to 2000 and possibly 3000 km (according to some palaeomagnetic data).

Oliverian-Grampian-Finnmarken deformations affected the north-western periphery of the Iapetus. The factor causing these is not wholly clear: it may have been a collision with the North American margin or of a volcanic arc arising on oceanic crust in the Late Cambrian or of a microcontinent concealed, in particular, under the Midland Valley.

The island arc stage of Iapetus development pertains to the Ordovician period. Several volcanic arcs arose, occupying various positions in different segments of the ocean basin and developing on different substrata—simatic or sialic. Volcanic arcs of Northern Wales and the Lake district of northern England (Fig. 6-4) are ensialic in Britain while hypothetical arcs of Cockburnland and Midland Valley are ensimatic. In Newfoundland (Fig. 6-5), the Appalachians (Northern but not Southern) and Scandinavia (Fig. 6-6) the existence of only an ensimatic arc (arcs) has been assumed.

At the end of the Middle and in the Late Ordovician, the second phase (epoch) of compressive deformation manifested in the Iapetus. This time the region of maximum manifestation was the Appalachians, especially its northern part. It is here that this epoch got its name Taconian. It was accompanied by a charrette of deepwater deposits of continental slope and rise and also oceanic crust of the Iapetus on formations of the outer shelf of

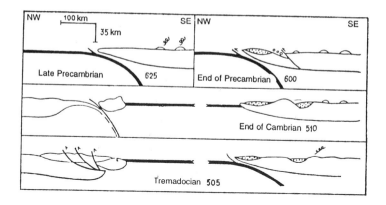

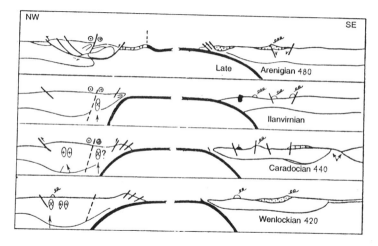

Fig. 6–4. Schematic palaeoprofile of British Caledonides (after A.H.G. Mitchell, 1984).

the North American margin. Regional metamorphism was accompanied by deformation. Thrusting of the thinned continental margin and oceanic crust of the Iapetus under the island arc in which intrusive and effusive magmatism manifested under the influence of this subduction is now regarded as the cause of this deformation process.

Taconian diastrophism is suggested at present for Newfoundland and the British isles. Moreover, it has been assumed (Hutton, 1987; Pluijin, 1987) that the main basin of the Iapetus was closed even at the end of the Ordovician while the deep basin existing in the Silurian behind the volcanic arc colliding with the North American continent represented a newly formed marginal sea between this arc and foreposts of Gondwana, i.e., Avalonia,

326

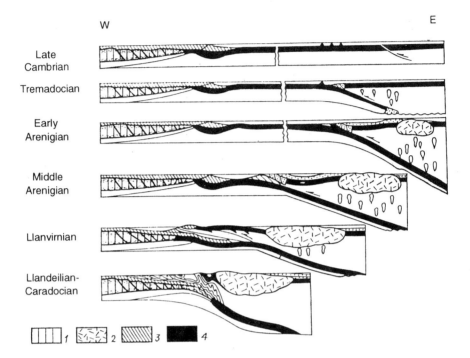

W   E

Late Cambrian

Tremadocian

Early Arenigian

Middle Arenigian

Llanvirnian

Llandeilian-Caradocian

Fig. 6–5. Schematic plate-tectonic reconstruction of evolution of the eastern margin of North America in western Newfoundland from the Late Cambrian to the Middle Ordovician (after J.F. Casey et al. In: Frasier and Schwimmer, 1987). Vertical magnification minimal.

1—continental crust of North America; 2—same, microcontinent; 3—post-rift sediments; 4—oceanic crust.

English Midland and Baltica (East Europe) in Scandinavia. Correspondingly, the new epoch of compression at the end of the Silurian to commencement of the Devonian-, the Scandian epoch, affecting Scandinavia and Britain and representing here the concluding cycle of Caledonian orogeny and the Acadian epoch in the Middle-Late Devonian, manifest in Newfoundland and the Appalachians and having also a decisive importance for the formation of their structure, apart from the Southern Appalachians, is regarded as the result of collision between the Ordovician island arc and the Gondwana and Baltica margins.

The non-uniformity of collision time in Scandinavia on the one hand, and in Britain and North America on the other, is explained by the occurrence of collision of North America with Baltica in the former region and North America with Armorica, i.e., Gondwana, in the second (Fig. 6–7). Baltica, i.e., the East European Craton, was separated by the south-eastern branch of the Caledonides from Gondwana in the Early Palaeozoic; these Caledonides

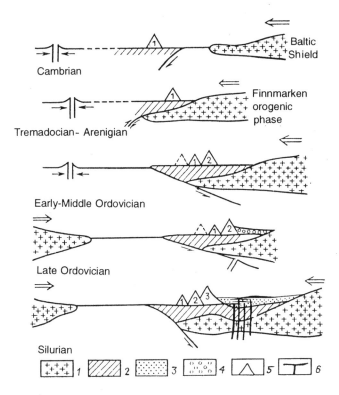

Fig. 6-6. Schematic illustration of plate-tectonic model of Cambrian-Silurian geological evolution of Balm region as a part of the Norwegian Caledonides (after Brekke and Furnes, 1984; and others).

1—continental crust; 2—obducted oceanic crust; 3—sediments; 4—coarse sediments; 5—island arcs (1, 2, 3—sequence of formations); 6—outpourings of mafic composition.

extend through the North Sea to the south-western margin of Baltica and farther, along the Teisseyre-Tornquist line.

The basin in which the German-Polish Caledonides were later formed is called the Tornquist sea or even ocean in contemporary literature (Iorsvik et al., 1990). It extended from Polish Pomorie and the coast of north-eastern Germany (Rügen island region) in the west in a latitudinal direction south of the south-western salient of the Baltic Shield, i.e., Ringköbing Fün arch in the southern part of Denmark, and farther to the well-known Caledonides of southern Belgium (Brabant) and south-eastern England. In the North Sea this basin should have joined with the Iapetus, giving birth to the British-Scandinavian Caledonides; some wells have indeed reached deformed and metamorphosed rocks of the Lower Palaeozoic here. However, some French

328

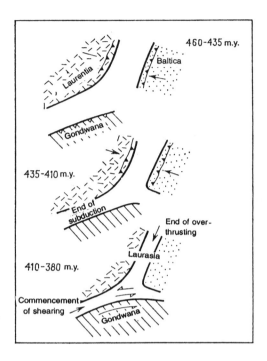

Fig. 6-7. Model of evolution of the Caledonides, indicating polarity of subduction zones and disposition of plates in the zone of triple junction in the region of the North Sea (after Hutton, 1987).

researchers (Paris and Robardet, 1990) assume the existence of a 'continental bridge' between Avalonia, the eastern projection of which is English Midland and Baltica; this assumption is dubious, however.

Lower Palaeozoic formations that accumulated in the Tornquist Sea underwent fold deformation commencing from the end of the Ordovician and a marginal trough was formed to the north of the fold system in southern Denmark. Concluding deformation in the German-Polish Caledonides system had already proceeded in the first half of the Early Devonian.

The Alleghanian epoch of diastrophism represented the concluding event in the development of the Iapetus. Commencement of this diastrophism lay in the Middle Carboniferous and its culmination probably in the Early Permian. Since processes of metamorphism and granitoid plutonism affected even the easternmost Piedmont zones in the Appalachians, regarded as the southern continuation of Avalonia, it must be assumed that Alleghanian diastrophism was associated with subduction to the west of a basin with oceanic crust situated between Avalonia and the main Gondwana massif. This basin was actually a continuation not of the Iapetus, but of the northern branch of the Palaeo-Tethys, and is often called the Rheicum.

Complete fusion of the Palaeo-Tethys with the southern tip of the Iapetus occurred at the level of Ouachita zone and its Mexican continuation. Volcanic island arcs existed in this segment of the Palaeo-Tethys possibly already in the Silurian or even the Ordovician and reliably in the Early and Middle Carboniferous while the total inversion in this system with flysch replaced by molasse occurred at the commencement of the Late Carboniferous; orogeny continued in the Early Permian. In the Mexican part of the system the main deformation occurred already in the Permian.

In the north the Iapetus joined with the Innuitian system and the Ural-Okhotsk belt. The Innuitian system embraced the structures of northern Alaska, Herald-Wrangel islands, Taimyr and Severnaya Zemlya as its further continuation (western in contemporary co-ordinates). The birth of all these mobile systems separating North America and Hyperborea in the east and Hyperborea and Siberia in the west occurred in the Riphean, most probably in its midperiod, judging from its exposed sequences in Taimyr-Severnaya Zemlya region, Wrangel Island and the northern Canadian Arctic archipelago (Ellesmere Island). Ophiolites present on Taimyr peninsula and Ellesmere Island suggest the oceanic nature of these regions. Data on the Taimyr peninsula indicates that the Proto-Arctic basin commenced closing in the middle of the Late Riphean, probably in the process of subduction of the Siberian plate under the Hyperborean plate. This closure was accompanied by formation of a folded mountain system most fully exposed in northern Taimyr and the southern archipelago of Severnaya Zemlya. In the Late Vendian-Cambrian repeated opening of a marine basin occurred south of this folded mountain structure. This Central Taimyr basin attained maximum width and depth in the Ordovician-Silurian. It should have been underlain by a crust of transitional to oceanic, if not by true oceanic (ophiolites are not known).

A very similar Hazen basin-trough appeared in the northern Canadian Arctic archipelago. From the north, however, it was bound by the Pearya volcanoplutonic belt, evidently a marginal belt of the Andean type on the margin of Hyperborea underlain by a zone of subduction inclined northward. Another interpretation of Pearya is also possible, as a volcanic arc on the southern margin of the ocean above a subduction zone inclined southward; in this case the Hazen basin represents a back-arc marginal sea.

At the end of the Silurian-Early Devonian, collision of Hyperborea and North America commenced in the Innuitian system in the northern part of Greenland; in the Middle-Late Devonian it extended into the Canadian part commencing from the Pearya arc; this represents the so-called Ellesmere orogeny reaching northern Alaska, in particular Brooks Range. In the west, on the Taimyr peninsula, similar events commenced only in the Middle

Carboniferous while already a sharp reduction in tectonic activity had taken place in Alaska and east of it. The Central Taimyr basin experienced closure and folding and the zone of subsidence was displaced southward into the modern Byrranga Range where a thick paralic coal-bearing clastic series intercalated with traps began to accumulate. Olistostrome detected in it by S.A. Kurenkov (1988) is proof of the preceding deformation of the Central Taimyr zone. The zone of the Byrranga Range itself underwent final deformation only at the end of the Triassic to commencement of the Jurassic in the Early Cimmerian tectonic epoch. It is evident that the Siberian continent here finally joined with Hyperborea only in this period. The intrusion of plutons of subalkaline granitoids with Permian-Triassic radiometric ages in the northern part of Taimyr is evidently proof of subduction of the Eurasian lithospheric plate under the Hyperborean plate, which continued during this period.

In our view, it is quite possible that the Central Taimyr basin extended westward through the present-day Kara Sea basin into the central part of Novaya Zemlya and later into the Barents Sea because similar relationships between shelf and thin deepwater formations of the Early and Middle Palaeozoic are noticed here while Devonian clinoforms detected in the Barents Sea basin by seismic survey point to a deep basin. According to Ustritsky (1990), this Middle and Late Palaeozoic deep basin, established geophysically as resting on a crust of oceanic or proximate type, could be a relict of the northern continuation of the Iapetus. Let us add that it could even be the Boreal basin from which Zechstein transgression extended southward at the beginning of the Late Permian.

The North Atlantic and Ural-Okhotsk belts were joined in the Barents-Kara region. Of the two, the Ural-Okhotsk belt is far wider and more complex in internal structure and history. Major blocks of ancient continental crust, microcontinents, have been preserved in the structure of the Ural-Okhotsk belt, the more prominent of them being Kazakhstan, possibly consisting of two independent terranes—Kokchetav and Ulytau-Northern Tien Shan—and Central Mongolia. The earliest stage of development of the belt encompassed the Middle and Late Riphean up to 850 m.y. ago. This stage was best manifest in the Yenisei range on the western margin of the Siberian Craton. It was also established here that an earlier epoch of diastrophism comparable to the Grenvillian at the border of 1000 m.y. preceded the main deformation and granite formation at the level of 850 m.y. (a similar scenario is noticed in northern Taimyr). Zaitsev (1984) detected manifestation of diastrophism of the same epoch, called Issedonian here, in the Central Kazakhstan microcontinent. Metamorphics aged 1000–850 m.y. (predominantly amphibolite facies) are also known in many parts of the Altay-Northern Mongolia region (Altukhov, 1986).

The next stage of development of the belt, i.e., Baikalian proper ending in the Vendian, is most fully represented in the Timan-Pechora region and in the Polar Urals. The respective formations were metamorphosed mainly into greenschist facies, reaching amphibolite facies only close to granite plutons. These formations have been traced in the north-west in the southern part of the Barents Sea outcropping in Rybachyi peninsula and in southern Novaya Zemlya; in the south-east the Baikalian complex extended along the other side of the Polar Urals into the basement of the West Siberian Platform where it comprised individual blocks in the form of microcontinents in the Palaeozoic ocean. The formations of this complex emerge also into the Altay-Sayan region and farther east up to Transbaikalia and Peri-Amuria inclusive.

The third stage, generally called the Early Caledonian or Salairian, is of extreme importance in the development of the Ural-Okhotsk belt. It commences with rifting in the second half of the Late Riphean (Karatavian) and spreading in the Vendian to Early Cambrian. The island arc stage of development of the corresponding system began in the Early and partly Middle Cambrian and encompassed much of the Altay-Sayan region and Northern Mongolia. The central zone of the Palaeo-Asian ocean was shifted to the Ob'-Zaisan system and Southern Mongolia with continuation into Greater Hinggan. Marginal seas on the eastern periphery of the ocean separated by island arcs were filled in the Late Cambrian to Ordovician with thick flyschoid terrigenous formations that experienced deformation in the Taconian epoch of the Late Ordovician. In the central parts of basins, however, a very similar sedimentation continued in the Silurian and transition to the orogenic stage occurred only at the end of the Silurian to commencement of the Devonian. This epoch of main Caledonian orogeny was accompanied by intense granite formation and metamorphism and was of decisive importance for the creation of a mature continental crust in the entire Altay-Northern Mongolia region, which has been reflected in the Tectonic Map of Northern Eurasia (1980).

Events in the western and south-western Ural-Okhotsk belt developed somewhat differently. Moderate uplift prevailed in the Urals in the Cambrian and destruction of continental crust existing at the end of the Baikalian stage occurred only at commencement of the Ordovician; an ocean type basin was formed that separated the East European continent from the Siberian continent and the Kazakhstan microcontinent (Figs. 6-8 and 6-9). The stages of opening of this basin, documented by ophiolites, continued up to the Middle Devonian but ensimatic island arcs had arisen in it already in the Silurian and subduction of crust eastward and a corresponding westward shift of the arc commenced. The first volcanic arc collision with the continental margin occurred at the verge of the Early and Middle Devonian and final closure of the ocean basin and collision of the Kazakhstan

332

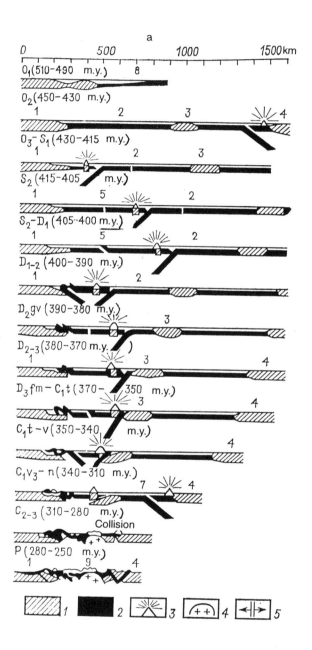

O_1(510-490 m.y.)   8

O_2(450-430 m.y.)
   1   2   3   4

O_3-S_1(430-415 m.y.)
   1   2   3

S_2(415-405 m.y.)
   1   5   2

S_2-D_1(405-400 m.y.)
   1   5   2

D_{1-2}(400-390 m.y.)
   1   2

D_2gv (390-380 m.y.)
       3

D_{2-3}(380-370 m.y.)
   1   3   4

D_3fm-C_1t (370-350 m.y.)
      3   4

C_1t-v (350-340 m.y.)
       4

C_1v_3-n (340-310 m.y.)
      7   4

C_{2-3}(310-280 m.y.)
Collision

P(280-250 m.y.)
   1   9   4

Fig. 6-8.

b

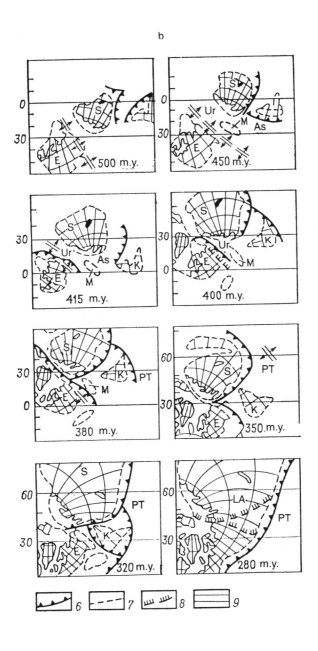

500 m.y.
450 m.y.
415 m.y.
400 m.y.
380 m.y.
350 m.y.
320 m.y.
280 m.y.

microcontinent with Laurussia in the course of the Late Palaeozoic, more accurately from the end of the Early Carboniferous to the Early Triassic inclusive in the far north and extreme south, i.e., in the classic Hercynian era of orogeny.

In Southern Tien Shan lying on the southern continuation of the eastern zones of the Urals, destruction of continental crust also commenced not later than the Ordovician but probably even in the Cambrian (Fig. 6–10). The main basin with an oceanic crust extended along the southern margin of the contemporary Ferghana basin between the Kazakhstan margin built up in the Caledonian stage and the Alay microcontinent. The existence is also possible of a more southern basin of such a type between the Alay and Afghan-Tadzhik microcontinents. These microcontinents lay on the continuation of the Tarim continent. The relatively quiescent development of the Southern Tien Shan system continued up to the middle of the Middle Carboniferous and the island arc stage here was weakly manifest, typical island arc volcanics being known only in the Devonian. The uplift of the Alay microcontinent and formation of gravity nappes of its carbonate cover transiting into olistoplaques and olistoliths of thick flysch-olistostrome formation of the Middle-Upper Carboniferous commenced in the Moscovian epoch. The orogenic stage of development of Southern Tien Shan with formation of molasse and granites had already set in in the Permian.

It was earlier thought that a basin with oceanic crust arose east of the Kazakhstan microcontinent in the Vendian-Early Cambrian. But all the modern datings of ophiolites, palaeontological as well as radiometric, point to the Late Cambrian-Ordovician, as in the case of the North Atlantic belt. There is nothing surprising about the fact that the youngest oceanic crust was preserved in both regions while the much older crust experienced subduction and entered the composition of island arcs, which were also manifest in the Ordovician in the eastern part of Central Kazakhstan and in Northern Tien Shan. In the Late Ordovician, however, a significant part of the region experienced intense deformation, some metamorphism and intrusion of numerous

---

Fig. 6–8. Scheme of development of the Southern Urals in the Palaeozoic (a—palinspastic sections, b—palinspastic reconstructions) (after L.P. Zonenshain In: *History of Development of Urals Palaeo-Ocean*, 1984).

1, 2—crust (1—continental, 2—oceanic); 3—volcanic arcs; 4—granite-gneiss domes and granites; 5, 6—plate boundaries (5—divergent, 6—convergent); 7—contours of continents; 8—collision zones; 9—sedimentary cover.

Figures and letters in reconstructions: 1—East Europe; 2—Uralian ocean; 3—Mugodzhary microcontinent; 4—Kazakhstan continent; 5—Sakmara marginal sea; 6—Magnitogorsk marginal sea; 7—Valeryanov arc; 8—Proto-Uralian ocean; 9—Uralian fold belt. E—Europe; S—Siberia; K—Kazakhstan; LA—Laurasia; M—Mugodzhary; As—Asian palaeo-ocean; Ur—Uralian palaeo-ocean; PT—Palaeo-Tethys.

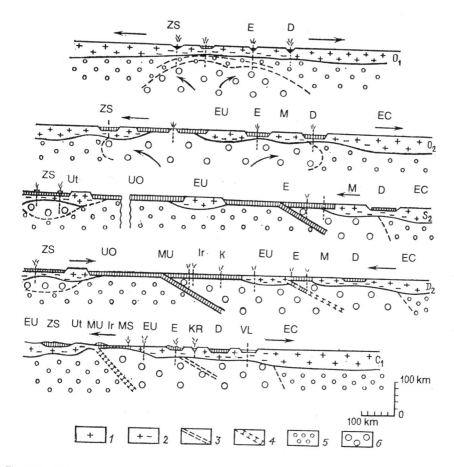

Fig. 6-9. Scheme of suggested development of the main structure of the Southern Urals in the Ordovician to the Carboniferous (after Ivanov et al., 1986).

1—continental pre-Ordovician Earth's crust; 2—same, deformed by extension and partly rebuilt in the Palaeozoic; 3—basaltoids and other rocks of oceanic crust and continental riftogenic depressions and also andesitoids of island arcs; 4—metamorphosed oceanic crust; 5 and 6—mantle (5—normal, 6—anomalous).

Legend: EE—East European continent; ZS—Zilair-Sakmara zone; Ut—Uraltau; UO—Uralian ocean; MU—Major deep fault of Urals; Ir—Irendyk; K—Karamalytash; MS—Magnitogorsk synclinorium; EU—East Uralian uplift; E—Eastern zone; KR—Karabutak rift; M—Mariiinsk zone; D—Denisov zone; Vl—Valeryanov zone; EC—Eastern continent.

granitoid plutons including the territory of the Kazakhstan microcontinent. The zone of sedimentation diminished in the Silurian while much of Central Kazakhstan as also Northern and Central Tien Shan were encompassed by orogeny at the end of the Silurian to the beginning of the Devonian. The

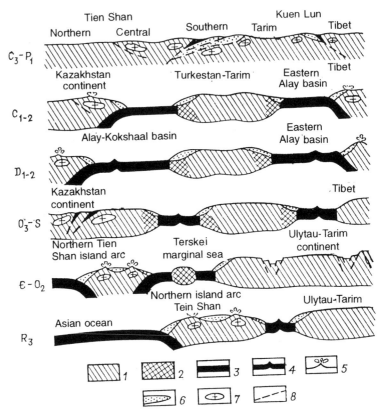

Fig. 6–10. Scheme of evolution of the Earth's crust in Tien Shan in the Late Riphean and Palaeozoic (after Zamaletdinov and Osmonbetov, 1988).

1—continental crust; 2—intermediate crust; 3—oceanic crust; 4—midoceanic ridges; 5—volcanic belts; 6—sedimentary formations; 7—intrusive formations; 8—faults.

sea receded into the Junggar-Balkhash basin in which continuous terrigenous sedimentation continued until the middle of the Viséan epoch. A mighty marginal volcanoplutonic belt manifested at the boundary of this basin and in the region of Caledonian cratonisation in the Early to Middle Devonian. A very similar belt was formed with displacement towards the centre of the Junggar-Balkhash region in the Late Palaeozoic in a period when it had already experienced the orogenic stage.

The Ob'-Zaisan system represents the youngest and central part of the Palaeo-Asian ocean. The Gobi-Hinggan zone and Amur-Okhotsk farther north-east represent the eastern continuation of the system. Its eastern branch with a latitudinal strike stretched through southern Dunbei into southern Russian Primorie (Maritime Region) and north-eastern part of the Korean peninsula (Tumangan trough). Spreading commenced in this part of the Palaeo-Asian ocean in the Silurian or even Ordovician and continued up to the Devonian. Subduction westward proceeded simultaneously under the Chingiz-Tarbagatai island arc separating the Ob'-Zaisan basin from Central Kazakhstan and later from Junggar-Balkhash; in the east subduction proceeded from the Devonian under the Rudnyi Altay marginal volcanoplutonic belt fringing the Altay-Sayan region entering its cratonisation stage. Subduction also proceeded in the east in the latitudinal segment in both directions: in the north under the Caledonian frame of the Siberian Craton and in the south under the marginal uplift of the Sino-Korea Craton. Closure of this axial strip of the Palaeo-Asian basin commenced from the Middle Carboniferous and its development concluded in the Late Permian when the Siberian continent joined with the Sino-Korea continent. The marginal north-eastern Amur-Okhotsk branch of the belt under consideration represents an exception: Hercynian diastrophism was barely manifested in it and deepwater marine sedimentation in the axial zone continued without interruption in the Late Palaeozoic and Early Mesozoic, up to the Middle Jurassic inclusive. From the Late Palaeozoic, this segment of the Ural-Okhotsk belt was actually transformed into an apophysis of the West Pacific belt, into a Pacific Ocean bay. Given this background, separation of the Amur-Okhotsk and Sikhote-Alin' zones is only tentative.

Riftogenic troughs filled with black shale formations developed in the eastern·and northern frame of the Ural-Okhotsk belt in the background of a subduction zone commencing from the Devonian and in the east throughout the rest of the Palaeozoic era. Volcanoplutonic belts were also formed with alkaline magmatism, especially in Northern Mongolia and eastern Transbaikalia.

The zone remaining now for consideration is the Mediterranean belt, i.e., the Tethys, separating the Laurasian group of continents and later the entire Laurasia and Gondwana. Evolution of the Tethys ocean began in the Late Riphean and is best documented by ophiolites of this age in Anti-Atlas and the Arabia-Nubian Shield. Within Europe, ophiolites of the northern and eastern framework of the Bohemian massif may be of Late Riphean age. Evidently two basins with oceanic crust separated by a central band of microcontinents existed in the west right from the beginning. Pre-Late Riphean crust is detected only with great difficulty within these microcontinents. On the whole, these basins constituted the Proto-Tethys ocean; it was separated in the north-west and west from the Proto-Iapetus (Celtic ocean) only by minor blocks of microcontinents (Armorican, Aquitaine-Iberian and hypothetical Senegalese). In the east, the junction of the Palaeo-Tethys with the Palaeo-Asian ocean in Central Asia was of similar character. The first phase of development of the Mediterranean belt, i.e., the phase of existence of the Proto-Tethys, encompasses the Late Riphean and ends in orogeny in the Vendian to the Cambrian, called the Cadomian in West Europe and Pan-african in Africa, and generally corresponding to Baikalian orogeny. Rocks involved in this orogeny have experienced, apart from folding, greenschist metamorphism and granitisation on a limited scale. As a result, cratonisation was complete in the south alone; as already mentioned, a fairly extensive Peri-Gondwana Platform arose here. The possibility cannot be excluded of spreading continuing uninterruptedly in the Vendian to Early Palaeozoic in the southern part of the basin whose formations are now included in the structure of the Alpine fold belt of Europe since ophiolites of this age are known in the Western Alps and Southern Carpathians. On the extreme eastern flank of the belt, Vendian-Cambrian ophiolites are known in Qilianshan and Qinling. But over much of the area of the Mediterranean belt, the Vendian and Cambrian represented a period of absolute or relative uplift and manifestation of riftogenic magmatism. Intense destruction commenced here only in the Ordovician and major deepwater basins with accumulation of predominantly fine clastic material and eruptions and intrusions of mafic magma appeared once again. Thick sandy-clay beds of continental slope and rise accumulated along the northern (Ardennes and Brabant) and eastern (margin of East European Craton) peripheries of the northern basin in the Ordovician and Silurian. Island arc volcanics are also known in Brabant region, pointing to northward subduction of the basin crust. Its eastern part simultaneously formed the south-eastern branch of the Iapetus and was set off from its main trunk by the Midland microcontinent and its North Sea continuation. In the Silurian and to commencement of the Devonian, the band from south-eastern England to western Ukraine and possibly Moldavia was encompassed by folding and uplift and accreted from the south the 'Continent of Old Red sandstone'. A similar process, but already in a more intense form, encompassed eastern Qilianshan and northern Qinling.

The more southern zones of the Palaeo-Tethys developed in an opposite direction in the Devonian. The marginal sea of the Rhenohercynian zone was formed here and its sediments are traced in the form of a broad arc from Southern Portugal to Silesia and Moravia. In the north they are partly superimposed on the Caledonian fold complex. Lizard ophiolites in southern Cornwall date the floor of this basin. In the east, in Central Europe, effusions and sills of diabases and keratophyres are known. The Rhenohercynian sea was separated from the more southern Saxothuringian by a non-volcanic arc (inherited from the Early Palaeozoic) of the Mitteldeutsche Schwelle adjoining the Armorican microcontinent in the west.

In the more southern basin of the European Palaeo-Tethys, much of its area falling in the Alpine fold belt, volcanic arcs developed in the Middle

Palaeozoic. Their formations outcrop in the Alps, Carpathians and Balkans and in the east in northern Caucasus and Badahshan.

Deformation and uplift accompanied by calc-alkaline volcanism were manifest in the Middle-Late Devonian in the central zone of the European Hercynides and roughly correspond to the Acadian orogeny of North America. At the verge of the Devonian and Carboniferous, the zone of uplift enlarged (Fig. 6–11), which caused replacement of the Devonian black schist formation by flysch or flyschoid Culm formation. A new impulse of diastrophism occurred in the Middle Viséan epoch; from this time deformation and uplift grew until commencement of the Middle Carboniferous; this is the Sudetan epoch of orogeny, the main one in the Hercynian period. It was manifest in the most energetic form not only in the European (Fig. 6–12), but also the Asian part of the Palaeo-Tethys and was accompanied by regional metamorphism and mighty granite formation. From this same period began the development of foredeeps and intermontane troughs filled with molasse, initially (C_{2+3}) mainly coal-bearing or salt-bearing paralic, and later (P) predominantly with continental, usually red coarse clastic molasse. This type of development persists surprisingly well over a vast area from West Europe to the eastern part of Asia.

Contrary tendencies began manifesting on the southern marginal strip of the Palaeo-Tethys in the Late Palaeozoic, especially in the Permian. Commencing from the south-western margin of the Carpathians and Dinarides in the west and continuing with the southern slope of Caucasus Major and the central Pamirs and right up to South-east Asia, marine conditions were preserved or even intensified in this strip in the Late Palaeozoic in the absence or weak manifestation of signs of Hercynian diastrophism. In some regions (Thailand and Vietnam-Lao) possibly destruction of continental crust was repeated. Traps of the Yunnan probably represented a sign of this destruction while traps of the Pir-Panjal in the north-western Himalayas point to the beginning of formation of the Indus-Tsangpo oceanic zone, i.e., the birth of the eastern part of the Neo-Tethys.

Thus there was a new set-up in the Mediterranean belt by the end of the Palaeozoic with much of the Palaeo-Tethys involved in orogeny and later adjoining the Laurasian supercontinent, which was formed at the same time. In the west, in the area from Central America to the Apennine peninsula, this process affected the entire zone of the Palaeo-Tethys, which led to the joining of Laurasia with Gondwana and to the formation of Pangaea II. The residual oceanic basin, already represented by the Meso-Tethys, formed an extensive bay opening into the Palaeo-Pacific and its axial zone ran considerably south of the axis of the Palaeo-Tethys through northern Anatolia, Transcaucasus, central Afghanistan, middle of Tibet-Qinghai plateau and farther towards South-east Asia with branching along the southern slope of Qinling (Sengör and Hsü, 1984). Simultaneously, a new oceanic basin, the

340

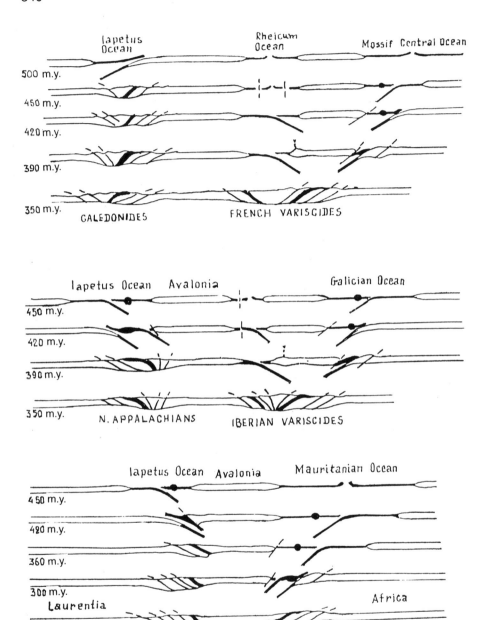

Fig. 6–11. Schematic plate-tectonic sections showing evolution of the Peri-Atlantic Palaeozoic belts between 500 and 270 m.y. (after Ph. Matte, 1991).

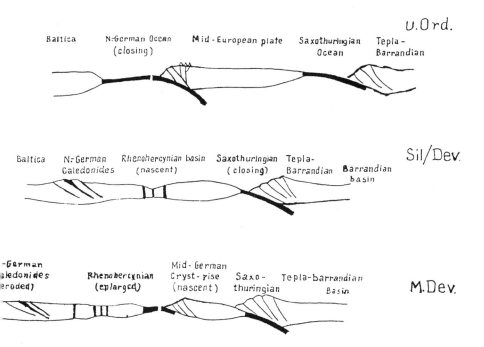

Fig. 6–12. Plate-tectonic model for the northern flank of the Variscan belt. Final stage of collision and northward subduction on the southern flank of the belt are not shown (after W. Franke, 1989).

Neo-Tethys, originated in the strip between Tibet and the Himalayas and was fully open by the Mesozoic.

6.3 TRANSGRESSIONS AND REGRESSIONS

The various views proposed concerning the problem of global transgressions and regressions have been examined in detail in a special article by Yanshin (1973). He concluded that the concept of the existence of worldwide transgressions and regressions in geological history is erroneous. As pointed out by Yanshin and other researchers, the problem could only be resolved on the basis of world palaeogeographic maps. So the authors analysed global qualitative data on the Phanerozoic and came to the conclusion of the reality of worldwide transgressions and regressions and their correspondence to certain stages of Caledonian, Hercynian and Alpine tectonic cycles of development of the Earth (Ronov et al., 1976).

Another approach to this problem was proposed at the end of the '70s with the development of seismostratigraphic studies and their application for identifying fluctuations in the level of the World ocean. The graph of such fluctuations for the entire Phanerozoic published by Vail and coauthors (1977) is the most well known. Vail's method was criticised for its failure to take into consideration palaeotectonic factors, inadequate representation of Palaeozoic data etc. (Cloetingh et al., 1985; Miall, 1986; Parkinson and Summerhayes, 1985; Watts, 1982). As an example, we reproduce here a diagram from Summerhayes' article (1986) which illustrates the degree of reliability of the correlation of regional seismostratigraphic data when plotting the general curve and isolating global cycles (Fig. 6–13). The number of regional sections was, however, somewhat larger than that shown in the figure but the diverse tendencies and the need for more statistical data are evident. However, in relation to Palaeozoic sequences, as shown by the scheme of data used by Vail and others (Fig. 6–14), the representativeness is clearly inadequate. Vast regions of Eurasia, South America, Australia etc. with extremely extensive development of Palaeozoic formations were not taken into consideration by Vail when plotting the first global curve for the Phanerozoic. Later, Vail's curve was improved when it reflected the eustatic fluctuations of sea level only for the Meso-Cenozoic history of the Earth. As a result, it must be acknowledged that, for the latest history of the Earth, Vail's method is comparatively reliable but does not provide good results for Palaeozoic periods.

To determine sea level changes in the Cambrian, Ordovician and Silurian, McKerrow (1979) and Leggett and coauthors (1981) used the analysis of remnants of fossil fauna and the distribution of black shales.

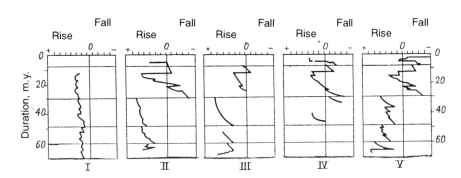

Fig. 6–13. Correlation of regional transgression schemes from which the scheme of global cycles of sea-level fluctuations was generated (after Vail et al., 1977).

I—Gippsland basin, Australia; II—North Sea; III—north-western Africa; IV—San Joaquin basin, California; V—global curve; vertical line—present sea level.

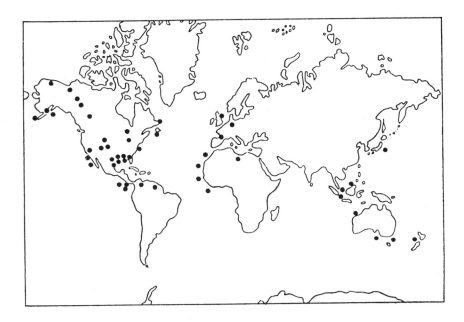

Fig. 6-14. Areas of regional seismostratigraphic research and drilling (shown by dots) from which Vail and coauthors (1977) plotted the graph of global fluctuations of sea level for the Phanerozoic.

They were able to gather detailed data only for some well-studied regions. The conclusion of these authors accords with ours concerning global transgressions. Further, there are several works in which the history of transgressions and regressions within certain sedimentary basins or very large areas over comparatively brief or rather prolonged periods has been analysed in detail. Examples of such studies are those of Hallam (1977) for the former USSR and North American territories, Ross and Ross (1985) for the Late Palaeozoic of north-eastern Europe, the Russian Platform and the Mississippi basin, Johnson and coworkers (1985) for Devonian Euramerica, Veevers and Powell (1987) for the Late Palaeozoic of all continents of the Southern Hemisphere and also Euramerica; in the latter work, the analysis includes palaeotectonic and other information. Almost all these researchers, however, resolved only regional problems.

A series of Palaeozoic lithological-palaeogeographic maps of continents published in 1984 provided new data on the development of transgressions and regressions in the Palaeozoic (Ronov et al., 1984). Data on quantitative analysis of these maps for the Early and Middle Palaeozoic was published (Seslavinsky, 1987) and information for the entire Palaeozoic era is given here for the first time. It must be pointed out that the discussion deals only with transgressions and regressions on continents but not global changes of sea level. Undoubtedly, epochs of global transgressions correspond to epochs of increased general sea level. However, Palaeozoic vertical sea level fluctuations are so far away in time from the common reference point, i.e., the present level of the ocean, that it would be desirable to refer to transgressions and regressions, all the more so since our analysis is based on palaeogeographic data on the areas of seas.

There is at present no doubt that global transgressions and regressions manifested repeatedly in the Phanerozoic. But how were they determined: by a change in volume of ocean water, or change in morphology of ocean floor, or general vertical movements of continents?

Judging from several geological studies (Simonov, 1978; Donovan and Jones, 1979; Holland, 1972), the mass of water in the World ocean remained almost unchanged during the Phanerozoic or perhaps rose steadily as a result of volcanic activity, excluding only the epochs of major continental glaciation. According to Timofeev and coauthors (1986), the amount of water on Earth rose sharply at the verge between the Palaeozoic and Mesozoic. This view is debatable and, moreover, we are interested here only in the Palaeozoic. Hence in the Palaeozoic the reason for global regression could be either a change of topography of the ocean floor (capacity of ocean basins) or the overall uplift of continents caused by tectonics. Donovan and Jones (1979) analysed this problem and regarded the change in volume of marine mountain ridges as the main reason while inundation of the continental crust and formation of deep trenches are treated as subordinate factors. Russel (1968), a little earlier, was the first to suggest a dependence between eustatic fluctuations of sea level and development of marine mountain ridges; this view was later well confirmed by data on the Cenozoic history of the Earth. There are, however, almost no data on the formation of the topography of the floor of Palaeozoic oceans—Iapetus, Palaeo-Tethys and Palaeo-Pacific. Thus the above question concerning the factors responsible for global transgressions for the Palaeozoic still awaits a solution. There is no doubt, however, about these factors being of a tectonic nature and their association with global changes in the development of the Earth. An analysis of Palaeozoic tectonic history of platforms and mobile belts reveals the nature of these changes and offers several suggestions.

When comparing the direction of transgressions and regressions in seven of the largest platforms (Fig. 6-15), the first to strike attention is the general chaotic trend almost all through the Palaeozoic. In most platforms only regressions coincide in the Late Silurian and Early Permian and transgressions in the Middle Devonian. There are, however, no examples of matching trends even for half of the Palaeozoic for any two platforms. Sharp changes in different platforms are usually of opposite signs and, as

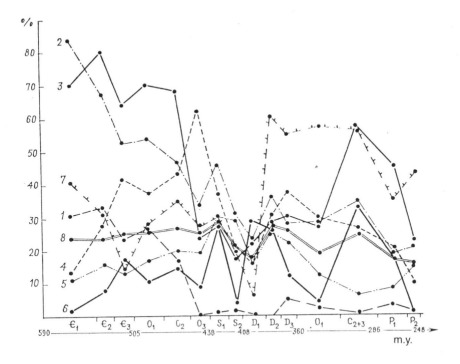

Fig. 6-15. Changes over time in area of platforms (as percentage of their total area) covered by sea (after Seslavinsky, 1987, with additions).

Platforms: 1—Australian; 2—Siberian; 3—Chinese; 4—North American; 5—African; 6—South American; 7—East European; 8—all platforms as a whole.

a result, the general curve of transgressions for platforms is rather smooth. The deepest regressions have been established in the Early Devonian, Early Carboniferous and Permian.

In general, the data for platforms points to rather small fluctuations of overall area covered by sea in all platformal structures. The changes ranged from 15 to 29% (Table 6.1).

The graph shown in Fig. 6-16 provides very general information on global transgressions and regressions. This graph depicts changes in the areas of marine basins on the continents as a whole. Information on the areas of Palaeozoic continents has been taken from the *Atlas of Lithological-Palaeogeographic Maps of the World* (Ronov et al., 1984). Platforms and regions with an orogenic regime of development with typical thick crust of the continental type form the basis of these areas.

Many zones in mobile belts were characterised by deepwater conditions and evidently a thinned crust of oceanic or transitional type. Such zones

Table 6.1. Area of sea on platforms (as percentage of their total area)

Platforms	ε_1	ε_2	ε_3	O_1	O_2	O_3	S_1
Australian	31	33	24	27	17		1
Siberian	84	67	53	54	47	34	46
Chinese	70	80	64	70	68	24	27
North American	13	28	42	37	43	62	37
African	11	16	13	17	20	19	30
South American	2	8	17	11	15	9	27
East European	41	31	14	28	35	27	30
All platforms as a whole	24	24	25	26	27	25	29

Platforms	S_2	D_1	D_2	D_3	C_1	C_{2+3}	P_1	P_2
Australian	2	—	—	5	2	—	3	1
Siberian	31	23	36	28	28	34	18	20
Chinese	17	22	29	30	27	57	45	22
North American	21	16	30	38	29	26	20	10
African	29	16	25	22	12	6	8	14
South American	4	29	26	12	4	33	17	—
East European	20	6	60	55	57	56	35	43
All platforms as a whole	21	17	27	26	19	25	17	15

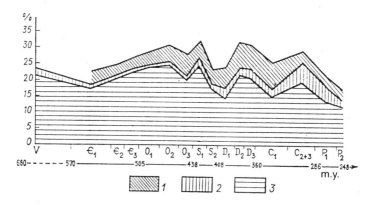

Fig. 6-16. Change over time in area covered by sea (as percentage of total area of continents) within continents (after Seslavinsky, 1987, with additions).

1—sea on platforms; 2—residual seas in orogenic regions; 3—shelf seas in mobile belts.

should not be included in the area of palaeocontinents. However, shallow-water shelf zones of mobile belts, according to all available data, generally possessed a fairly thick continental crust and are considered here as having taken part in the area of palaeocontinents of the Earth. The interpretation of such zones for Precambrian history involves difficulties and is tentative. Their area for the Riphean and Vendian has therefore not been calculated and our figure does not contain information on the areas of shelf seas of mobile zones for the Vendian.

The upper curve of the graph reflects the course of global transgressions and regressions from the Early Cambrian onwards while the two lower ones show the contribution of platform basins, residual marine basins in orogenic regions and shelf basins of mobile belts to the total area of seas on continents. Detailed data (up to the level of epochs) worked out from original lithological-palaeogeographic maps helped identify only comparatively major fluctuations with a periodicity of not less than 10–15 m.y.

Thus two transgressive-regressive cycles occurred on the continents in the Palaeozoic. The area of seas decreased after the Vendian (towards the beginning of the Cambrian) and progressively increased up to 30% in the Middle Ordovician (Table 6.2). In the Late Ordovician it decreased by 2.5%, which can positively be associated with a drop in ocean level as a result of continental glaciation (McKerrow, 1979) encompassing extensive areas of the African Platform at the end of the Caradocian. In the Early Silurian the area of seas again increased, reaching almost 32% (the highest value known so far). This was followed by a major, sharp regression as a result of which the area of the sea decreased by 8%. The first transgressive-regressive cycle concluded with this regression and its time frame distinctly coincided with the Caledonian tectonic cycle of development of the Earth.

Table 6.2. Areas of continental seas on platforms, in orogenic regions and shelf zones of mobile belts (as percentage of total area)

Area of seas (as percentage of total area of continents)	V	ε_1	ε_2	ε_3	O_1	O_2	O_3	S_1
On platforms	21.6	17.3	20.8	22.1	24.2	24.5	20.4	24.9
In orogenic regions	1.5	1.2	0.4	0.7	0.4	0.3	0.3	1.6
In shelf zones of mobile belts	—	4.2	3.5	3.6	4.4	5.5	6.7	5.4

Area of seas (as percentage of total area of continents)	S_2	D_1	D_2	D_3	C_1	C_{2+3}	P_1	P_2
On platforms	17.4	14.2	22.0	20.5	15.0	19.7	13.5	11.9
In orogenic regions	1.5	3.4	2.0	2.7	2.3	6.3	5.0	2.2
In shelf zones of mobile belts	4.6	6.4	7.9	7.5	8.1	3.0	2.8	3.4

The factors responsible for such a global transgressive-regressive cycle should be sought in the general tectonic history of development of mobile belts. In the early stage of the Caledonian cycle, at the end of the Vendian and in the Early Cambrian, deepwater sediments were more extensively developed (Seslavinsky, 1987) and evidently many basins (deep trenches) deepened, thus generating additional capacity in the Palaeozoic oceans and causing regression on continents.

Later, with increasing intensification of collision and accretionary processes in the middle period of the Caledonian cycle, these deep basins were filled with sediments and volcanic products and were ultimately transformed into folded structures and orogenic regions. As a result, the volume of ocean basins may have decreased and continents were gradually inundated with an interruption, as pointed out above, by partial glacial regression in the Late Ordovician.

The Silurian period, representing the concluding stage of the Caledonian cycle, is characterised by a generally sharp intensification of the scale of manifestation of orogenic processes. The total area of orogenic regions more than tripled from the Late Ordovician to the Late Silurian (Seslavinsky, 1987). Further, a montane topography was formed in many regions and the tendencies of the predominance of uplift extended to much of the territory of the platforms. All this led to Late Silurian global regression.

The Early Devonian occupies an intermediate position between the Caledonian and Hercynian tectonic cycles. Palaeotectonic data tends to favour manifestation of several new deep troughs in this period in mobile belts, which probably again led to an increase in the capacity of oceans. Such an assumption is confirmed by the vast spread of deepwater formations in the Early Devonian compared to Late Silurian (Seslavinsky, 1987). As a result, the generally regressive conditions on the continents were maintained. However, the tendencies of the development of different types of structures differed for the first time. Platforms still experienced uplift and regression propagated at almost the same rate as before. Within the marginal parts of continents, however, strong transgression commenced in the shelf basins of mobile belts. Subsidence intensified even in troughs in regions with an orogenic regime.

An extensive general transgression occurred in the Middle Devonian and the area of epicontinental seas again reached almost 32% of the total area of continents and was maintained at that level up to the end of the Devonian. Here our data accords with the viewpoint of Dineley (1980) about the exclusiveness of the Devonian period in the Palaeozoic in relation to the area of marine basins on continents.

By the Early Carboniferous sharp regression encompassed most basins on the North American, South American and African platforms (see Fig. 6-15). This epoch was regressive even for the small Australian and

Sino-Korea platforms, which together led to a perceptible general reduction in area of seas. The tendency changed in the Middle Carboniferous. Extensive transgression on the South American, Siberian and Sino-Korea platforms and a significant increase in area of marine troughs in orogenic regions again raised the level of the sea cover of continents by almost 30%. In the Late Carboniferous, judging from regional geological data, regression again intensified. This, however, could not be reflected in our graph since data for the Middle as well as Late Carboniferous were grouped together, due to the peculiarity of basic data, and only the Middle Carboniferous tendency towards transgression was manifest.

A particularly deep global regression was established in the Early Permian period as a result of reduction in area of marine basins on almost all platforms and in orogenic regions. One more factor intensifying regression at the end of the Carboniferous to the Early Permian is the well-known Gondwana continental glaciation. Its importance and detailed chronology were recently studied by Veevers and Powell (1987). Regression deepened even more in the Late Permian and the area of seas on continents was the least for the entire Palaeozoic (about 18%). Late Permian reduction of sea area occurred as a result of the inversion of most residual troughs in orogens, notwithstanding perceptible transgression on the East European and Arabian platforms and disappearance of continental glaciation.

The scale of Permian regression was even higher than that of the Late Silurian. A sharp reactivation of all forms of tectonic activity concluding the Hercynian cycle of evolution and associated with the formation of Pangaea was the main feature in the tectonic development of the Earth in the Late Carboniferous and Permian. Processes of granitisation, metamorphism and diverse tectonic deformations were very extensively manifest; the area of orogenic regions enlarged by one-and-a-half times compared to the Early Carboniferous. All this led to a general uplift of continents and global regression which concluded the second transgressive-regressive cycle in the Palaeozoic.

The co-ordinated manner of changes in the area of seas on platforms, in orogenic regions and shelf zones of mobile belts confirms their global character and, too, that the main causative factor should be sought in the most general features of the tectonic evolution of the Earth and primarily its ocean regions. This is supported by an exception: mismatching of tendencies for platforms and mobile belts in the Early Devonian indicates an arrival of transgressions with lag of one epoch from the side of oceans. In the Meso-Cenozoic history of the Earth, a relationship can be traced between the rate of processes of spreading and level variations. Flemming and Roberts (1973) were the earliest scientists to note such a relationship. This relationship could have prevailed in the Palaeozoic too.

Thus in the Palaeozoic the maximum area of seas on continents was attained in the Early Silurian and Middle Devonian and the minimum in the Early Cambrian, Late Silurian and Late Permian. During this period two transgressive-regressive cycles were manifest. These are the Cambrian-Silurian and Devonian-Permian, corresponding to the Caledonian and Hercynian tectonic cycles of development of the Earth.

6.4 MAGMATISM AND METAMORPHISM

6.4.1 Ophiolite Magmatism

The oldest group of ophiolites of interest to us here belongs to the latter half of the Late Riphean-Vendian-Early Cambrian. Ophiolites of the Mauritanides in western Africa (pers. comm. from J. Sougy), Anti-Atlas (El Graara) and Arabian Shield are reliably of Late Riphean age (about 850 m.y.). In the latter two regions, they correspond to the opening of the Proto-Tethys and mark its southern-south-western (in present-day co-ordinates) margin. Ophiolites outcropping along the boundary of the Bohemian massif with the Saxothuringian zone of the Central European Hercynides and within the latter from the Münchberg massif to Gory Sowie may be of similar age or slightly younger. On their western extension lie the ophiolites of Audierne bay in the extreme western part of the Armorican massif and distinctly allochthonic nappes of the Central Iberian zone in the extreme north-western part of the Iberian peninsula. Ophiolites of the same age are evidently developed in the Balkans, in the lower portions of the 'diabase-phyllitoid formation', in the Southern Carpathians (Corbu series in Iron Gates) and possibly in the Dinarides (Vardar zone). Ophiolites of the eastern part of the Altay-Sayan region (Western and Eastern Sayan) and adjoining region of western Mongolia belong to this same group. Such an age is geologically (not radiometrically) more probable. Pre-Ordovician age ophiolites are found in the Chersky range in north-eastern Russia. Ophiolites of West Tasmania and the Lachlan system of eastern Australia likewise belong to this group.

It can be seen from the above list that processes of destruction of continental and newly formed crust of the oceanic type were already widely prevalent in the Late Riphean to the Early Cambrian and were manifest in the Mediterranean (Proto-Tethys and Central European ocean), Ural-Okhotsk (Sayan-Mongolia region) and West Pacific (Verkhoyansk-Kolyma and Tasmania regions) belts. Soon after their formation, ophiolites of this group in Europe experienced in most cases Baikalian (Cadomian) folding and in Asia a somewhat later Salairian folding.

The next age group of ophiolites, even more widespread and generally better dated, arose at the end of the Cambrian to the Early and Middle Ordovician, predominantly within the future Caledonides. Foremost in this group are ophiolites of the Appalachians, British and Scandinavian

Caledonides, i.e., rocks of the Iapetus ocean floor. It is interesting that, although its opening should have commenced in the Early Cambrian on the basis of general geological grounds, nearly all radiometric age determinations point to the Early Ordovician, Tremadocian-Arenigian, except for one of Late Ordovician (Dunning and Pedersen, 1988). According to geological data, opening of a small marginal sea in Norway is also possible in the Silurian. A similar picture, i.e., a decisive predominance of Ordovician (O_{1-2}) ophiolites on the basis of palaeontological data (conodonts), is noticed even in central Kazakhstan (Cambrian ophiolites, too, are present, however); in Northern Tien Shan, based on the same data, the age of ophiolites is Late Cambrian to Early Ordovician. The situation is less clear in Southern Tien Shan: island arc volcanics are manifest already in the Ordovician and Silurian, indicating subduction of oceanic crust but the latter is encountered only in isolated fragments and has not been dated radiometrically. Ophiolites of the very beginning of the Ordovician (496 ± 6 m.y.) were recently established in the Western Alps (Ménot et al., 1988). In the West Pacific belt Penzhina ophiolites (north-western Kamchatka) fall in this group; in the East Pacific belt, these are Oregon and eastern Sierra Nevada in the Cordillera, Precordillera of Argentina and possibly Cordillera Coastal of Chile in the Andes. Predominance everywhere of Ordovician ophiolites over Cambrian in this group may be explained by subduction of much older ophiolites to a greater extent than much younger ones.

The third age group, partly overlapping the preceding group, is weakly manifest and more modest in scale. It includes ophiolites of the end of the Ordovician and Silurian. Ophiolites of this group belong predominantly to Hercynian fold systems: Urals (in : *History of Development of Urals' Palaeoocean*, 1984), Ob'-Zaisan (Chara zone) and Gobi-Hinggan systems and also the Nippon-Sakhalin system. In the West Pacific belt, ophiolites of this age are not known (yet ?) but have been detected in the East Pacific belt in the North American Cordillera, in Oregon.

The fourth group is distributed slightly more extensively and comprises ophiolites of Mid-Devonian age. This, like the preceding group, belongs to Hercynian systems and includes a new generation of tholeiite-basalts of the Urals. This group also includes ophiolites of the Rhenohercynian (Cape Lizard in Cornwall, 375 m.y.) and Southern Portugal zones of the European Hercynides, the northern Innuitian system, Southern Tien Shan and Pamirs and the Yunnan-Malaya system. Apart from these, it also includes Early Devonian ophiolites of the Mongolia-Amur and Gobi-Hinggan (Amur-Okhotsk zone) systems, similar formations of the Nippon-Sakhalin system formed even from the Silurian to the end of the Mid-Devonian (Faure and Charvet, 1987; Khanchuk et al., 1988) and New England system of Eastern Australia. Their presence is also possible, tentatively, in the West Antarctic region.

At the end of the Devonian and Early Carboniferous, an interesting and deep gap occurred in the formation of ophiolites and hence of oceanic crust whose subduction nevertheless continued at least in the periphery of the Pacific Ocean, judging from the distribution of glaucophane schists (see below). Dobretsov (1988) distinctly depicted this gap in his diagram.

The formation process of oceanic crust continued in the Late Palaeozoic but on a significantly smaller scale and led to opening of basins that had experienced closure already in the Mesozoic, mainly Early Cimmerian, partly Late Cimmerian epochs. Late Palaeozoic ophiolites are known on both sides of the Pacific Ocean, in the west in Japan (Faure and Charvet, 1987) and New Zealand and in the east in the Cordillera of North America, in Oregon, and southern Sierra Nevada. Further, such ophiolites were detected on Greater Lyakhovsky Island in the Arctic (296 m.y., as communicated by S.S. Drachev). Their development occurred probably in the central Pamirs (Shvol'man, 1980), in central Afghanistan and in Tibet. Late Palaeozoic ophiolites are known in the Yunnan-Malaya system in the region of Uttaradit in Thailand (Helmcke, 1985) and are possibly present in the Shongda zone in northern Vietnam (if these are ophiolites at all). All these regions represented in the Late Palaeozoic apophyses of the Pacific Ocean.

6.4.2 Island Arc Volcanism

The formation of island arc systems played an important role in the development of all Palaeozoic mobile belts. It led to material redistribution in the mantle-crust system with a general oxidation of crustal material and an increase in its volume. Further, a complex and prolonged evolution of the magmatism of island arcs occurred commencing from the period of their origin as a result of extension in the early stage of development of mobile belts being replaced by compression (commencement of subduction of lithospheric plates) up to transformation into complexly deformed belts with a thick crust of the continental type. The main manifestation of this magmatism was the large-scale volcanism of predominantly a calc-alkaline type. The early stage of evolution of arcs was preceded by tholeiitic and the concluding stage by subalkaline and alkaline types of volcanism (Bogatikov and Tsvetkov, 1988). All these are depicted in Fig. 6-17 by a common sign of island arc volcanism.

Fig. 6-17. Scheme depicting correlation between tectonic, magmatic and metamorphic processes in mobile belts of the Earth for the Palaeozoic era.

1—granitisation; 2—regional metamorphism; 3—folding; 4—ophiolite complexes; 5—island arc complexes; 6—complexes of marginal-continental volcanic belts; 7—bimodal complexes; 8—alkaline volcanic complexes; 9—age intervals of manifestation of volcanism (broken lines = assumed); 10—flysch complexes; 11—molasse; 12—manifestation of orogenic-tectonic regime; 13—manifestation of platformal tectonic regime.

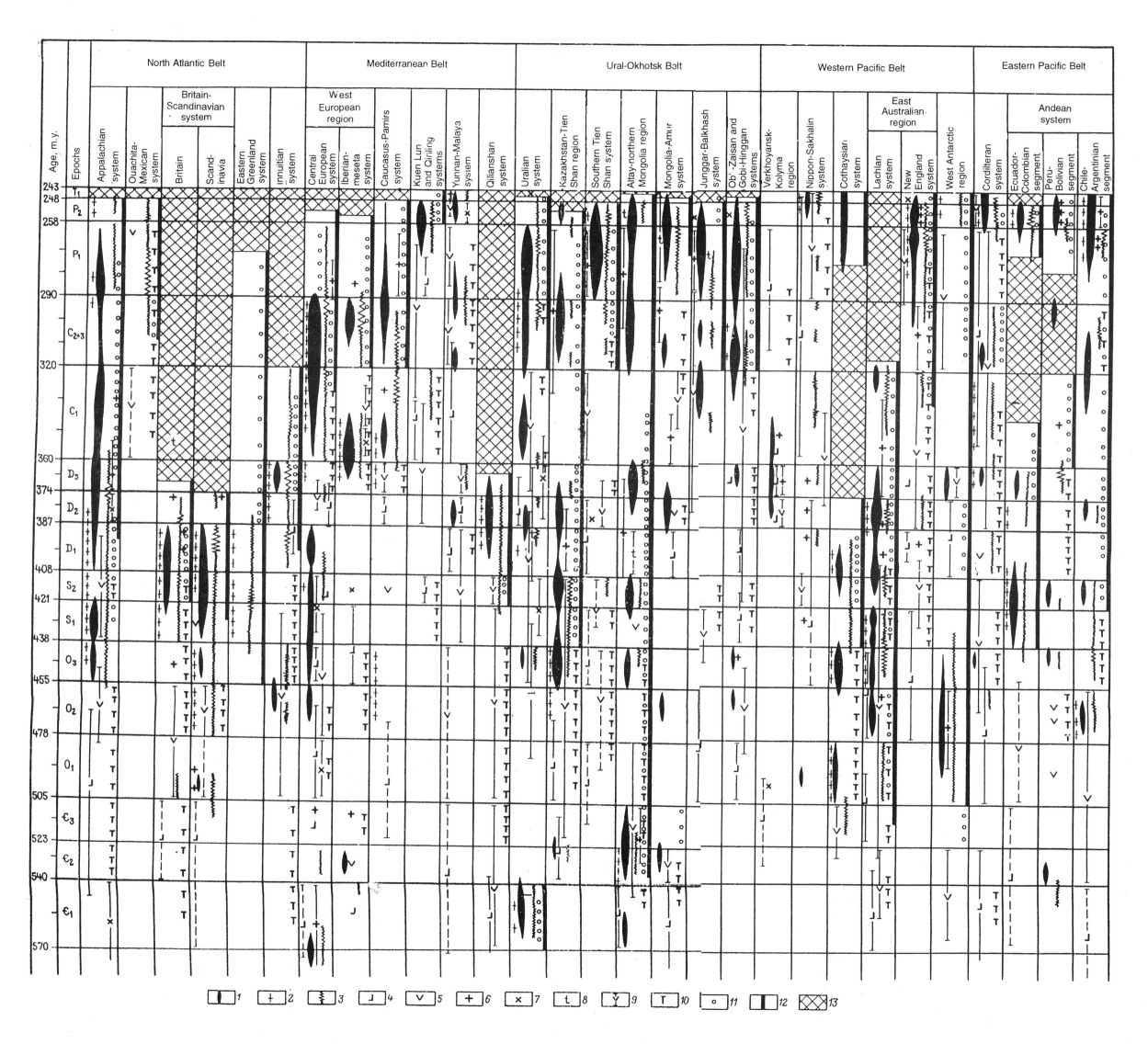

The duration and magnitude of such manifestation naturally differed in various mobile belts, systems and zones. Sometimes they were local and brief outbursts as, for example in the Early Ordovician in the Peru-Bolivian system of the Andes or the Middle Cambrian in the Ossa-Morena zone of the Iberian peninsula. But more often the duration of formation of island arc systems ran into tens of millions of years and their extent corresponded to the entire length of mobile structures. A typical example of the prolonged development of island arc volcanism is the British-Scandinavian system where this complex was formed in the Early and Middle Ordovician over a period of 40–50 m.y.

However, Palaeozoic history provides examples of complexes with even more prolonged periods of development. Thus in the East Australian region, in the Lachlan system, volcanic eruptions of the calc-alkaline type continued from the end of the Middle Cambrian to the beginning of the Early Devonian (about 120 m.y.) and in the eastern parts of the Kazakhstan-Tien Shan region from Late Cambrian to the end of the Silurian (about 110 m.y.); the last Palaeozoic cycle in the Cordillera continued from Middle Devonian to the end of the Early Permian (about 130 m.y.). Detailed palaeotectonic analysis shows that almost continuous (over time) and echelon-like (in space) alternation of generation of island arcs, the evolution of each extending for 40–60 m.y., occurred during such prolonged cycles (Veevers, 1986). The overall regional palaeotectonic setting of compression and subduction remained unchanged, however.

Wide variation in the duration of manifestation of island arc volcanism is also characteristic of the Meso-Cenozoic history of the development of the Earth (Bogatikov and Tsvetkov, 1988): 8 m.y. for the Southern Sandwich arc and 125–150 m.y. for Kamchatka. Based on examples of present-day island arcs, a definite relationship has been recognised between the type of magmatism and rate and angle of subduction of lithospheric plates under them, ratio of compressive stress and extension within seismofocal zones, their form and other parameters of subduction zones and also relative thickness and composition of the Earth's crust, lithosphere and asthenosphere in the region of mantle wedge. Having interpreted the alternation over time of types of manifestation of Palaeozoic volcanism, variation in rates of subduction regimes and of other geodynamic characteristics for that period could be recognised.

It must be pointed out that great diversity was manifest in the evolution of island arc volcanism in actual structures and deviations from the standard scheme (tholeiite—calc-alkaline—subalkaline—alkaline) are encountered almost as often as standard examples themselves. Alkaline volcanism interrupted volcanism of other types or developed simultaneously; tholeiite and calc-alkaline eruptions could have occurred simultaneously at some time; other combinations are also possible (see Fig. 6–17). Evidently an explanation for such diversity should be sought in the superposition of local or regional processes of extension—rifting and other processes occurring at depth—on compression regimes more prevalent in island arcs.

The relationship between island arc volcanism and subduction settling helped in considering this type of volcanic activity a good indicator of corresponding geodynamic conditions of the geological past. It is evident that a global analysis of the distribution and evolution of this volcanism over time provides data for judging the general pattern of development of the lithosphere, for evaluating its tectonic state and for resolving problems of manifestation of global phases of compression and extension.

Our data on the development of island arc volcanism in Palaeozoic mobile belts, shown in Fig. 6-17, serves as a basis for identifying such very common patterns. This data was summed up and not only the number of structures in which island arc volcanism occurred in a given epoch, but also the magnitude of this process in each system was calculated. Local manifestation, such as the Early Ordovician in the Peru-Bolivian system of the Andes was assigned 1 ball, usual volcanism with formation of developed island arcs 2 balls if the given interval of time was not wholly covered by these processes and 3 balls when they occurred throughout the interval. In several structures (for example, Yunnan-Malaya), island arc volcanism was identified tentatively and evaluated as 1 ball. This gradation is, of course, extremely provisional and sometimes even subjective but the use of such a method is wholly justified for global semi-quantitative evaluations and the results obtained can be regarded as fairly reliable, all the more since they accord in general with the results of analysis of changes in the intensity of andesite volcanism carried out by one of the authors by altogether different methods (Seslavinsky, 1987).

So we plotted a histogram of the changes of global distribution of manifestation of island arc volcanism during the Palaeozoic (Fig. 6-18). All the time intervals in it are proximate in duration (about 10 m.y.) with amplitude of changes ranging from 6 to 15 m.y. Such a detailed analysis enables detection of general mean frequency variations within a period of a few tens of millions of years.

An analysis of the histogram revealed the following very general characteristics. Peaks of manifestation of island arc volcanism separated by stable dips corresponding to epochs of decrease of such activity are distinctly seen in it. The large amplitude of these fluctuations excludes possible errors due to inaccuracies in identifying the type of volcanism even in 2 or 3 mobile structures.

Another important feature is the absence of total cessation of island arc volcanism in the Palaeozoic era, i.e., these processes are always manifest in one or another region but vary significantly in overall magnitude.

The first and weakest peak is confined to the end of the Middle Cambrian (Maya age). At this time the scale of island arc volcanism was comparatively

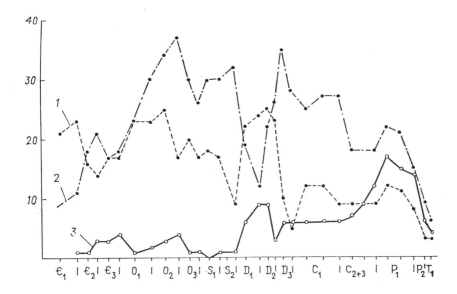

Fig. 6-18. Variation in distribution of ophiolites, island arc volcanics and marginal volcanoplutonic belts in the Palaeozoic.

1—ophiolites; 2—island arc volcanics; 3—marginal volcanoplutonic belts.

broad in the Altay-Northern Mongolia region, in the Lachlan system of Australia and some other areas.

The second peak, maximal in the Palaeozoic, pertains to the Llandeilian stage of the Ordovician. At this time island arc volcanic complexes were formed in 13 of 29 mobile structures; the North Atlantic and East Pacific belt in which volcanic arcs developed throughout the period are of primary interest.

Later, the overall extent of this volcanism decreased up to the Ashgillian and later grew again up to the third peak in the Late Silurian. In this epoch the main regions of this type of activity were the structures of the Ural-Okhotsk belt—Uralian system and Kazakhstan-Tien Shan region—and also the Appalachian system of the North Atlantic belt, Lachlan system of the West Pacific belt etc.

Subsequent reduction in activity of island arc volcanism continued up to the Emsian stage of the Devonian and was deeper than in the Ashgillian stage. At the end of the Early Devonian, these processes manifested on an extensive scale only in the Uralian system. The new burst in activity attained a peak, fourth in order, in the Frasnian age and its main area again underwent a change. Now the centre of island arc volcanism shifted into mobile structures of the Mediterranean belt: Central European, Caucasus-Pamirs and Yunnan-Malaya regions. However, high activity was maintained in the Uralian system also.

The Frasnian and subsequent peaks are not divided by very deep minima. In the Late Palaeozoic there was gradual reduction of scale of island arc volcanism to minimal levels in the Early Triassic at the beginning of the Mesozoic when it was manifest only in the Nippon-Sakhalin and Cordilleran systems. Some increase is noticed in the Namurian and beginning of the Early Permian: the first was associated with island arc volcanism in the Ural-Okhotsk belt and the second mainly with processes in the West Pacific belt. The general Late Palaeozoic reduction of the role of island arc volcanism was accompanied by a growth of subaerial eruptions on continental margins where extended volcanic belts were often formed. These processes were manifest most extensively in the Early Permian in the Mediterranean and Ural-Okhotsk belts.

Thus a distinct rhythm is noticed in the development of island arc volcanism and is manifest in the alternation of periods of its general intensification and weakening. The period of rhythms was 40–50 m.y. The main ranges of volcanism changed over time. At the end of the Middle Cambrian it was the Ural-Okhotsk belt and at the end of the Middle Ordovician the North Atlantic and Eastern Pacific belts; in the Middle Silurian again the Ural-Okhotsk belt, at the beginning of the Late Devonian the Mediterranean belt, at the end of the Early Carboniferous once again the Ural-Okhotsk belt and at the beginning of the Early Permian the West Pacific belt. The foregoing reactivation epochs of island arc volcanism correspond to global phases of the predominance of compression and subduction. Of them, the Middle Ordovician epoch of reactivation was the most intense.

Attention must be drawn to shift over time of these phases of predominance of compression relative to phases of preponderance of extension and spreading which were manifest in the enlargement of scale of ophiolite formation. A Middle Cambrian peak followed the Early Cambrian phase of ophiolite formation, Llandeilian followed the Llanvirnian, Late Silurian followed the Early Silurian (very weak), Frasnian followed the Eifelian and Namurian followed the Viséan. Thus the period of shift was usually about 10–15 m.y.

6.4.3 Granite Formation and Regional Metamorphism

The most important processes in the tectonic development of the lithosphere are granite formation and regional metamorphism. They played a significant role in the formation of the Earth's continental crust and were concentrated mainly in mobile belts. The affinity of granite formation and regional metamorphism to the concluding (orogenic) stages of development of mobile zones and the direct association of the magnitude of their manifestation with the changes in thermal regime of the Earth as a whole and the lower parts of the Earth's crust and upper mantle in particular are well known. Therefore, data on the development of these processes over time provides the most important information on the evolution of the Earth's interior; its comparison with other palaeotectonic data reveals several important characteristics of the formation of the structure of the Earth's crust on the continents.

Fig. 6–17 gives data on the periods and scale of manifestation of intrusive magmatism in Palaeozoic mobile belts. The main form of such activity was the intrusion of granitoids, which is generally reflected in our diagram without subdivision into petrochemical types. There was no need for this since the object was to determine the more general characteristics and special phenomena; for example, local intrusions of plagiogranites or gabbro in the early and mature stages of development of geosynclinal structures were not taken into consideration. The bulk of the intrusions studied consisted of such associations as granite-leucogranite, leucogranite-alaskite, monzonite-granodiorite-syenite, granodiorite-granite, granite-migmatite and gabbro-granite. Intensification of the scale of granite formation is shown in the diagram by thickening the corresponding sign and, while analysing the overall situation, these bold signs were evaluated on a 3-ball scale from minimum (1 ball) to maximum (3 balls) without taking into consideration the zero values for those time periods when no such magmatism was manifest. The results of total data summation are shown in a histogram (Fig. 6–19).

The duration of phases of granite formation in different structures differed but two characteristic intervals—about 20 and about 60 m.y.—have been distinguished on the basis of this feature. Examples of the first could be the Middle Devonian phase of granitisation of the Kazakhstan-Tien Shan region, Late Ordovician of the Cathaysian system. Examples of the second interval are the Carboniferous Central European region or Middle Carboniferous to Early Permian Uralian system. There were also extremely brief outbursts (as, for example, the Famennian in the Cordilleran system) and the longest period of granite formation in the Appalachian system encompassed the period from commencement of the Devonian to the Early Permian (about 150 m.y.). There is a striking close correlation between the duration of typical phases of granite formation with the most prevalent duration of the formation of island arc complexes (about 50 m.y.).

The period of intensification of the overall activity of intrusive magmatism in each of the five mobile belts did not coincide in time. In the North Atlantic belt such periods were the Silurian and Early Devonian and in the Mediterranean belt the Carboniferous. In the Ural-Okhotsk belt two such periods are noticed in the Palaeozoic: Middle Devonian and the period of greatest activity, Late Carboniferous to Early Permian. In the West Pacific belt such periods were manifest less distinctly but here, too, there were two: in the Devonian and Late Permian to Early Triassic. In the East Pacific belt

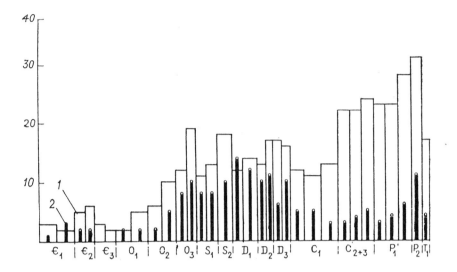

Fig. 6-19. Change of intensity of granite formation and regional metamorphism in the Palaeozoic.

1—intensity of granite formation; 2—intensity of regional metamorphism.

the Late Permian period was distinctly manifest while the Late Silurian was poorly manifest.

In the histogram (Fig. 6-19) summarising the entire data on phases of granitisation in mobile belts, the first aspect to strike the eye is the rising tendency of magnitude of the process analysed from the beginning of the Cambrian to the end of the Permian. It is possible, however, that this general trend is partly due to poor coverage of magmatic complexes of increasingly older age with decreasing overall areas of outcrops of older geological formations compared to younger ones because of the superposition on the former by the latter complexes or their subsequent reworking by some geological processes.

In addition to the above-mentioned general trend in the evolution of Palaeozoic granitisation, three global periods of its intensification are distinctly discernible. In the first (Middle Cambrian) period intrusion of granitoids occurred in the Atlas zone, Polar Urals, Kazakhstan-Tien Shan, Altay-Northern Mongolia regions and the Peru-Bolivian system of the Andes.

In the second period, more prolonged and covering the Late Ordovician to the end of the Carboniferous, three particularly active phases have been distinguished. The first is the Ashgillian when granite formation occurred in the Innuitian system, West European, Kazakhstan-Tien Shan and Altay-Northern Mongolia regions, Uralian, Cathaysian and Lachlan systems,

West Antarctic region and the Peru-Bolivian system. The magnitude of manifestation in the aforesaid structures was different and most intense in the Kazakhstan-Tien Shan, Cathaysian and Lachlan systems. The second phase is the Late Silurian when granitoids intruded into British and Scandinavian zones, West European, Kazakhstan-Tien Shan and Altay-Northern Mongolia regions, Cathaysian and Lachlan systems and also into the Ecuador-Colombian, Peru-Bolivian and Chile-Argentinian segments of the Andes. Maximum magnitude here is noticed in the Kazakhstan-Tien Shan region. The third phase is confined to the Givetian stage of the Devonian period. This phase is distinguished from the preceding ones by a distinct reduction with a very small intermediate increase in the Emsian. At the end of the Middle Devonian, granite formation occurred in the Appalachian, Yunnan-Malaya and Qilianshan systems, Kazakhstan-Tien Shan and Altay-Northern Mongolia regions, Mongolia-Amur, Verkhoyansk-Kolyma, Lachlan and Chile-Argentinian systems. In this phase the process was most intense in the Appalachian and Lachlan systems.

The third period continued throughout the Middle-Late Carboniferous and Permian. There were two phases of intensification of granitisation in it, one each in the Late Carboniferous and the Late Permian, being maximal in the latter phase for the entire Palaeozoic. In the Late Carboniferous granitoids intruded into the Appalachian system, West European, Caucasus-Pamirs and Yunnan-Malaya regions, Uralian and Southern Tien Shan systems, Kazakhstan-Tien Shan and Altay-Northern Mongolia regions, Ob'-Zaisan, Gobi-Hinggan and New England systems and also in the Peru-Bolivian segment of the Andean system. The magnitude of granitisation was maximal in the West European region. In the Late Permian granitisation manifested in most of those structures in which a geosynclinal or orogenic regime was preserved by the end of the Palaeozoic, i.e., in 16 of 23 structures. This process was most intense in Central Asia—in the Southern Tien Shan system and Altay-Northern Mongolia region.

Thus the three Palaeozoic periods of intensification of granite formation can be put 120 m.y. apart while much smaller phases follow one another at intervals of 40 m.y. These assessments are tentative, however, and require to be checked with more extensive data. Nevertheless, they draw attention to the matching of this periodicity with the general rhythm of manifestation of island arc processes (40–50 m.y.) or global Caledonian and Hercynian tectonic cycles. Enlargement of granitisation in the Middle Palaeozoic corresponds in time to conclusion of the Caledonian cycle and Permian maximum to the end of the Hercynian.

Between the periods and phases of intensification of granitisation and epochs of intensification of island arc volcanism there is some correlation only for the first half of the Palaeozoic. This is associated with a change in general tectonic setting in the mobile belts. In the Early Palaeozoic granitisation was more often a consequence of the manifestation of global and regional phases of subduction and roughly coincided with the peaks of island arc volcanism. At the end of the Palaeozoic granite formation occurred under conditions of convergence, i.e., collision of continental masses and formation of Pangaea.

Manifestation of regional metamorphism in the Palaeozoic mobile structures was predominantly localised or in the form of belts and confined to the same regions and palaeotectonic conditions as magmatism. In our generalisations we considered only the phenomena with maximal magnitude which covered considerable areas and were often accompanied by the build-up of granite-granodiorite plutons (see Fig. 6–17). The duration of metamorphism ranged from a few million years to almost 90 m.y. in the Appalachian system but, as a rule, was shorter than the periods of manifestation of granitisation and just a few tens of millions of years. It is important to note that development of metamorphism in each given region and its relationship to magmatic and tectonic processes were peculiar. Further, metamorphism in many cases anticipated granitisation and corresponded in time to the stage of transition from a geosynclinal to an orogenic regime.

Evaluation of the general scale of regional metamorphism shown in Fig. 6–19 is quite tentative. Development of a quantitative method of determination and detailed analysis of all manifestations of metamorphism with reliable determination of composition and age—these are important problems beckoning the attention of scientists in future. The present preliminary evaluation took into consideration only well-known facts. If metamorphic processes continued throughout the period under analysis, their intensity was assigned 2 balls and cases of incomplete duration in a given interval 1 ball.

Thus the histogram of changes in the global scale of metamorphism reveals three periods of increase, coinciding in general, as anticipated, with the periods of intensification of granitisation. Of these, the Late Ordovician-Devonian corresponding in time to epochs of the conclusion of the Caledonian tectonic cycle of development of the Earth is most distinctly manifest. In its maximal Early Devonian phase, processes of regional metamorphism developed in all structures of the North Atlantic belt, Qilianshan and Uralian systems, Kazakhstan-Tien Shan region, Cathaysian and Lachlan systems and the Ecuador-Colombian segment of the Andes. According to palaeomagnetic data, convergence of continental blocks of Laurasia intensified during this period.

The second period is marked at the end of the Permian but its magnitude is somewhat less although, as pointed out above, granitisation was most intensely manifest in the Late Permian. Early Cambrian intensification of metamorphism is discernible with difficulty and is clearly a secondary phenomenon.

6.4.4 High Pressure and Low Temperature Metamorphism

This type of metamorphism is characteristic of the ancient zones of subduction and collision. Glaucophane schists are widespread in most of the fold systems formed in the Palaeozoic. Since geological and radiometric data on the age of glaucophane schists is generally inadequate, their division should be restricted to three age groups: Baikalian or Early Palaeozoic, Caledonian or Middle Palaeozoic and Hercynian or Late Palaeozoic.

Only the blueschists of Anglesey Island with phengite, barroisite and crossite dated 560–550 m.y. by the Ar-Ar method (Caridroit et al., 1987), i.e., falling in the Early Cambrian, can be reliably placed in the first, B a i k a l i a n, group. These schists evidently mark the closure of the Proto-Iapetus or Celtic sea. According to a review by Dobretsov (1988), some more glaucophane schists of the Cordillera—Sierra Nevada and coastal ranges of Oregon and Washington, New Zealand, Eastern Alps and Dinarides, i.e., the West Pacific and Mediterranean belts, and also the Northern and Polar Urals, Southern Urals (Maksyutov complex) and Tien Shan (Makbal and Atbashy), fall in the interval 620–550 m.y. In China, glaucophane schists aged 1100 and 900–700 m.y. are respectively lying in the northern frame of the Tarim and Yangtze cratons unconformably under the Sinian and above the Early Precambrian crystalline basement. Very similar schists aged 806 m.y. are known in western Yunnan under the Ordovician; a reference has also been made to the existence of Late 'Precambrian glaucophane schists in the western part of an ancient massif at the boundary of Heilongjian province and Russian Primorie (Dong Shenbao, 1988).

Glaucophane schists of Norway, the British Isles, Northern Tien Shan, Altay, Sayan-Mongolia region (Gornyi Altay, Western Sayan and Western Transbaikalia), i.e., the eight classic districts of the Caledonides, fall in the second, C a l e d o n i a n, group. Glaucophane schists and eclogites of the same age are also known in much younger fold systems, however. Thus a glaucophane schist belt aged 430–400 m.y. has been traced in the European Hercynides from the Iberian Meseta through the Armorican and Central to the Bohemian massif (House et al., 1978). It is associated everywhere with outcrops of ophiolites and corresponds, according to Matte (1991), to the early phase of obduction pertaining already to the Hercynian cycle of orogeny. Another example is provided by high-pressure metamorphics aged 447 ± 9 m.y. in the eastern Klamath mountains of California (Schenk, 1972), i.e., in the Mesozoic fold system of the Cordillera (N.L. Dobretsov, 1988, places the metamorphics of Oregon and Sierra Nevada in the same group); New Zealand too was mentioned.

Chinese geologists (Dong Shenbao, 1988) provide information about glaucophane schists of Caledonian age in Qilianshan (448 ± 1 m.y.), Inner Mongolia (426 ± 15 m.y.) and south-western Junggar (schists in the Ordovician and pebbles in the Silurian).

The third group, H e r c y n i a n, of glaucophane schist belts includes those prevalent in all the main mobile belts which remained active in the second half of the Palaeozoic. Glaucophane schist belts in the Ural-Okhotsk mobile belt include primarily an extremely extended (2000 km) belt traced from the Polar Urals to the Southern Urals. High-pressure metamorphism (eclogite-glaucophane) was superposed here not only on, and even not so much on the Palaeozoic, as on the Riphean formations; the Maksyutov complex of the Southern Urals is particularly well known in this respect. On this basis it is possible to presume the existence in the Urals not only of Palaeozoic, but also Riphean high-pressure metamorphics. However, radiometric data invariably records Late Palaeozoic age including even the Maksyutov complex. Nevertheless, as pointed out before, N.L. Dobretsov (1988) assigned Late Precambrian age for this complex.

An eclogite-glaucophane schist belt of considerable extent has been established in the Southern Tien Shan system not only in Kirghizian, but also Chinese (Dong Shenbao, 1988) parts of it. In Uzbekistan territory, it has been traced in the north-west up to Sultanuizdag, i.e., up to the junction with the Uralian system. The next belt extends from northern Peri-Balkhash to eastern Junggar. Another belt has been traced from Chara zone of the Ob'-Zaisan system to Mongolia.

In the West Pacific mobile belt manifestation of Late Palaeozoic high-pressure and low-temperature metamorphism is known from western Koryakia (Penzhina ridge) and Japan to Eastern Australia and New Zealand. The Sangun belt in the interior zone of south-western Japan is most interesting, being one of the classic belts of this type. According to present data, it is of Permian age. In the East Pacific belt, glaucophane schists of Late Palaeozoic age have been established in the Cordillera, in Washington, Oregon and California (Sierra Nevada), and in the Andes in the Coastal zone of central and southern Chile, for which ages of 341 ± 10 and 310 ± 11 m.y. respectively have been reported.

6.5 EPOCHS OF HIGH TECTONIC ACTIVITY

The analysis of palaeotectonic development of continents and mobile belts given in the preceding chapters confirms the well-known concept of the continuously discontinuous course of tectonic processes. Tectonic movements encompass much of the Earth's volume, proceed continuously and steadily and exhibit in some epochs a sharp increase in intensity, leading to significant qualitative changes and restructuring of the Earth's crust and evidently of the mantle.

Periodicity of tectonic movements was noted as early as the first half of the nineteenth century when the palaeontologist A. d'Orbigny, a disciple of G. Cuvier, established that the Earth has suffered 32 catastrophes in its history on the basis of distribution of hiatuses and angular unconformities in stratigraphic sequences and differences in the composition of fossil fauna. Subsequently, many researchers, adopting essentially the same method of identifying widely persistent hiatuses and angular unconformities, not only improved this list but also supplemented it. The most prominent of such efforts is the work of the German structural geologist H. Stille, first published in the form of a compendium in 1924 *List of Orogenic Phases* (actually phases of folding).

At the end of the nineteenth century, the works of the French geologist M. Bertrand were published in which another problem closely related to tectonic periodicity, i.e., the problem of tectonic cycles, was examined. Based on a study of the succession of characteristic sedimentary complexes (formations), Bertrand demonstrated that the Alpine geosyncline underwent definite stages of development in the Mesozoic and Cenozoic and that the same stages of development were experienced by Palaeozoic fold systems of Europe and Late Precambrian systems of North America. Thus the Huronian, Caledonian, Hercynian (Variscan) and Alpine tectonic cycles were established. Improvement of these concepts was associated once again with the works of Stille who differentiated the corresponding tectonic era (Huronian was replaced by Assynthian) and, within these eras, several orogenic phases classified as initial, main and concluding phases.

Later, the concept of orogenic phases (folding phase) as well as tectonic cycles encountered serious objections. While Stille and his followers regarded the phases of folding as almost instantaneous and manifest strictly simultaneously in extremely divergent regions and separated by protracted periods of tectonic rest, critics have pointed out that folding could proceed over an extremely prolonged period and simultaneous with subsidence and accumulation of sediments and be reflected in changes in their composition and thickness (synsedimentary folding) and not necessarily in angular unconformities. Critics have also pointed out that the concept of a strict synchronism of orogenic phases and their alternation with prolonged epochs of tectonic rest is based on inaccurate stratigraphic dating of unconformities and that such non-orogenic epochs did not occur in the Earth's history. Thus, in their opinion, there is no need to identify special tectonic (orogenic) phases. This viewpoint subsequently led to the replacement of concepts concerning periodicity, intermittent and discontinuous folding (most distinct manifestation of tectogeny) by an opposite concept, its continuous nature.

Long drawn-out discussions have demonstrated that both these concepts are one-sided. A.A. Pronin carried out a quantitative treatment of data on stratigraphic and angular unconformities in a large number of regional stratigraphic sequences of Phanerozoic formations of all continents. His studies and the results of other investigators have shown that, in fact, all types of tectonic movements including movements responsible for folding reveal a definite periodicity, sometimes attenuating (but not ceasing altogether) and sometimes intensifying. The growth of folds is a prolonged process and some of them proceed even in the process of accumulation of sedimentary complexes in the margins of continents. This is confirmed by data on seismostratigraphy of the sedimentary cover of present-day shelf zones of continents. However, the final configuration of linear fold zones is determined by comparatively brief epochs of fold formation under stress generated in collision zones against a background of uplift in the initial phase of the orogenic stage of development of mobile systems. These tectonic phases are fairly simultaneous only within individual major structural units but a concentration of such phases in certain intervals of geological history (for example, Late Silurian, Late Permian etc.) is noticed on a global scale.

As noted above, the concept of tectonic cycle also underwent (and is undergoing) criticism mainly on the following grounds: 1) the term cycle implies a strict repetition of events which in reality is not noticed; 2) when studying individual mobile zones and systems, the matching of repetition of periods of accumulation of different sedimentary and volcanic complexes and periods of orogeny and folding are far less important than the main trend of evolution, with complication of the general structure and accretion under conditions of different types of continental margins.

In spite of the correctness of these premises, we regard them as inadequate for contradicting cyclicity. From the viewpoint of dialectical logic, cycles need nowhere be wholly identical. On the contrary, since development does not proceed in a closed circle, cycles represent only stages of even larger cycles. Correspondingly, each subsequent cycle differs systematically from the preceding one and a combination of repetition and developmental trend is observed. Repetition is manifest only in extremely general features: predominance of extension conditions is replaced by predominance of compression, ophiolite complexes are replaced by island arc and volcanoplutonic continental complexes, deep sedimentation is replaced by shelf and continental sedimentation and basic magmatism by felsic. Collision processes are usually confined to the end of the cycle. These processes include intense folding and regional metamorphism and enlargement of regions with an orogenic regime. However, some of these features may be lacking or the degree of their manifestation may vary. Thus from cycle to cycle the role of continental sediments increases at the expense of marine sediments and subaerial volcanism at the expense of submarine volcanism. In some mobile belts some cycles may not manifest at all, having merged with the adjoining ones, but they can be recognised statistically on a planetary scale. Their reality is confirmed by quantitative analysis of change in the proportion of terrigenous and carbonate rocks, rates of sedimentation, intensity and composition of volcanism etc., as demonstrated by A.B. Ronov and V.E. Khain in studies on lithological complexes of the World for Hercynian and Alpine cycles and K.B. Seslavinsky for the Caledonian cycle.

There is thus no basis for contradicting the concepts concerning global tectonic cycles and, together with cycles of duration 150–250 m.y. (Bertrand cycles), megacycles (Wilson cycles) with a longer (500–600 m.y.) duration in which the former cycles play the role of individual stages should be distinguished.

So, processes of tectogeny including folding proceed continuously and steadily, revealing a periodic and fairly sharp increase in intensity, leading to significant qualitative changes, reorganisation of structural plan and internal structure of the lithosphere. The much smaller of these maxima of tectonic activity manifest within relatively restricted areas (zones) in the course of not more than 1–2 m.y. are called tectonic phases and the much larger ones represent a group of convergent (in time) phases of duration not exceeding 15–20 m.y. and revealing relative simultaneity on a planetary scale, tectonic epochs. A group of tectonic epochs constitutes a tectonic era up to 250 m.y. in duration, equivalent to tectonic cycles.

Given these preliminary observations, let us now briefly review the Palaeozoic tectonic history of the Earth to identify in its history the main tectonic epochs and eras, i.e., classify tectonic periods to recognise the more common (global) regularities of development of the tectosphere of the Earth. This review is based on the scheme of correlation of the main tectonic, magmatic and metamorphic processes manifest in all mobile belts (see Fig. 6-17) discussed in the preceding sections.

Aside from other information, the scheme shows the duration and magnitude of folding in 29 mobile structures, covering nearly all the main elements of Palaeozoic mobile belts. The amplitude of zigzags of the sign representing folding correspond approximately to the intensity of fold formation processes. This data was summarised taking into consideration also the number of structures in which folding manifested in one or the other epochs and the magnitude of such manifestation was determined on a 4-ball system. A minor folding of duration less than that of a given time interval was assigned 1 ball. An example could be the folding at the end of the Early Cambrian in Anti-Atlas. If such a folding continued for the entire duration of the interval (epoch), as in the Early Permian in the Appalachians, it was assigned 2 balls. Folding of moderate intensity throughout the interval (Late Silurian in Scandinavia) was put at 3 balls and intense (major) folding, also for the entire duration (Late Permian in the Ecuador-Colombian system) at 4 balls. Manifestation of intense and moderate folding but incomplete in time lowered the evaluation by 1 ball. This system of evaluation was, without doubt, schematic, quite approximate, and probably subjective in many cases. Unfortunately, no other approach is available to date for global evaluation of intensity of folding and we hope that this first semi-quantitative estimation will reveal the most common regularities. This view is supported by agreement of the above data with results of quantitative analysis of another palaeotectonic feature, i.e., the average rate of sedimentation, carried out by one of the authors based on independent material and methodology.

The results of calculations are depicted in the form of a histogram showing the changes in general intensity of folding in the Palaeozoic (Fig. 6-20). This figure also shows a curve for changes in average rates of sedimentation in mobile belts (geosynclines). This data was taken from the work of Seslavinsky (1987) and supplemented with data for the Carboniferous and Permian. It may be recalled that determination of average rates of sedimentation is based on the volumetric method developed by A.B. Ronov. This data was supplemented to enable taking into consideration the influence of the antiquity of sediments and the duration of age intervals under comparison. It should be borne in mind that the division of time intervals for curves of sedimentation rate and for a histogram differ, being much smaller for the latter, and all the 32 intervals are proximate in duration (6–15 m.y.). This reveals the average frequency variation of the total intensity of folding with a period of a few tens of millions of years, i.e., planetary

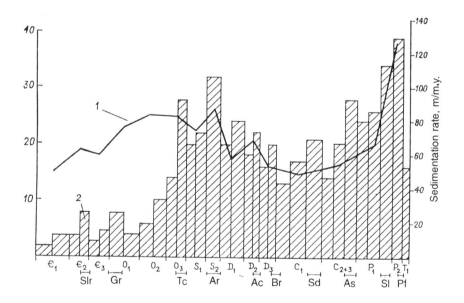

Fig. 6-20. Change of sedimentation rate in mobile belts and intensity of folding in the Palaeozoic.

1—rate of sedimentation; 2—intensity of folding. *Epochs of folding:* Slr—Salairian; Gr—Grampian; Tc—Taconian; Ar—Ardennian; Ac—Acadian; Br—Bretonian; Sd—Sudetan; As—Asturian; Sl—Saalian; Pf—Pfalzian.

tectonic epochs and much longer tectonic eras, but does not help in recognising tectonic phases of duration extending for a few million years. Data on average sedimentation rates is less detailed and helps determine reliably only the tectonic eras and the most prominent tectonic epochs.

The histogram showing changes of general intensity of folding draws attention foremost to the build-up of this process over time from the commencement of the Palaeozoic to its end. We have already noted a similar tendency when analysing granite formation and expressed the view that such a tendency could be partly apparent since its association with some objective phenomena is evident. Firstly, the extent of coverage of studies on much older geological bodies is invariably poorer than on younger ones and this renders dating of older phases of folding difficult. Secondly, on a contemporary surface, the areas of outcrops of geological bodies of a given age are much less, the older these formations. This is due to the subsequent partial reworking by much later geological processes or superposition, for example, by a cover of platform deposits etc.

So far, we have no means of evaluating quantitatively the effect of such objective phenomena on the results of our analysis. We can only tentatively, therefore, refer to the most common trend of development of folding deformation and tectonic activity for the Palaeozoic as a whole. For a final solution to this problem, similar data should be drawn at least for the Mesozoic and Cenozoic and desirably for the Late Precambrian. Attention should be drawn, however, to the position of the Late Palaeozoic in the Late Riphean-Palaeozoic megacycle of tectonic development of the Earth reflected in the transformations of structure of the lithosphere from Pangaea I to Pangaea II. This stage concludes with the Late Palaeozoic with a build-up of collision processes and it is wholly possible that the scale of folding deformation actually enlarged.

In the background of general build-up, the histogram reveals three periods of intensification of tectonic activity coinciding with the concluding stages of tectonic eras and several epochs of orogeny concentrated in these periods.

The first period covers the Middle Cambrian, which includes the Maya age representing culmination of the Salairian epoch. Folding occurred on the largest scale in this epoch in the Kazakhstan-Tien Shan and Altay-Northern Mongolia regions. It may be recalled that distinct peaks of granite formation and formation of island arc complexes are associated with the end of the Middle Cambrian (Fig. 6–19). Thus there is no doubt about the global character of the Salairian epoch of orogeny marking intense compression in the tectonic history of the Earth. The Middle Cambrian tectonic reactivation is confirmed by data on the average rates of sedimentation in geosynclines. This data points to the subsequent fall in tectonic activity at the verge between the Cambrian and Ordovician. For the commencement of the Ordovician, the sum total of our data points to a preponderance of extension in the lithosphere of the Earth. The highest peak of ophiolite formation, which occurred in one-half of all the mobile structures, is confined to the Tremadocian.

The second period of tectonic activity intensification continued from the end of the Early Ordovician to the Devonian and its magnitude is significantly higher. It included several epochs of orogeny proximate in time. The first, the Grampian, manifested in the Arenigian. Folding then occurred in Britain, Scandinavia and the Lachlan system of Australia. The second, the Taconian, reached maximum in the Ashgillian. This was a period of intense deformation in the Appalachian and British-Scandinavian systems, in the Kazakhstan-Tien Shan region, in the Lachlan system, the Chile-Argentinian segment of the Andes and many others. A sharp intensification of granite formation and metamorphism was also established in this epoch. Only in the Mediterranean belt has the Taconian orogeny almost not been detected.

The next epoch of orogeny, Ardennian or Scandian, is confined to the Late Silurian. It concludes the Caledonian era of tectonic development and is manifest extremely well. Folding has been established at this time in all mobile belts but was manifest most intensely in the British-Scandinavian and Lachlan systems and in the Kazakhstan-Tien Shan region. Peaks of granite formation and island arc volcanism and also a sharp minimum in the formation of ophiolites are likewise associated with the Late Silurian (see Figs. 6–18 and 6–19). This was a period of increase in average rates of sedimentation and manifestation of global regression (see Figs. 6–16 and 6–20). The sum total of the aforesaid characteristics points to the exceptionally important role of the Ardennian-Scandian epoch in the overall tectonic development of the Earth; undoubtedly, it had the nature of a planetary epoch of compression.

Commencement of the Early Devonian, compared to the Late Silurian, was a period of relative tectonic quiescence. Fold formation processes ceased in all the structures of the Ural-Okhotsk belt and the magnitude of granitisation and sedimentation rates decreased in many others; however, intense ophiolite formation and subaerial volcanic activity commenced in the form of marginal continental belts. An exception here is regional folding in the Central European system, involving Caledonian volcanosedimentary complexes at the end of the Gedinnian age. Palaeotectonic criteria for the end of the Early Devonian (Emsian age) are contradictory. Some intensification of fold formation and granitisation helps identify tentatively in the Emsian yet one more epoch of orogeny. But, on the other hand, this was a period of extremely steep drop in the activity of island arc processes and increase in ophiolite formation. The nature of such a combination remains as yet an enigma. There was perhaps some error in the initial data.

The next epoch of orogeny, Acadian, is more distinctly manifest with a peak in the Givetian. Fold formation occurred in the Appalachians, Scandinavia, Central Europe, New England and other regions. In this epoch granite formation intensified and the development of island arc complexes increased. This was the last epoch in the second period of overall intensification of tectonic processes. Later, some secondary variations occurred in the overall intensity of fold formation and the commencement of the Carboniferous is characterised by a fall in tectonic activity. It is interesting that the Bretonian epoch of orogeny is traditionally associated with the Tournaisian stage. However, this orogeny was only regional and manifested most intensely only in the West European region.

The third period of overall intensification of tectonic activity extends in time from the Viséan to the end of the Permian. The former, as well as the latter, includes several epochs and reveals gradual increase with a maximum in the Late Permian. Since detailed information on sedimentation rates for the Late Palaeozoic is not available, different tectonic epochs cannot be identified, but the data accords well with the overall trend of fold formation (see Fig. 6–20).

In the Viséan age, the Sudetan epoch of orogeny, the first in this period, was manifest. The end of the Early Carboniferous was a period of intense fold formation in many parts of the Mediterranean belt, in the East Australian region, Cordilleran system and others and also intensification of granite formation. The magnitude of island arc processes and ophiolite formation diminished and this tendency held until the end of the Palaeozoic, reflecting a gradual change in the form of tectonic processes as a result of growing magnitude of orogeny, caused by collision of continents grouping in the Pangaea.

The Late Carboniferous Asturian epoch was more energetic than the Sudetan. This was a period of major folding in the Variscides of the West European region, in the Uralian system and in other structures. Moreover, at the end of the Carboniferous, granitisation processes enlarged and were manifest on a different scale in all mobile belts but maximally in the Mediterranean and Ural-Okhotsk belts.

In the Palaeozoic era, the latter orogenies (Saalian and Pfalzian) were confined to the second half of the Permian with maxima in the Artinskian and Late Permian. Folding occurred in most mobile structures at the end of the Permian. In the Cordilleran system in the Ecuador-Colombian segment of the Andes, the Ob'-Zaisan and Gobi-Hinggan systems, linear folding due to general piling up occurred. In other systems, for example the Appalachian, folding of this age was less intense. In the Late Permian the average rate of sedimentation in geosynclines rose sharply. At the end of the Early Permian and especially in the Late Permian epoch, granitisation attained maximum importance for the Palaeozoic and metamorphic processes intensified. All this tends to favour a predominance of compressive stress and collisional orogeny and agrees with palinspastic reconstructions reflecting the evolution of Pangaea towards the end of the Palaeozoic. There is no doubt that the Late Permian represented a global epoch of tectogeny in spite of a sharp reduction in intensity of all forms of volcanic activity.

Thus, in the Palaeozoic there was not even a single period of total quiescence in respect of tectonic activity. Fold formation, granitisation, metamorphism and volcanic processes were manifest continuously on one scale or another in one or the other regions of the Earth. It may be assumed that some overall increase in their magnitude occurred by the end of the Palaeozoic.

Three periods of very sharp increase in tectonic activity may be distinguished and there is no doubt about their global character. These are: Middle Cambrian, Late Ordovician-Silurian and Late Carboniferous-Permian periods of intensification of compression and collision processes. They are separated by periods of relative predominance of extension whose course was not smooth, but was disturbed by secondary epochs of orogeny. The duration of the first period was 120 m.y. and of the second 150 m.y. These periods correspond to tectonic eras, i.e., Caledonian and Hercynian cycles, in the development of the Earth.

The Late Silurian and Early Permian represented epochs of maximum tectonic activity or 'major' epochs of tectogeny in the Palaeozoic. This has been established by a complex of independent features, such as folding, sedimentation rate, granite formation, regional metamorphism and island arc processes. The Late Cambrian to Early Ordovician and commencement of the Early Carboniferous represented epochs of similar rank with minimum activity and predominance of extension.

Conclusions

The leitmotif of tectonic development of the Earth at the end of the Precambrian and in the Palaeozoic, was the transition from Pangaea I to Pangaea II. Deciphering this brief formula, it may be stated that the Late Precambrian to Palaeozoic evolution of the Earth included destruction and disintegration of the supercontinent Pangaea I formed in the Middle Proterozoic, followed by formation and accretion of a new supercontinent, Pangaea II. The first stage (destruction) could very roughly be regarded as extending from the middle of the Late Riphean to the Middle Devonian (850–380 m.y. ago) and the second stage from the Middle Devonian to the Middle Triassic (380–230 m.y. ago), i.e., partly covering the Early Mesozoic. The destructive nature of the first stage is emphasised by the practically continuous manifestation of ophiolite magmatism, i.e., ocean formation, while the accretionary and constructive nature of the second stage is established by the growing rate of folding right from the Late Ordovician, metamorphism, granite and molasse formation. It is necessary to state, however, that this division is quite tentative and is justified only on a global scale. Ophiolite (oceanic crust) formation prevailed, although on a relatively modest magnitude, even in the Late Palaeozoic but was confined to the Pacific Ocean fringe of Pangaea and to apophyses of the Pacific Ocean[1] that penetrated from the east (in present-day geography). Granites and molasse were formed already in the Early Palaeozoic but on a limited scale, their manifestation being largest in the second half of the Palaeozoic. This is also true of marginal volcanoplutonic belts, which acquired a perceptible role in the structure of the Earth's crust commencing only from the Devonian. Island arc volcanism and regional metamorphism were most distinct in the Middle Palaeozoic, the former in the Middle Ordovician and Early Carboniferous and the latter in the Late Ordovician to Late Devonian. However paradoxical this might appear at first glance, continental rifting and magmatism under conditions of accretion and consolidation of Pangaea II reveal distinct growth in the latter half of the Palaeozoic commencing from the Devonian. This can be explained only by assuming that they somehow anticipated the

---

[1] The Ancient Pacific Ocean refers here to the oceanic hemisphere of the Earth, Panthalassa, of which the present-day Pacific Ocean is a relict.

disintegration of Pangaea II in the Mesozoic: accretion continued along the periphery while a reverse tendency began to manifest at the centre.

All these changes fall within the framework of the concept that the interval of the history of the Earth under study represents one of the megacycles corresponding on a global scale to the so-called Wilson cycle. At the same time, it corresponds to commencement of one of the major stages of development of the tectonic life of the Earth, i.e., the stage of formation of present-day type of manifestation of plate tectonics as distinguished from 'tectonics of minor plates' of the Early Proterozoic (Khain and Bozhko, 1988). The onset of this stage is supported by very marked manifestation of ophiolites and glaucophane schists, extensive development of island arc volcanics, large-scale formation of marginal volcanoplutonic belts and accumulation of molasse. Although all these formations, with the possible exception of glaucophane schists, are known even in the much earlier stages of development of the Earth, they manifested together and at such magnitude for the first time in this stage; some of them, for example volcanoplutonic belts, were considerably developed only in the second half of the Palaeozoic. Separation of ancient platforms (cratons) and mobile (geosynclinal) belts fringing or dividing them occurred in this same period.

It must also be said that the diversity of organic life on the Earth began with this megastage. This reveals a clear, albeit indirect association with the development of full-scale plate tectonics, i.e., with disintegration of Pangaea into individual continents with an extended and dissected coastline and formation of extensive shelf seas with well-warmed waters and upwelling at their boundaries with deepwater parts of basins.

In the general background of progressive development of the Earth from the Late Precambrian to the Late Palaeozoic, features of relatively brief changes, i.e., cyclicity of a higher order stand out prominently in graphs (see Figs. 6-18, 6-19 and 6-20). This is the same cyclicity which is well known to geologists from the time of Marcel Bertrand; such cycles are called Bertrand cycles as distinct from Wilson cycles (megacycles). Two such main cycles are noticed within the frame of the Palaeozoic proper. These are the Caledonian and Hercynian which border the Middle and Late Devonian. These are most distinctly manifest in a change of sedimentation rate in mobile belts with two maxima, one in the Ordovician-Silurian and the other in the Permian; in a change in intensity of fold-overthrust deformation with peaks in the Late Ordovician, Late Silurian, Late Carboniferous and Late Permian (see Fig. 6-20); degree of development of island arc complexes, maxima in the Middle to Late Ordovician, end of Silurian to beginning of the Devonian and in Late Devonian to Early Carboniferous (see Fig. 6-18) and volcanoplutonic belts, maxima in Devonian and Late Palaeozoic; partly in the magnitude of granite formation most extensively manifest from the Late Ordovician to the Late Devonian and later with a relative fall in the

Early Carboniferous and Middle Carboniferous to Late Permian; and regional metamorphism in the Middle Ordovician to Early Carboniferous and Late Permian (see Fig. 6–19).

Since the starting point of our graphs is the verge between the Vendian and Cambrian, they do not reflect manifestation of the Baikalian (Cadomian, Brazilian, Pan-African) cycle of orogeny while a small increase in granite formation in the Middle Cambrian and the fact that independent peaks of ophiolite formation and continental rifting occurred in the Early Cambrian (and Vendian) reflect to some extent the Salairian cycle of a higher order. However, considering the Late Precambrian-Palaeozoic megacycles as a whole, three, not two, Bertrand cycles are recognised, viz. Baikalian, Caledonian and Hercynian, each occupying a different position in the megacycle and hence of varying importance. The Baikalian cycle was preparatory; nevertheless, it led to the final consolidation of Western Gondwana (South America plus Africa) and then of Gondwana as a whole, to the formation of an extensive Epibaikalian Platform along the western, northern and northeastern peripheries of Gondwana which, however, later experienced partial destruction and, even more to the north, development of a similar Barents-Pechora Platform. The Caledonian cycle occupies a central place in the megacycle; its commencement was marked by highly significant destruction (abundance of ophiolites!) while the end was accompanied by extensive mountain building and granite formation, and the appearance of a major generation of volcanoplutonic belts, the first in the Phanerozoic. Island arc volcanism flourished in the Caledonian cycle and at the very beginning of the Hercynian cycle. Large regions of much earlier cratonisation adjoining epigeosynclinal orogens were involved in mountain formation from the end of this Caledonian cycle in the Devonian.

The Hercynian cycle plays the role of a conclusive cycle in the megacycle under consideration (its complementary character relative to the Caledonian was noticed by N.S. Shatsky). New formation of oceanic crust, its starting point, occurred in the Ordovician-Silurian-first half of the Devonian, i.e., essentially coincides with the Caledonian cycle. Folding, regional metamorphism, granite formation and mountain building (molasse) attain maximum in the Hercynian cycle that concluded the formation of Pangaea II. Simultaneous with the latter half of this cycle, however, a new period of ocean formation commenced. This period concluded in folding and mountain building even in the Mesozoic, i.e., the initial and concluding stages of adjoining cycles are also mutually overlapping each other at this time.

Alternation of epochs (phases) of high tectonic activity with epochs of relative (but not total) decrease serves as manifestation of cyclicity of an even higher order than the Bertrand cycles. In the megacycle under consideration there were 11 such phases for about 470 m.y.; almost all were predicted in Stille's 'List' except the Grampian (Finnmarken, Sardian and

Delamerian) in the Early Ordovician, which differed from the Middle Cambrian Salairian phase. These phases repeat roughly at 40–45 m.y. intervals, which generally corresponds to the cyclicity noticed by many researchers based on data of analysis of sedimentary sequences. Almost all the phases shown in Fig. 6–20 are of genuine global significance because manifested in different regions, often widely separated. Yet not one of them covered all the mobile belts of corresponding epochs, let alone all segments of the same belt. Moreover, there was evidently no strict synchronism in the manifestation of individual phases. A detailed study of Rheinische Schiefergebirge by West German scientists suggests a gradual migration of folding across the strike of mobile systems which, in our view, does not eliminate the uneven course of this process.

The association between fold-overthrust deformation, regional metamorphism and granite formation remains omnipresent. Extra-geosynclinal granites, apart from marginal volcanoplutonic belts, are also encountered in epiplatformal, secondary orogens (Baikalian region, Sierra Pampeanas and Northern Patagonia massif in South America). High-pressure and low-temperature metamorphism usually occurs, as pointed out by Dobretsov (1988), immediately after the formation of ophiolites.

According to A.S. Monin and O.G. Sorokhtin, alternation of major periods of formation and disintegration of the supercontinents, the Pangaeas, in the history of the Earth uniting the entire mass of continental crust formed until then and concentrated in the Southern Hemisphere, is explained by a cardinal reorganisation of mantle convection, i.e., replacement of bicellular by monocellular convection. Such is the explanation for megacyclicity; in so far as Bertrand cyclicity is concerned, it can be explained by a change of plan in convective currents only in the upper mantle. Reorganisation of plate kinematics of a still smaller scale may serve as a factor of periodic reactivation of tectonic deformations, granitic magmatism and regional metamorphism. It is quite evident that similar changes in the intensity of heat flow from the interior of the Earth correspond to this cyclicity of the third order: long periods of predominant accumulation of heat should alternate with rather brief phases of intense heat liberation in the phase of reorganisation of a convective regime. In particular, as already pointed out in the literature, the accumulation of heat under supercontinents with a thick and poorly conducting crust and lithosphere may serve as a precondition for their subsequent disintegration.

It is quite possible that the periodic change of the Earth's radius, i.e., pulsation of its volume, is associated with the periodic heating of the interior and 'discharge' of heat in the phase of high tectonic activity. In this context a striking feature, as already noted, is the near total cessation of ophiolite formation and, perhaps, spreading too in the Late Devonian to Early Carboniferous, i.e., at the commencement of the Hercynian tectonic cycle, with island arc volcanism, an index of subduction, continuing simultaneously. It could be assumed that intense general compression of the Earth took place at this time, but such a conclusion does not find confirmation in the history of tectonic deformations, since their intensity was not particularly high in this epoch. Further, rifting continued in the same epoch in the central parts of continents. Thus there is no basis for presuming alternation of epochs of general compression and overall extension; there could only have been relative or local predominance of one or the other process.

Literature Cited

Aceñolaza, F.G. (1973). Sobre la presencia de trilobites en las cuarcitas del Grupo Meson, de Potrerillos, Provincia de Salta. *Rev. Asoc. Geol. Arg.*, vol. 28, no. 3.

Aceñolaza, F.G. (1982). El sistema Ordovicico en Sudamerica. *Acta Geologica Lilloana*, 16 (1): 77–91.

Adamia, Sh.A. (1984). Pre-Alpine basement of Caucasus, its composition, structure and formation. In: *Tectonics and Metallogeny of Caucasus*, pp. 3–104. Metsniereba, Tbilisi.

Aitchison, J.C., Ireland, T.R., Blake, M.C., Jr. and Flood, P.G. (1992). 530 m.y. zircon age for ophiolite from the New England orogen: oldest rocks known from eastern Australia. *Geology*, vol. 20, pp. 125–128.

*Akhmedzhanov, M.A., Abdullaev, R.N. and Borisov, O.M. (1979). *Lower Palaeozoic of Central and Southern Tien Shan*. FAN, Tashkent.

*Aliev, M.M., Vysotsky, V.I. and Golenkova, N.P. (1979). *Geological Structure and Oil and Gas Fields of Northern Africa, Near and Middle East*. ELM, Baku.

*Altukhov, E.N. (1986). *Tectonics and Metallogeny of Southern Siberia*. Nedra, Moscow.

Amos, A.J. (1981). Correlación de las formaciones carbónicas y pérmicas de Argentina. *An. Acad. Brasil. Ciênc.*, 53 (2): 347–356.

Anderton, R. (1982). Dalradian deposition and the Late Precambrian-Cambrian history of the North Atlantic region: A review of the early evolution of the Iapetus Ocean. *J. Geol. Soc.* 139 (4): 421–431.

*Aplonov, S.V. (1987). *Geodynamics of Early Mesozoic Ob' Palaeo-ocean*. Institute of Oceanology, Moscow.

Atlas of the Palaeogeography of China. (1985). Beijing, China.

Autran, A. (1980). Evolution structurale du Protérozoïque aux distensions posthercyniènnes. *Mém. BRGM*, no. 107, pp. 10–17.

Autran, A. and Cogné, J. (1980). La zone interne de l'orogène varisque dans l'Ouest de la France et sa place dans le développement de la chaîne hercyniènne. In: *Colloque C 6. Géologie de l'Europe*, pp. 90–111. 26 CGI. Villeneuve d'Ascq.

Asterisked works are in Russian—General Editor.

Avcin, M.J. and Koch, D.I. (1979). The Mississippian and Pennsylvanian (Carboniferous) in the United States. *Geol. Surv. Profess. Pap.*, no. 1110 M-DD, pp. 1–13.

Azcuy, C.L. (1985). Late Carboniferous palaeogeography and stratigraphy of Argentina. In: *C.R. 10ème Congr. Int. Stratigr. et Géol. Carbonifère*, vol. 4, pp. 281–293. Madrid.

Babale, H.A. (1987). Palaeogeographic and tectonic implications of the Golconda allochthon, southern Toiyabe Range, Nevada. *Bull. Geol. Soc. Amer.*, 99 (2): 231–243.

Babin, C., Arnaud, A. and Blaise, J. (1976). The Ordovician of the Armorican massif (France). In: *The Ordovician System*, Bassett, M.G. (ed.). Univ. Wales Press and Nat. Mus. Wales, Cardiff pp. 359–385.

Bahafzallah, A., Jux, P. and Omara, S. (1981). Stratigraphy and facies of the Devonian Jauf Formation, Saudi Arabia. *Neues Jahrb. Geol. und Paläontol. Monatsh.*, no. 1, pp. 1–18.

Bahlburg, H. (1987). Sedimentology, petrology and geotectonic significance of the Palaeozoic flysch in the Coastal Cordillera of northern Chile. *Neues Jahrb. Geol. und Paläontol. Monatsh.*, no. 9, pp. 527–559.

Bahlburg, H., Breitkreuz, Ch. and Zeil, W. (1987). Palaeozoic basin development in northern Chile (21°–27° S). *Geol. Rdsch.*, 76 (2): 633–646.

Barber, A.J. and Max, M.D. (1979). A new look at the Mona complex (Anglesey, North Wales). *J. Geol. Soc.*, 136 (4): 407–432.

Bard, J.P., Burg, J.P. and Matte, Ph. (1980). La chaîne hercyniènne d'Europe occidentale en terme de tectonique de plaques. *Mém. BRGM*, no. 108, pp. 233–242.

Barr, S.M. and Macdonald, A.S. (1987). Nan River suture zone, northern Thailand. *Geology*, 15 (10): 907–910.

Bassett, M.G. (1984). Lower Palaeozoic Wales—a review of studies in the past 25 years. *Proc. Geol. Assoc.*, 55 (4): 291–311.

Beauchamp, J. and Izert, A. (1987). Early Carboniferous basins of the Atlas Meseta domain (Morocco): Sedimentary model and geodynamic evolution. *Geology*, 15 (9): 797–800.

Bebien, J., Garny, Cl. and Rocci, G. (1980). La place du volcanisme dévono-dinantien dans l'évolution magmatique et structurale de l'Europe moyenne varisque au Paléozoïque. *Mém. BRGM*, no. 108, pp. 213–225.

Behr, J.-H., Walliser, O.H. and Weber, K. (1980). The development of the Rhenohercynian and Saxo-Thuringian zones of the mid-European Variscides, *Mém. BRGM*, no. 108, pp. 77–89.

Behr, J.-H., Engel, W. and Franke, W. (1984). The Variscan Belt in Central Europe: Main structures, geodynamic implications, questions. *Tectonophys.*, 109 (1–2): 15–40.

*Bekker, Yu.R. (1982). Precambrian Molasse of Europe. Author's abstract of Doctoral Dissertation, All-Union Geological Research Institute (VSEGEI), Leningrad.

*Bekker, Yu.R. (1987). Geological map of the basement, Pre-Vendian cover and fold framing of the Russian platform. *Sov. Geologiya*, no. 8, pp. 63–71.

*Belichenko, V.G. (1977). *Caledonides of the Baikalian Mountain Region.* Nauka, Novosibirsk.

*Belousov, A.F., Lapin, B.N. and Polyakov, G.V. (1978). Volcanic complexes of the Upper Precambrian and Palaeozoic in the Altay-Sayan region. *Tr. VSEGEI*, no. 270, pp. 97–133.

*Belov, A.A. (1981). History of tectonic development of the Alpine fold region in the Palaeozoic. *Tr. GIN AN SSSR*, no. 347.

*Belyakov, L.N., Dembovsky, B.Ya. and Kishka, N.P. (1985). Nappe structure of Pay-Khoy. *Dokl. AN SSSR*, 282 (1): 151–153.

*Belyakova, L.T. (1985). Structural-formational zoning of the Baikalides of Timan-Ural region. In: *Problems of Precambrian and Lower Palaeozoic Geology of the Urals*, pp. 21–24. Nauka, Moscow.

Bender, F. (1975). Geology of the Arabian Peninsula. Jordan. *Geol. Surv. Profess. Pap.*, no. 560–561, 136 pp.

Bevins, R.E., Kokelaar, B.P. and Dunkley, P.N. (1984). Petrology and geochemistry of Lower to Middle Ordovician igneous rocks in Wales: A volcanic arc to marginal basin transition. *Proc. Geol. Assoc., London*, 95 (4): 337–347.

Bluck, B.J. (1985). The Scottish paratectonic Caledonides. *Scot. J. Geol.*, 21 (4): 437–464.

*Bogatikov, O.A. and Tsvetkov, A.A. (1988). *Magmatic Evolution of Island Arcs.* Nauka, Moscow.

*Bonchev, E. (1985). Principal moments in Caledonian and Hercynian development of the Balkan peninsula. *Geol. Balcan*, no. 8, pp. 61–82.

Bond, G.C., Nickeson, P.A. and Kominz, M.A. (1984). Break-up of a supercontinent between 625 m.y. and 555 m.y.: New evidence and implications for continental histories. *Earth Planet Sci. Lett.*, 70 (2): 325–345.

Bond, G.C., Christi-Blick, N. and Kominz, M.A. (1985). An Early Cambrian rift to post-rift transition in the Cordillera of western North America. *Nature*, 315 (6022): 742–746.

Borch, C.C. (1980). Evolution of Late Proterozoic to Early Palaeozoic Adelaide fold belt, Australia: Comparison with Post-Permian rifts and passive margins. *Tectonophys.*, vol. 70, pp. 115–134.

*Borisenok, V.I., Gerasimova, N.A. and Zaitsev, Yu.A. (1979). Stratigraphy of Lower Palaeozoic volcanic-siliceous beds of central Kazakhstan. *Byull. MOIP, Otd. Geol.*, 54 (4): 54–66.

Borrello, A.V. (1969). Los Geosinclinales de la Argentina. *Dir. Nac. Geol. Miner., Anal., Buenos Aires*, no. 24, pp. 17-73.

Boucot, A.G. (1992). Comment on E.M. Moores and J.W.D. Dalziel papers. *Geology*, 20 (1): 87-88.

Boulin, J. (1991). Structures in Southwest Asia and evolution of the Eastern Tethys. *Tectonophysics*, vol. 196, pp. 211-268.

Bourrouilh, R., Coccoza, T. and Demange, M. (1980). Essai sur l'évolution paléogéographique, structurale et métamorphique du Paléozoïque du Sud de la France et de l'Ouest de la Méditerranée, In: *Colloque C 6 Géologie de l'Europe*, pp. 159-188. 26 CGI, Villeneuve d'Ascq.

Boyer, C., Brillanceau, A. and Brousse, R. (1975). Volcanisme paléozoïque sousmarin à pillow-lavas en Vendée. *C.R. Acad. Sci.*, 282 (10): 973-975.

*Bozhko, N.A. (1984). *Late Precambrian of Gondwana*. Nedra, Moscow.

Bradshaw, M.A. (1991). The Devonian Pacific margin of Antarctica. In: *Geological Evolution of Antarctica*, pp. 193-197. M.R.A. Thomson et al. (eds.). Cambridge University Press, Cambridge.

Breemen, O. van, Aftalion, M. and Bowes, D.R. (1982). Geochronological studies of the Bohemian massif, Czechoslovakia, and their significance in the evolution of Central Europe. *Trans. Roy. Soc. Edinburgh, Earth Sciences*, vol. 73, pp. 89-108.

Brekke, H. and Furnes H. (1984). Lower Palaeozoic convergent plate margin volcanism on Bamlø (Balm), South Norway, and its bearing on the tectonic environments of the Norwegian Caledonides *J. Geol. Soc.*, 141 (6): 1015-1032.

*Brodskaya, N.G., Gavrilov, V.K. and Solovyeva, N.A. (1979). Sedimentation and volcanism in the Late Palaeozoic-Early Mesozoic Sakhalin basin. In: *Sedimentation and Volcanism in Geosynclinal Basins*. Tr. GIN AN SSSR no. 337, pp. 82-129.

Buggisch, W., Marzela, C. and Hügel, Ph. (1979). Die fazielle und paläogeographische Entwicklung der infrakembrischen bis ordovizischen Sedimentes im Mittleren Antiatlas um Agdz (S. Marokko). *Geol. Rdsch.*, 68 (1): 195-224.

Burchett, R.R. (1979). The Mississippian and Pennsylvanian (Carboniferous) System in the United States—Nebraska. *Geol. Surv. Profess. Pap.*, no. 1110 M-DD, pp. 1-15.

*Butov, Yu.P. (1985). Some problems of Pre-Mesozoic stratification of the Sayan-Baikalian mountain Region. Article 1. Crisis in the traditional stratigraphic scheme. *Byull. MOIP, Otd. Geol.*, 60 (6): 40-53.

Caputo, M.V. (1985). Late Devonian glaciation in South America. *Palaeogeogr., Palaeoclimat., Palaeoecol.*, 51 (1-4): 291-317.

Caputo, M.V. and Crowell, J.C. (1985). Migration of glacial centres across Gondwana during the Palaeozoic era. *Geol. Soc. Amer. Bull.*, 96 (8): 1020-1036.

Caridroit, M., Faure, M. and Charvet, J. (1987). Nouvelles données stratigraphique et structurales sur le Paléozoïque supérieur des zones internes du Japon sud-ouest. Un essai sur l'orogenèse permienne. *Bull. Soc. Géol. Fr.*, 8 (4): 683–691.

Carlier, G., Grandin, G. and Laubacher, G. (1982). Present knowledge of the magmatic evolution of the Eastern Cordillera of Peru. *Earth-Sci. Rev.*, 18 (3–4): 253–283.

Cas, R. (1983). Palaeogeographic and tectonic development of the Lachlan fold belt, Southeastern Australia. *Geol. Soc. Austral.*, Spec. Publ., no. 10, 104 pp.

Cas, R.A.F. and Jones, J.G. (1979). Palaeozoic interarc basin in Eastern Australia and a modern New Zealand analogue. *N.Z.J. Geol. and Geophys.*, 22 (1): 71–85.

Casey, J.F., Elthon, D.L. and Siroky, F.X. (1985). Geochemical and geological evidence bearing on the origin of the Bay of Islands and Coastal Complex ophiolites of Western Newfoundland. *Tectonophys.*, vol. 116, pp. 1–40.

Castanos, A. and Rodrigo, L.A. (1980). Palaeozoico superior de Bolivia. *An. Acad. Brasil. Ciênc.*, 52 (4): 851–866.

Castellarin, A. and Vai, G.B. (1981). Importance of Hercynian tectonics within the framework of the Southern Alps. *J. Struct. Geol.*, 3 (4): 477–486.

Chaloupsky, J. (1978). The Precambrian tectonogenesis in the Bohemian massif. *Geol. Rdsch.*, 67 (1): 72–90.

Chang, W.T. (1980). A review of the Cambrian of China. *J. Geol. Soc. Austr.*, 27 (1–2): 137–150.

Chi-ching, H. (1978). An outline of the tectonic characteristics of China. *Eclog. Geol. Helv.*, 71 (3): 611–635.

Chronic, J. (1979). The Mississippian and Pennsylvanian (Carboniferous) Systems in the United States—Colorado. *Geol. Surv. Profess. Pap.*, no. 1110 M-DD, pp. 1–25.

*Chumakov, N.M. (1984). Major glacial events of the past and their geological significance. *Izv. AN SSSR, Ser. Geol.*, no. 7, pp. 35–53.

Churkin, M. and McKee, E.H. (1974). Thin and layered subcontinental crust of the Great Basin of western North America inherited from Palaeozoic marginal ocean basins? *Tectonophys.*, 23 (1/2): 1–15.

Churkin, M. and Eberlein, G.D. (1977). Ancient borderland terranes of the North American Cordillera: Correlation and microplate tectonics. *Geol. Soc. Amer Bull.*, vol. 88, pp. 769–786.

Churkin, M., Carter, C. and Trexler, J. (1980). Collision-deformed Palaeozoic continental margin of Alaska—foundation for microplate accretion. *Bull. Geol. Soc. Amer.*, 91 (1): 648–654.

Churkin, M., Jr., Poster, H.L. and Chapman, R.M. (1982). Terranes and suture zones in East Central Alaska. *J. Geophys. Res.*, 87 (B5): 3718–3730.

Cloetingh, S., McQueen, H. and Lambeck, K. (1985). On a tectonic mechanism for regional sea level variations. *Earth Planet. Sci. Lett.*, vol. 75, pp. 157–166.

Cloud, P. A. and Glaessner, M. F. (1982). The Ediacaran period and system: Metazoa inherit the Earth. *Science*, vol. 217, pp. 783–792.

Cloud, P., Wright, L.A. and Williams, E.G. (1974). Giant stromatolites and associated vertical tubes from the Upper Proterozoic Noonday dolomite, Death Valley region, eastern California. *Geol. Soc. Amer. Bull.*, vol. 85, pp. 1869–1882.

Cogné, J. and Wright, A.E. (1980). L'orogène cadomien. In: *Colloque C 6. Géologie de l'Europe*, pp. 29–55. 26 CGI. Villeneuve d'Ascq.

Coira, B., Davidson, J. and Mpodozis, C. (1982). Tectonic and magmatic evolution of the Andes of northern Argentina and Chile. *Earth-Sci. Rev.*, 18 (3–4): 303–332.

Collinson, J.W. (1991). The Palaeo-Pacific margin as seen from East Antarctica. In: *Geological Evolution of Antarctica*, pp. 199–204. M.R.A. Thomson et al. (eds.). Cambridge University Press, Cambridge.

Coney, P.J. et al. (1990). The regional tectonics of the Tasman orogenic system, eastern Australia. *J. Struct. Geol.*, vol. 12, pp. 519–543.

Conrad, J., Massa, D. and Weyant, M. (1986). Late Devonian regression and Early Carboniferous transgression on the northern African platform. *Ann. Soc. Geol. Belg.*, 109 (1): 113–122.

Cook, F.A. (1983). Some consequences of palinspastic reconstructions in the Southern Appalachians. *Geology*, 11 (2): 86–89.

Corbett, K.D., Reid, K.O. and Corbett, E.B. (1974). The Mount Read volcanism and Cambrian-Ordovician relationships at Queenstown, Tasmania. *J. Geol. Soc. Austral.*, vol. 21, pt. 2, pp. 173–186.

Crespo-Blanc, A. and Orozco, M. (1988). The Southern Iberian shear zone: A major boundary in the Hercynian folded belt. *Tectonophys.*, pp. 221–227.

Crowell, J.C. (1983). Ice ages recorded on Gondwana continents. *Trans. Geol. Soc. S. Afr.*, 86 (3): 238–261.

Cuerda, A.J. (1974). Stratigraphie des Altpaläozoikums in Argentinien. *Geol. Rdsch.*, vol. 63, pp. 1261–1277.

Cwojdzinski, S. (1979). A simplified mobilistic model of evolution of Polish Variscides. *Bull. Acad. Pol. Sci., Ser. Sci. Terre*, 27 (1–2): 1–16.

Dalla Salda, L. (1982). Nama La Tinta vel inicio de Gondwana. *Acta Geologica Lilloana*, 16 (1): 23–38.

Dallmeyer, R.D. and Gibbons, W. (1987). The age of the blue schist metamorphism in Anglesey, North Wales: Evidence from $^{40}Ar/^{39}Ar$

mineral dates of the Pennynydd schists. *J. Geol. Soc. Lond.*, vol. 144, pp. 843–852.

Dallmeyer, R.D., Wright, J.E. and Secor, D.T. (1986). Character of the Alleghanian orogeny in the Southern Appalachians. Pt. II: Geochronological constraints of the tectonothermal evolution of the eastern Piedmont in South Carolina. *Geol. Soc. Amer. Bull.*, 97 (11): 1329–1344.

Dallmeyer, R.D.. Caen-Vachette, M. and Villeneuve, M. (1987). Emplacement age of post-tectonic granites in southern Guinea (West Africa) and the peninsular Florida subsurface: Implications for origins of Southern Appalachian exotic terranes. *Geol. Soc. Amer. Bull.*, 99 (7): 87–93.

Dalmayrac, B., Laubacher, G. and Marocco, R. (1980a). *Géologie des Andes Péruviennes.* ORSTOM, Peru, 501 pp.

Dalmayrac, B., Laubacher, G. and Marocco, R. (1980b). La chaîne hercynienne d'Amerique du Sud: Structure et évolution d'un orogéne intracratonique. *Geol. Rdsch.*, 69 (1): 1–21.

Dalziel, J.W.D. (1991). Pacific margin of Laurentia and east Antarctica-Australia as a conjugate rift pair: Evidence and implications for an Eocambrian supercontinent. *Geology*, vol. 19, pp. 598–602.

Davoudzadeh, M. and Weber-Diefenbach, K. (1987). Contribution to the palaeogeography, stratigraphy and tectonics of the Upper Palaeozoic of Iran. *N. Jb. Geol. Paläontol. Abh.*, 175 (2): 121–146.

De Cserna, L., Grat, J.L. and Ortega-Gutierres, F. (1977). Aloctono del Palaeozoico inferior en la region de Ciudad Victoria, estado de Tamaulipas. *Rev. Inst. Geol.*, 1 (1): 33–43.

Degeling, P.R., Gilligan, L.B. and Scheibner, E. (1986). Metallogeny and tectonic development of the Tasman fold belt in New South Wales. *Ore Geol. Rev.*, vol. 1, pp. 259–313.

Dergunov, A.B., Lavsandanzan, B. and Pavlenko, V.S. (1980). Tectonics of Western Mongolia. *Tr. Sovm. Sov-Mongol'sk. N.-I. Geol. Eksped.*, No. 31.

Devlin, W.J. and Bond, G.C. (1988). The initiation of the Early Palaeozoic Cordilleran miogeocline: Evidence from the uppermost Proterozoic-Lower Cambrian Hamill group of south-eastern British Columbia. *Can. J. Earth Sci.*, 25 (1): 1–19.

Dewey, J.F. and Shackleton, R.M. (1984). A model for the evolution of the Grampian tract in the early Caledonides and Appalachians. *Nature*, 84 (312): pp. 115–121.

Deynoux, M., Dia, O. and Sougy, J. (1972). La Glaciation "Fini-Ordovicienne" en Afrique de l'Ouest. *Bull. Soc. Géol. Minéral. Bretagne*, (C), IV (1): 9–16.

Dineley, D.L. (1980). Tectonic setting of Devonian sedimentation. *Spec. Pap. Palaeontol.*, no. 23, pp. 49–63.

*Dobretsov, N.L. (1988). Regular periodicity of formation of glaucophane schists and ophiolites as an index of the periodicity of geological processes. *Dokl. AN SSSR*, 300 (2): 427–431.

Dobretsov, N.L., Coleman, R.G. and Liou, J.G. (1987). Blueschist belts in Asia and possible periodicity of blueschist facies metamorphism. *Ofioliti*, 12 (3): 445–456.

*Dodin, A.L. (1979). *Geology and Minerageny of Southern Siberia*. Nedra, Moscow.

Dong Shenbao (1988). *The General Features and Distribution of the Glaucophane Schist Belts of China*. Peking.

Donovan, D.T. and Jones, E.J.W. (1979). Causes of world-wide changes in sea level. *J. Geol. Soc. Lond.*, vol. 136, pp. 187–192.

Dunning, G.R. and Krogh, T.E. (1985). Geochronology of ophiolites of the Newfoundland Appalachians. *Can. J. Earth Sci.*, vol. 22, pp. 1659–1670.

Dunning, G.R. and Pedersen, R.B. (1988). U/Pb ages of ophiolites and arc-related plutons of the Norwegian Caledonides: Implications for the development of Iapetus. *Contrib. Mineral Petrol.*, 98 (1): 13–23.

Dutro, J. (1979). The Mississippian and Pennsylvanian (Carboniferous) Systems in the United States—Alaska. *Geol. Surv. Profess. Pap.*, no. 1110 M-DD, pp. 1–16.

Ebanks, W.J., Jr., Brady, L.L. and Heckel, Ph.H. (1979). The Mississippian and Pennsylvanian (Carboniferous) Systems in the United States—Kansas. *Geol. Surv. Profess. Pap.*, no. 1110 M-DD, pp. 1–30.

Eisbacher, G.H. (1985). Late Proterozoic rifting, glacial sedimentation, and sedimentary cycles in the light of Windermere deposition, Western Canada. *Palaeogeograph., Palaeoclimat., Palaeoecol.*, vol. 51, pp. 231–254.

*Explanatory Note to the Tectonic Map of Antarctica on Scale 1 : 10,000,000 (1980). Nedra, Leningrad.

Fabre, J. (1988). Les séries Paléozoïques d'Afrique: une approche *J. Afr. Earth Sci.*, 7 (1): 1–40.

Farooq, A.Sh. (1986). Depositional environments of the Triassic system in central Saudi Arabia. *Geol. J.*, 21 (4): 403–420.

Farrar, S.S. (1985). Tectonic evolution of the easternmost Piedmont, North Carolina. *Geol. Soc. Amer. Bull.*, 96 (3): 362–380.

Faure, M. and Charvet, J. (1987). Late Permian-Early Triassic orogeny in Japan: Piling up of nappes, transverse lineation and continental subduction of the Honshu block. *Earth and Planet. Sci. Lett.*, 84 (2–3): 295–308.

Fay, R.O., Friedman, S.A. and Johnson, K.S. (1979). The Mississippian and Pennsylvanian (Carboniferous) Systems in the United States—Oklahoma. *Geol. Surv. Profess. Pap.*, no. 1110 M-DD, pp. 1–35.

384

Feist, R. (1978). Das Altpaläozoikum Südfrankreichs. *Schriftenr. Erdwiss. Kommis. Österr. Akad. Wiss.*, vol. 3, pp. 191–200.

Fichter, L.S. and Diecchio, R.J. (1986). Stratigraphic model for timing the opening of the Brito-Atlantic Ocean in northern Virginia. *Geology*, 14 (4): 307–309.

Flemming, N.S. and Roberts, D.G. (1973). Tectonoeustatic changes in sea level and sea floor spreading. *Nature*, 243 (1): 19–22.

Floch, J.-P., Grolier, J. and Guillot, P.-L. (1977). Essai de correlations lithostratigraphiques dans le socle de l'ouest Massif Central. Evolution tectonique et métamorphique. *5e Réun. Annu. Sci. Terre*, Rennes, p. 29.

Forsythe, R.D. (1982). The Late Palaeozoic to Early Mesozoic evolution of southern South America: A plate tectonic interpretation. *J. Geol. Soc. Lond.*, vol. 139, pp. 671–682.

Franke, D. (1978). Entwicklung und Bau der Paläozoiden im nördlichen Mitteleuropa. Part 1: Paläogeographisch-paläotektonisĉhe Entwicklung des Prädevon, *Leitŝchr. Geol. Wiss. Jahrg.*, 6 (1): 5–32.

Franke, W. (1989). Tectonostratigraphic units in the Variscan belt of Central Europe. *Geol. Soc. America*, Special Paper 230, pp. 67–90.

Frasier, W.J. and Schwimmer, D.R. (1987). *Regional Stratigraphy of North America.* Plenum Press, New York and London, 719 pp.

*Gamkrelidze, I.P. (1984). Tectonic structure and alpine geodynamics of Caucasus. In: *Tectonics and Metallogeny of Caucasus.* pp. 105–184. Metoniereba, Tbilisi.

*Garris, M.A. (1977). *Stages of Magmatism and Metamorphism in Pre-Jurassic History of Urals and Peri-Urals.* Nauka, Moscow.

Garzanti, E., Casnedi, R. and Jadoul, F. (1986). Sedimentary evidence of a Cambro-Ordovician orogenic event in the northwest Himalaya. *Sediment. Geol.*, 48 (3–4): 237–265.

*Gatinsky, Yu.G. (1986). *Lateral Structure-Formational Analysis.* Nedra, Moscow.

Gee, D.G. and Sturt, E.A. (eds.) (1985). *The Caledonide Orogen—Scandinavia and Related Areas.* J. Wiley & Sons, Chichester.

Geological Structure of the USSR and Pattern of Location of Mineral Deposits. Vol. 1. Russian Craton. Nedra, Leningrad (1985).

Geological Structure of the USSR and Pattern of Location of Mineral Deposits. Vol. 4. Siberian Craton. Nedra, Leningrad (1987).

*Geology and stratigraphy of Novaya Zemlya. *Sb. Nauchn. Tr. NII Geol. Arktiki, Leningrad*, no. 113 (1979).

Geology of Poland. Vol. IV. Tectonics. Wydawnictwa Geologiczna, Warsaw, 718 pp (1977).

Geology of Yakutsk ASSR. Nedra, Moscow (1981)

Gilligan, L.B. and Scheibner, E. (1978). Lachlan fold belt in New South Wales, *Tectonophys.*, 84 (3–4): 217–266.

*Gordienko, I.V. (1987). *Palaeozoic Magmatism and Geodynamics of Central Asiatic Fold Belt*. Nauka, Moscow.

Gray, G.C. (1986). Native terranes of the central Klamath mountains, California. *Tectonics*, 5 (7): 1043–1054.

Greenough, J.D., McCutcheon, S.R. and Paperik, Y.S. (1985). Petrology and geochemistry of Cambrian volcanic rocks from the Avalon zone in New Brunswick. *Can. J. Earth Sci.*, 22 (6): 881–892.

Grocholski, A. (1986). Proterozoic and Palaeozoic of southwestern Poland in the light of new data. *Biul. Inst. Geol.*, 255 (1): 7–29.

Grocholski, A. (1987). Facies differentiation of Palaeozoic rocks in southwestern Poland. *Bull. Pol. Ac. Sci.*, 35 (3): 209–214.

*Gryaznov, O.N., Dushin, V.A. and Makarov, A.B. (1986). Geological formations and history of development of the montane part of the Polar Urals. *Byull. MOIP, Otd. Geol.*, 61 (4): 39–60.

Guy Tamain, A.L. (1978). L'évolution calédono-varisque des Hesperides *Pap. Geol. Surv. Can.*, no. 78B, pp. 183–212.

Hahn, L., Koch, K.-E. and Wittekindt, H. (1986). Outline of the geology and the mineral potential of Thailand. *Geol. Jahrb.*, Reihe B, Hf. 59, 49 pp.

Haley, B.R., Glick, E.E. and Caplan, W.M. (1979). The Mississippian and Pennsylvanian (Carboniferous) Systems in the United States—Arkansas. *Geol. Surv. Profess. Pap.*, no. 1110 M-DD, pp. 1–14.

Hallam, A. (1977). Secular changes in marine inundation of USSR and North America through the Phanerozoic. *Nature*, vol. 269, pp. 768–772.

Hambrey, M.J. (1985). The Late Ordovician-Early Silurian glacial period. *Palaeogeogr., Palaeoclimat., Palaeoecol.*, vol. 51, pp. 273–289.

Hammann, W. (1976). The Ordovician of the Iberian peninsula—a review. In: *The Ordovician System*, pp. 367–408. M.G. Bassett (ed.). Univ. Wales Press and Nat. Mus. Wales, Cardiff.

Hannah, J.I. and Moores, E.M. (1986). Age relationships and depositional environments of Palaeozoic strata, northern Sierra Nevada. *Geol. Soc. Amer. Bull.*, 97 (7): 787–797.

Haq, B.U., Hardenbol, J. and Vail, P.R. (1987). Chronology of fluctuating sea levels since the Triassic (250 m.y. to present). *Science*, vol. 235, pp. 1156–1167.

Harland, W.B. (1972). The Ordovician ice-age, essay review. *Geol. Mag.*, 109 (5): pp. 96–103.

Harland, W.B. (1979). A stratigraphic outline of Greenland. *Geol. Mag.*, 116 (2): 145–153.

Harland, W.B., Armstrong, R.L., Cox, V., Craig, L.E., Smith, A.G. and Smith, D.G. (1990). *Geological Time Scale, 1989*. Cambridge Univ. Press, London-New York.

Harris, A.L., Rathbone, P.A. and Watson, J. (1980). The Pre-Caledonian evolution of the British Isles. In: *Colloque C 6. Geology of Europe from*

Precambrian to the Post-Hercynian Sedimentary Basins, pp. 22–28. Villeneuve d'Ascq.

Harrison, S.M. and Piercy, B.A. (1991). Basement gneisses in north-west Palmer Land: Further evidence for Pre-Mesozoic rocks in Lesser Antarctica. In: *Geological Evolution of Antarctica*, pp. 341–344. M.R.A. Thomson et al. (eds.). Cambridge Univ. Press, Cambridge.

Havlicek, V. (1971). Stratigraphy of the Cambrian of Central Bohemia. *Sborn. Geol. Věd. Geologie Rada G. Praha*, no. 20, pp. 7–52.

Heidlauf, D.T., Hsui, A.T. and Klein, G. de V. (1986). Tectonic subsidence analysis of the Illinois basin. *J. Geol.*, 94 (6): 779–794.

Helmcke, D. (1985). The Permo-Triassic "Palaeotethys" in mainland Southeast Asia and adjacent parts of China. *Geol. Rdsch.*, 74 (2): 215–228.

Henriksen, N. (1978). East Greenland Caledonian fold belt. *Pap. Geol. Surv., Can.*, no. 78B, pp. 105–109.

Hervé, F. (1988). Late Palaeozoic subduction and accretion in Southern Chile. *Episodes*, 11 (3): 183–188.

Hervé, F., Davidson, J. and Godoy, E. (1981). The Late Palaeozoic in Chile: Stratigraphy, structure and possible tectonic framework. *An. Acad. Brasil. Ciênc.*, 53 (2): 361–373.

Higgins, M.W. (1972). Age, origin, regional relations and nomenclature of Glenarm series, Central Appalachian Piedmont: A reinterpretation. *Geol. Soc. Amer. Bull.*, 83 (4): 989–1026.

History of Development of Urals Palaeo-ocean. Institute of Oceanology, Moscow (1984).

Hobday, D.K. and Tankard, A.J. (1978). Transgressive-barrier and shallow-shelf interpretation of the Lower Palaeozoic Peninsula formation, South Africa. *Geol. Soc. Amer. Bull.*, vol. 89, pp. 504–513.

Hoffman, J. and Paech, H.-J. (1980). Zum Strukturgeologischen Bau am Westrand der Ostanarktischen Tafel. *Z. Geol. Wiss.*, 8 (4): 425–437.

Hoffman, P.F. (1991). Did the breakout of Laurentia turn Gondwanaland inside out? *Science*, vol. 252, pp. 1409–1412.

Holland, H.D. (1972). The geological history of sea water: An attempt to solve the problem. *Geochim. Cosmochim. Acta*, vol. 36, pp. 637–651.

House, M.R., Richardson, J.B. and Chaloner, W.G. (1978). A correlation of the Devonian rocks in the British Isles. *Geol. Soc. Spec. Rept.*, no. 8, pp. 1–110.

Hurst, V.J. (1973). Geology of the Southern Blue Ridge belt. *Amer. J. Sci.*, vol. 273, pp. 643–670.

Huszeldine, R.S. (1984). Carboniferous North Atlantic palaeogeography: Stratigraphic evidence for rifting, not megashear or subduction. *Geol. Mag.*, 121 (5): 443–463.

Hutton, D.H.W. (1987). Strike-slip terranes and a model for evolution of British and Irish Caledonides. *Geol. Mag.*, 124 (5): 405–425.

*Igolkina, N.S., Kirikov, V.P. and Kochin, G.G. (1981). Geological formations of the sedimentary cover of the Russian Platform. *Tr. Vses. N.-I. Geol. In-ta*, no. 296, pp. 3–154.

*Ilyin, A.V. (1986). Tectonics of southern China. *Geotektonika*, no. 1, pp. 32–46.

Isaacson, P.E. (1975). Evidence for a Western Extracontinental Land Source during the Devonian period in the Central Andes. *Geol. Soc. Amer. Bull.*, 86 (1): pp. 36–46.

*Ivanov, S.N. (1979). Baikalian Ural and the nature of metamorphic series in eugeosynclinal framework (preprint). Urals Scientific Centre, AN SSSR, Sverdlovsk.

*Ivanov, S.N. et al. (1986). *Formation of the Earth's Crust in the Urals*. Nauka, Moscow.

Jacobi, R.D. and Wasowski, J.J. (1985). Geochemistry and plate-tectonic significance of the volcanic rocks of the Summerford Group, north-central Newfoundland. *Geology*, 13 (2): 126–130.

Jaeger, H., Bonnefous, J. and Massa, D. (1975). Le Silurien en Tunisie: ses rélations avec le Silurien de Libye nord-occidentale. *Bull. Soc. Géol. Fr.*, 17 (1): 68–76.

Jain, A.K., Geol, R.K. and Nair, N.G.K. (1980). Implications of Pre-Mesozoic orogeny in the geological evolution of the Himalaya and Indo-Gangetic plains. *Tectonophys.*, 62 (1): 67–86.

Jelinek, E. (1984). Geochemistry of a dismembered megaophiolite complex, Letovice, Czechoslovakia. *Trans. Roy. Soc. Edinburgh, Earth Sciences*, 75 (1): 37–48.

Jezek, P., Willner, A.P. and Aceňolaza, F.G. (1985). The Puncoviscana trough—a large basin of Late Precambrian to Early Cambrian age on the Pacific edge of the Brazilian shield. *Geol. Rdsch.*, 74 (3): pp. 573–584.

Jin-Lu, L., Fuller, M. and Zhang, W.-X. (1985). Palaeogeography of the North and South China blocks during the Cambrian. *J. Geodyn.*, 2 (2–3): 91–114.

Johnson, J.G., Klapper, G. and Sandberg, C.A. (1985). Devonian eustatic fluctuations in Euramerica. *Geol. Soc. Amer. Bull.*, vol. 96, pp. 567–587.

Julivert, M., Martinez, F.-J. and Ribeiro, A. (1980). The Iberian segment of the European Hercynian fold belt. *Mém. BRGM*, no. 108, pp. 123–158.

Kapoor, H.M. and Shah, S.C. (1979). Lower Permian in Kashmir-Himalaya—A discussion. *Geol. Surv. India Misc. Publ.* no. 41, pt. 1, pp. 97–113.

Katzung, G. and Krull, P. (1986). Zur tektonischen Entwicklung Mittel-und Nordwesteuropas während des Jungpaläozoikums. *Wiss.-Techn. Informations-dienst*, A27 (2): 24–33.

*Kazakov, A.V. (1939). *Phosphate Facies. Origin of Phosphates and Geological Factors of Formation of Their Deposits*. Nedra, Moscow.

*Kazantseva, T.T. and Kamaletdinov, M.A. (1986). Geosynclinal development of the Urals. *Dokl. AN SSSR*, 288 (6): 1449–1453.

*Kaz'min, V.G. (1988). Tectonic development of Mozambique belt: from accretion to collision. *Geotektonika*, no. 3, pp. 26–34.

*Khain, V.E. (1984). *Regional Geotectonics. Alpine Mediterranean Belt.* Nedra, Moscow.

*Khain, V.E. and Bozhko, N.A. (1988). *Historical Geotectonics. Precambrian.* Nedra, Moscow.

*Khanchuk, A.I., Panchenko, I.V. and Nemkin, I.V. (1988). *Geodynamic Evolution of Sikhote-Alin' and Sakhalin in the Palaeozoic and Mesozoic.* Vladivostok.

*Kheraskov, N.N. (1975). Formation and stages of geosynclinal development of Western Sayan. *Geotektonika*, no. 1, pp. 35–53.

*Kheraskova, T.N. (1979). Siliceous formations of the Lower Palaeozoic of Central Kazakhstan. In: *Sedimentation and Volcanism in Geosynclinal Basins*, pp. 43–51. Nauka, Moscow.

*Kheraskova, T.N. (1986a). Vendian-Cambrian Caledonides formations of Asia. *Tr. GIN AN SSSR*, no. 386.

*Kheraskova, T.N. (1986b), Vendian-Cambrian Caledonides formations of Kazakhstan, Altay-Sayan region and western Mongolia. *Geotektonika*, no. 6, pp. 69–84.

*Kheraskova, T.N. and Dergunov, A.B. (1986). Vendian-Early Palaeozoic marine formations of the Kazakhstan-Mongolia region. *Byull. MOIP, Otd. Geol.*, 61 (5): 29–36.

Kier, R.S., Brown, L.F. and McBride, E.F. (1979). The Mississippian and Pennsylvanian (Carboniferous) Systems in the United States—Texas. *Geol. Surv. Profess. Pap.*, no. 1110 M-DD, pp. 1–45.

Kimura, T.I. (1986). The Chichibu geosyncline. The development of the Japanese islands. *Proc. Jap. Acad.*, B62 (10): 385–387.

King, Ph.B. (1975). Ancient southern margin of North America. *Geology*, 3 (12): 732–734.

King, Ph.B. (1978). Tectonics of the North American Cordillera near the Fortieth Parallel. *Tectonophys.*, vol. 47, pp. 275–294.

Klein, G. de V. (1987). Current aspects in basin analysis. *Sedimentary Geology*, vol. 50, pp. 95–118.

Klitzsch, E. (1978). Geologische Bearbeitung Südwest—Agyptens. *Geol. Rdsch.*, 67 (2): 507–520.

*Klyuzhina, M.L. (1985). *Palaeogeography of the Urals in the Ordovician Period.* Nauka, Moscow.

Kobayashi, T. (1968). The Ordovician palaeogeography of Eastern Asia, Turaky Dzaccu. *J. Geogr.*, 77 (6): 313–328.

Kokelaar, B.P., Howells, M.F. and Bevins, R.E. (1984). The Ordovician marginal basin of Wales. In: *Marginal Basin Geology*, pp. 245–269. Geol. Soc. Spec. Publ., no. 16.

Kominz, M.A. (1984). Oceanic ridge volumes and sea level change—an error analysis. *Am. Assoc. Petr. Geol. Mem.*, vol. 36, pp. 108–128.

Krebs, W. (1974). Devonian carbonate complexes of Central Europe. *Soc. Econ. Palaeontol. and Miner.*, Spec. Publ. no. 18, pp. 155–208.

Kröner, A. (1981). Precambrian crustal evolution and continental drift. *Geol. Rdsch.*, 70 (2): 412–428.

Kröner, A., Byerly, G. and Lowe, D.R. (1991). Chronology of early Arabian granite-greenstone evolution in the Barberton Mountain Land, South Africa, based on precise dating by single zircon evaporation. *Earth Planet. Sci. Lett.*, 103 (1): 41–54.

*Kurenkov, S.A. (1988). Geodynamic environments of the formation of the Mesozoic traps of Eastern Siberia and Tamyr. In: Current Problems of Tectonics in the USSR.

*Kuz'min, M.I. and Filippova, I.B. (1979). History of the development of Mongolia-Okhotsk belt in the Middle-Late Palaeozoic and Mesozoic. In: *Structure of Lithospheric Plates. Interaction between Plates and Resultant Structures of the Earth's Crust*, pp. 189–226. Inst. Oceanology, Moscow.

Laird, M.G. (1981). Lower Palaeozoic rocks in the Ross Sea area and their significance in the Gondwana context. *J. Roy. Soc. N.Z.*, 11 (4): 425–438.

Laird, M.G. (1991). Lower and Middle Palaeozoic sedimentation and tectonic pattern on the Palaeo-Pacific margin of Antarctica. In: *Geological Evolution of Antarctica*, pp. 177–185. M.R.A. Thomson et al. (eds.). Cambridge Univ. Press, Cambridge.

Lapierre, H., Brouxel, M. and Martin, Ph. (1986). The Palaeozoic and Mesozoic geodynamic evolution of the Eastern Klamath Mountains (North California) inferred from the geochemical characteristics of its magmatism. *Bull. Soc. Géol. Fr.*, 2 (6): 969–980.

Larson, E.E., Patterson, P.E. and Curtis, G. (1985). Petrologic, palaeomagnetic and structural evidence of a Palaeozoic rift system in Oklahoma, New Mexico, Colorado and Utah. *Geol. Soc. Amer. Bull.*, vol. 96, pp. 1364–1372.

Larson, E.R. and Langenheim, R.L. (1979). The Mississippian and Pennsylvanian (Carboniferous) systems in the United States—Nevada. *Geol. Surv. Profess. Pap.*, no. 1110 M-DD, pp. 1–19.

Leggett, J.K. (1980). The sedimentological evolution of a Lower Palaeozoic accretionary fore-arc in the Southern Uplands of Scotland. *Sedimentology*, vol. 27, pp. 401–417.

Leggett, J.K., McKerrow, W.S. and Cock, L.R. (1981). Periodicity in the Early Palaeozoic marine realm. *J. Geol. Soc.*, vol. 138, pp. 167–176.

Legrand, Ph. (1974). Essai sur la paléogéographie de l'Ordovicien au Sahara algérien. *Not. et Mém.*, no. 11, pp. 121–136.

Legrand, Ph. (1981). Essai sur la paléogéographie du Silurien au Sahara algérien. *Not. et Mém.*, no. 16, pp. 9–24.

Legrand, Ph. (1985). Réflexions sur la transgression silurienne au Sahara algérien, In: *Act. 110ᵉ Congr. Nat. Soc. Savant.*, pp. 233–244. Montpellier. Sec. Sci. Fasc. 6, Paris.

*Lennykh, V.I. (1984). Pre-Uralides in the confluence zone of the Eastern European Craton and Urals. In: *Metamorphism and Tectonics of Western Zones of the Urals*, pp. 21–42. Sverdlovsk.

*Leont'ev, A.N., Litvinovsky, B.A. and Gavrilova, S.P. (1981). *Palaeozoic Granitoid Magmatism of the Central Asiatic Fold Belt.* Nauka, Novosibirsk.

Lin Baogu (1979). Silurian system in China. *Acta Geol. Sinica*, 53 (3): 173–191 (in Chinese).

Lindsay, J.F., Korsch, R.J. and Wilford, J.R. (1987). Timing the breakup of a Proterozoic supercontinent: Evidence from Australian intracratonic basin. *Geology*, 15 (11): 1061–1064.

Liu Hung-yun, Sha Chiang-an and Hu Shih-ling (1973). The Sinian system in Southern China. *Scientia Sinica*, 16 (2): 266–278.

Lorenz, V. and Nicholls, J.A. (1984). Plate and intraplate processes of Hercynian Europe during the Late Palaeozoic. *Tectonophys*, 107 (1–2): 25–56.

Lutzens, H. (1980). Ein Beitrag zur geologischen Entwicklung des Harzes unter den besonderen Bedingungen des paläotektonischen Regimes während des variszischen Flyschetappe. *Veröff. Zentralinst. Phys. Erde*, no. 58, pp. 1–22.

*Makarychev, G.I., Shtreis, N.A. and Morkovkina, V.F. (1981). Formational characteristics of the continental Earth's crust in Tien Shan. In: *Relationships between Geological Processes in Palaeozoic Fold Structures of Central Asia*, pp. 19–30. Frunze.

Mason, R. (1988). Did the Iapetus Ocean really exist? *Geology*, 16 (9): 823–826.

Matte, Ph. (1986). Tectonics and plate tectonics model for the Variscan belt of Europe. *Tectonophys.*, vol. 126, pp. 329–374.

Matte, Ph. (1991). Accretionary history and crustal evolution of the Variscan belt in Western Europe. *Tectonophys.*, vol. 196, pp. 309–337.

Matthews, S.C., Chauvel, J.J. and Robardet, M. (1980). Variscan geology of Northwestern Europe. In: *Colloque C 6.Geology of Europe from Precambrian to the Post-Hercynian Sedimentary Basins*, pp. 69–76. Villeneuve d'Ascq.

Max, M.D. and Long, C.B. (1985). Pre-Caledonian basement in Ireland and its cover relationships. *Geol. J.*, 20 (4): 341–366.

*Mazarovich, A.O. and Richter, A.V. (1987). Palaeozoic and Mesozoic history of development of the Far East. *Tr. GIN AN SSSR*, no. 417, pp. 178–197.

McCann, A.M. and Kennedy, M.J. (1974). A probable glacio-marine deposit of Late Ordovician-Early Silurian age from the north central Newfoundland Appalachian belt. *Geol. Mag.*, 111 (6): 549–564.

McClure, H.A. (1980). Permian-Carboniferous glaciation in the Arabian Peninsula. *Geol. Soc. Amer. Bull.*, pt. 1, 91 (12): 707–712.

McGill, P. (1979). The stratigraphy and structure of the Vendon Fiord area. *Bull. Can. Petrol. Geol.*, 22 (4): 361–386.

McKerrow, W.S. (1979). Ordovician and Silurian changes in sea level. *J. Geol. Soc. Lond.*, 136 (2): 137–145.

Mégard, F. (1978). Etude géologique des Andes du Pérou central. Contribution à l'étude géologique des Andes, N 1. *Mém. ORSTOM*, no. 86, 310 pp.

Mellen, F.F. (1977). Cambrian system in Black Warrior basin. *Bull. Amer. Assoc. Petrol. Geol.*, 61 (10): 1897–1900.

Ménot, R.-P., Peucat, J.J. and Scarenzi, D. (1988). 496 m.y. age of plagiogranites in the Chamrousse ophiolite complex (external crystalline massifs in the French Alps): Evidence of a Lower Palaeozoic oceanisation. *Earth Planet Sci. Lett.*, 88 (1): 82–92.

Mesolella, K.J. (1978). Palaeogeography of some Silurian and Devonian reef trends, Central Appalachian basin. *Bull. Amer. Assoc. Petrol. Geol.*, 62 (9): 1607–1644.

Metcalfe, I. (1984–1985). Stratigraphy, palaeontology and palaeogeography of the Carboniferous of Southeast Asia. *Mem. Soc. Geol. Fr.*, no. 147, pp. 107–118.

Miall, A.D. (1986). Eustatic sea level changes interpreted from seismic stratigraphy: A critic of methodology with particular reference to the North Sea Jurassic record. *Bull. Amer. Ass. Petrol. Geol.*, 70 (2): 131–137.

Michard, A. (1976). Elements de géologie Marocaine. *Not. et mém. Serv. Géol. Maroc*, no. 252, 408 pp.

Miller, M.M. (1987). Dispersed remnants of a north-west Pacific fringing arc: Upper Palaeozoic terranes of Permian McCloud faunal affinity, Western US. *Tectonics*, 6 (6): 807–830.

Milne, A.J. and Millar, J.L. (1991). Mid-Palaeozoic basement in eastern Graham Land and its relation to the Pacific margin of Gondwana. In: *Geological Evolution of Antarctica*, pp. 335–340. M.R.A. Thomson et al. (eds.). Cambridge Univ. Press, Cambridge.

Minato, M., Hunahashi, M. and Watanabe, J. (eds.) (1979). *Variscan Geohistory of Northern Japan: The Abean Orogeny*. Hokkaido University, Sapporo, Japan. Tokai University Press, 427 pp.

*Mishina, A.V. (1979). Late Palaeozoic and Mesozoic of Indonesia-Philippine region. *Izv. Vuzov Geol. i Razvedka*, no. 4, pp. 46–57.

Mitchell, A.H.G. (1984). The British Caledonides: Interpretations from Cenozoic analogues. *Geol. Mag.*, 121 (1): 35–46.

Moores, E.M. (1991). South-west US-east Antarctic (SWEAT) connection: A hypothesis. *Geology*, vol. 19, pp. 425–428.

Moran, Z.D.J. (1986). Breve revision sobre la evolucion tectonica de Mexico. *Geofis. Int.*, vol. 25, no. 1; Dyn. and Evol. Lithos—Results and Perspects. *Geophys. Res. Mexico*, pt. A, pp. 9–38.

Morris, R.C. (1974). Sedimentary and tectonic history of the Ouachita Mountains. *Soc. Econ. Paleontol. and Miner.*, Spec. Publ., no. 22, pp. 120–142.

*Mossakovsky, A.A. and Dergunov, A.B. (1983). Caledonides of Kazakhstan and Central Asia (tectonic structure, history of development and palaeotectonic set-up). *Geotektonika*, no. 2, pp. 16–33.

Murphy, J.B., Cameron, K. and Dostal, J. (1985). Cambrian volcanism in Nova Scotia, Canada. *Can. J. Earth Sci.*, 22 (4): 599–606.

Murray, C.G. and Kirkegaard, A.G. (1978). The Thomson orogen of the Tasman orogenic zone. *Tectonophys.*, 48 (3/4): 299–326.

Nance, R.D. (1987). Model for the Precambrian evolution of the Avalon terrane in southern New Brunswick, Canada. *Geology*, 15 (8): 753–756.

Nedjari, A. (1981). Un bassin paralique pérmocarbonifère à charbon et évaporites en bordure de la plate-forme saharienne: le Mezarif, Sud-Ouest Oranais, Algérie. *CR Acad. Sci.*, Sér. 2, vol. 292, pp. 847–850.

Nelson, K.D., McBride, J.H. and Arnow, J.A. (1985). A new COCORP profiling in the south-eastern United States. Part II. Brunswick and East coast magnetic anomalies, opening of the north-central Atlantic Ocean. *Geology*, 13 (10): 718–721.

*Nenashev, Yu.P. and Petrovsky, A.D. (1981). Stratigraphy of the Upper Proterozoic formations of the south-western part of the Sahara Platform. *Izv. AN SSSR, Ser. Geol.*, no. 1, pp. 61–73.

*Nikolaeva, I.V., Vakulenko, L.G. and Yadrenkina, A.G. (1968). *Lower Ordovician of South-eastern Siberian Platform (Lithology and Facies)*. Nauka, Novosibirsk.

Novaya Zemlya in Early Stages of Geological Development. Sb. Nauchn. Tr. PGO Sevmorgeologiya, Nedra, Leningrad (1984).

Nur, A. and Ben Avraham, Z. (1982). Oceanic plateaus, the fragmentation of continents, and mountain building. *J. Geoph. Res.*, 87 (B5): 3644–3661.

O'Brien, S.J., Werdle, R.J. and King, A.F. (1983). The Avalon zone: A Pan-African terrane in the Appalachian orogen of Canada. *Geol. Journ.*, vol. 18, pp. 195–222.

Oftedahl, Chr. (1980). Geology of Norway. *Norg. Geol. Unders.*, no. 356, pp. 3–114.

Okulitsch, A.V. (1985). Palaeozoic plutonism in Southeastern British Columbia. *Can. J. Earth Sci.*, vol. 22, pp. 1409–1424.

Opdyke, N.D., et al. (1987). Florida as exotic terrane: Palaeomagnetic and geochronologic investigations of Lower Palaeozoic rocks from the sub-surface of Florida. *Geology*, 15 (10): 900–903.

* *Ordovician Stratigraphy of the Siberian Platform*, No. 200. Tr. In-ta Geol. i Geofiz. SO AN SSSR, Nauka, Novosibirsk (1975).

Osberg, Ph.H. (1978). Synthesis of the geology of the Northeastern Appalachians, USA. *Pap. Geol. Surv. Can.*, no. 78B, pp. 137–147.

Paech, H.-J. (1981). Bemerkungen zur paläotektonischen Entwicklung Antarktikas. *Z. Geol. Wiss.*, 9 (10): pp. 1107–1111.

*Pai, V.M. (1992). Tectonic nature of Early Carboniferous-Middle Jurassic volcanics of Zaalay range. *Vestn. Mosk. Un-ta, Geol.*, no. 2, pp. 13–25.

* *Palaeogeography and Lithology of the Vendian and Cambrian of the Western Part of the East European Craton* (1980). Nauka, Moscow.

*Palei, I.P. (1981). *Structure of the Basement of the Palaeozoic structures of Mongolia. Problems of Tectonics of the Earth's Crust*, pp. 159–166. Nauka, Moscow.

Paris, F. and Robardet, M. (1990). Early Palaeozoic palaeogeography of the Variscan region. *Tectonophys.*, vol. 177, pp. 193–213.

Paris, F., Morzadec, P. and Herric, A. (1986). Late Devonian-Early Carbonif-erous events in the Armorican massif (western France): A review. *Ann. Soc. Geol. Belg.*, vol. 109, pp. 187–195.

Parkinson, N. and Summerhayes, C. (1985). Synchronous global sequence boundaries. *Bull. Amer. Assoc. Petrol. Geol.*, vol. 69, pp. 685–687.

Pedersen, S.A.S. (1986). A transverse, thin-skinned, thrust-fault belt in the Palaeozoic North Greenland fold belt. *Geol. Soc. Amer. Bull.*, 97 (12): 1442–1455.

Peiffer-Rangin, F. and Perez, A. (1980). Sur la présence d'Ordovicien supérieur à Graptolites dans le nord-ouest du Méxique. *CR Ac. Sc. Paris*, vol. 290, sér. D, pp. 13–16.

* *Phanerozoic of Siberia*. Vol. 1. *Vendian. Palaeozoic*. Tr. In-ta Geol. i Geofiz. SO AN SSSR, no. 595 (1984).

Phipps, S.P. (1988). Deep rifts as sources for alkaline intraplate magmatism in eastern North America. *Nature*, 334 (1): 27–31.

Picha, F. and Gibson, R.S. (1985). Cordilleran hingeline: Late Precambrian rifted margin of the North American craton and its impact on the depositional and structural history—Utah and Nevada. *Geology*, 13 (7): 465–468.

Pierce, W.H. (1979). The Mississippian and Pennsylvanian (Carboniferous) systems in the United States—Arizona. *Geol. Surv. Profess. Pap.*, no. 1110 M-DD, pp. 1–20.

Pin, C., Majerowicz, A. and Wojciechowska, I. (1988). Upper Palaeozoic oceanic crust in the Polish Sudetes: Nd-Sr isotope and trace element evidence. *Lithos*, vol. 21, pp. 195–209.

Piqué, A. (1976). Evolution sedimentaire du Nord-Ouest de la Meseta marocaine au cours du Carbonifère. Les étapes du comblement du bassin. *CR Acad. Sci.*, 282 (10): pp. 957–960.

Pluijin, B.A. van der (1987). Timing and spatial distribution of deformation in the Newfoundland Appalachians: A "multistage collision" history. *Tectonophys.*, 135 (1–3): 15–24.

Plumb, K.A. (1979). The tectonic evolution of Australia. *Earth Sci. Revs.*, vol. 14, pp. 205–249.

*Postel'nikov, E.S. (1980). *Geosynclinal Development of the Yenisei Range in the Late Precambrian*. Tr. GIN AN SSSR, no. 341.

*Preiss, V.V. and Krylov, I.N. (1980). Stratigraphy and vegetative remnants of the Upper Precambrian in southern Australia. *Izv. AN SSSR, Ser. Geol.*, no. 7, pp. 61–74.

*Pronin, A.A., 1960. *Caledonian Cycle in the Tectonic History of the Earth*. Nauka, Leningrad.

*Pyzhyanov, I.V., Sonin, I.I. and Karapetov, S.S. (1980a). Palaeozoic of Afghanistan. Article I (Vendian-Silurian). *Izv. Vuzov Geol. i Razvedka*, no. 4, pp. 26–32.

*Pyzhyanov, I.V., Sonin, I.I. and Karapetov, S.S. (1980b). Palaeozoic of Afghanistan. Article II (Devonian). *Izv. Vuzov Geol. i Razvedka*, no. 6, pp. 31–37.

*Raaben, M.E. (1975). *Upper Riphean as a Unit of the General Stratigraphic Scale*. Nauka, Moscow, 241 pp.

Ramos, V.A. (1988). Late Proterozoic-Early Palaeozoic of South America—a collisional history. *Episodes*, 11 (3): 168–173.

Ramsbottom, W.H. (1979). Rates of transgression and regression in the Carboniferous of NW Europe. *J. Geol. Soc. Lond.*, 136 (1): 147–153.

Rast, N., Skehan, S.J. and James, W. (1983). The evolution of the Avalonian plate. *Tectonophys.*, 100 (1–3): 257–285.

Read, J.F. (1980). Carbonate ramp-to-basin transition and foreland basin evolution, Middle Ordovician, Virginia Appalachians. *Bull. Amer. Assoc. Petrol. Geol.*, 64 (10): pp. 1575–1612.

Ren Jishun, Chen Tingyu and Liu Zhingan (1984). Some problems concerning the geotectonic zoning of Eastern China. *Int. Geol. Rev.*, 30 (4): 382–385.

*Repina, L.N. (1985). Palaeogeography of Early Cambrian seas based on trilobites. In: *Biostratigraphy and Biogeography of the Palaeozoic of Siberia*, pp. 5–15. Nauka, Novosibirsk.

Restrepo, J.J. and Toussaint, J.F. (1988). Terranes and continental accretion in the Colombian Andes. *Episodes*, 11 (3): 189–193.

Rice, A.H.N. (1987). A tectonic model for the evolution of the Finnmarkian Caledonides of North Norway. *Can. J. Earth Sci.*, vol. 24, pp.602–616.

Robardet, M. and Doré, F. (1988). The Late Ordovician diamictic formations from Southwestern Europe: north Gondwana glacio-marine deposits. *Palaeogeogr., Palaeoclimat., Palaeoecol.*, 56 (1): 19–31.

Roberts, D., Grenne, T. and Ryan, P.D. (1984). Ordovician marginal basin development in the Central Norwegian Caledonides. In: *Marginal Basin Geology*, pp. 233–244. Oxford and others.

Roberts, J. and Endel, B.A. (1987). Depositional and tectonic history of the southern New England orogen. *Austr. J. Earth Sci.*, 34 (1): 1–20.

*Ronov, A.B. Khain, V.E., Balukhovsky, A.N. and Seslavinsky, K.B. (1976). Change in the distribution, volumes and accumulation rates of sediments and volcanic deposits in the Phanerozoic (within present-day mainland boundaries). *Izv. AN SSSR, Ser. Geol.*, no. 12, pp. 5–12.

*Ronov, A.B., Khain, V.E. and Seslavinsky, K.B. (1984). *Atlas of Lithological-Palaeogeographic Maps of the World. Late Precambrian and Palaeozoic of Continents.* Nedra, Leningrad.

Ross, Ch.A. (1986). Palaeozoic evolution of the southern margin of the Permian basin. *Geol. Soc. Amer. Bull.*, 97 (5): 536–554.

Ross, Ch.A. and Ross, J.R.P. (1985). Late Palaeozoic depositional sequences are synchronous and worldwide. *Geology*, 13 (3): 194–197.

Ross, R.J. (1976). Ordovician sedimentation in the Western United States. In: *Ordovician System*, pp. 73–105. Univ. Wales Press and Nat. Mus. Wales.

Rowell, A.J., Rees, M.N. and Suczek, C.A. (1979). Margin of the North American continent in Nevada during Late Cambrian time. *Amer. J. Sci.*, 279 (1): 1–18.

*Rozanov, V.I. (1986). Development of the Late Permian basin on the northern Russian platform. *Izv. AN SSSR, Ser. Geol.*, no. 2, pp. 59–65.

Ruitenberg, A.A., Fytte, L.R. and McCutcheon, S.R. (1977). Evolution of Precambrian tectonostratigraphic zones in the New Brunswick Appalachians. *Geosci. Can.*, 4 (4): 171–181.

Russel, K.L. (1968). Oceanic ridges and eustatic changes in sea level. *Nature*, vol. 218, pp. 861–862.

*Ruzhentsev, S.V., Rozman, Kh.S. and Manzhin, Ch. (1991). Formation period of the Southern Mongolia palaeo-ocean. *Dokl. AN SSSR*, 319 (2): 451–455.

*Ruzhentsev, S.V., Pospelov, I.I. and Badarch, G. (1992). Ophiolite sutures of Inner Mongolia. *Dokl. AN SSSR*, 322 (5): 953–958.

*Ryazantsev, A.V., German, L.L. and Degtyarev, K.E. (1987). Lower Palaeozoic chaotic complexes in eastern Yerementau (central Kazakhstan). *Dokl. AN SSSR*, 296 (2): 406–410.

Sano, H. and Kanmera, K. (1988). Palaeogeographic reconstruction of accreted oceanic rocks, Akiyoshi, southwest Japan. *Geology*, 16 (7): 600–603.

Saul, R.B., Bowen, O.E. and Stevens, C.H. (1979). The Mississippian and Pennsylvanian (Carboniferous) systems in the United States—California, Oregon, and Washington. *Geol. Surv. Profess. Pap.*, no. 1110 M-DD, pp. 1–5.

Scheibner, E. (1978). Tasman fold belt system in New South Wales—general description. *Tectonophys.*, 86 (3–4): 207–216.

Scheibner, E. (1985). Suspect terranes in the Tasman fold belt system, Eastern Australia. In: *Tectonostratigraphic Terranes in the Circum-Pacific*, pp. 493–514. Council for Energy and Miner. Resources, Earth Sciences, Ser., No. 1.

Scheibner, E. (1987). Palaeozoic tectonic development of Eastern Australia in relation to the Pacific region. In: *Circum-Pacific Orogenic Belts and Evolution of the Pacific Ocean Basin*, pp. 133–165. J.W.H. Monger, and J. Francheteau (eds.). Geodynamics Series, vol. 18, Boulder, Washington.

Schenk, P.E. (1972). Possible Ordovician glaciation of Nova Scotia, *Can. J. Earth Sci.*, 9 (95): 95–107.

Schoon, R.A. (1979). The Mississippian and Pennsylvanian (Carboniferous) systems in the United States—South Dakota. *Geol. Surv. Profess. Pap.*, no. 1110 M-DD, pp. 1–10.

Schutz, W., Ebeneth, J. and Meyer, K.-D. (1987). Trondhjemites, tonalites and diorites in the South Portuguese zone and their relations to the volcanites and mineral deposits of the Iberian pyrite belt. *Geol. Rdsch.*, 76 (1): 201–212.

Schweikert, R.A. and Lahren, M.M. (1987). Continuation of Antler and Sonoma orogenic belts to the eastern Sierra Nevada, California and Late Triassic thrusting in a compressional arc. *Geology*, 15 (3): 270–273.

Secor, D.T., Snoke, A.W., and Bramlett, K.W. (1986). Character of the Alleghanian orogeny in the Southern Appalachians: Pt. 1. Alleghanian deformation in the eastern Piedmont of South Carolina. *Geol. Soc. Amer. Bull.*, 97 (11): 1319–1328.

Sengör, C.A.M. and Hsü, K.J. (1984). The Cimmerides of Eastern Asia: History of the eastern end of Palaeo-Tethys. *Mem. Soc. Géol. Fr.*, N.S., no. 147, pp. 139–167.

Sengör, C.A.M., Yilmaz, Y. and Sungurlu, O. (1985). Tectonics of the Mediterranean Cimmerides: Nature and evolution of the western termination of Palaeo-Tethys. In: *The Geological Evolution of the Eastern Mediterranean*, p. 77. Spec. Publ. Geol. Soc., No. 17, Oxford: Blackwell Scient. Publ.

*Sennikov, V.M. (1977). Developmental history of the structure of the southern part of the Altay-Sayan fold region in the Ordovician, *Tr. SNIIGGIMS*, Barnaul, no. 201.

*Sergeeva, L.A., Nochev, N.K. and Malyakov, I.G. (1979). V"rkhya paleozoiskata v"zrast na metamorfitite v Strandzha. *Spisanie B"lg. Geol. Druzhestva*, 40 (1): 10–17.

*Seslavinsky, K.B. (1987). *Caledonian Sedimentation and Volcanism in the History of the Earth*. Nedra, Moscow.

Shibata, K., Ishihara, Sh. and Ulriksen, C.E. (1984). Rb-Sr ages and initial $^{87}Sr/^{86}Sr$ ratios of Late Palaeozoic granitic rocks from Northern Chile. *Bull. Geol. Surv. Jap.*, 35 (11): 537–545.

Shubert, D.H. and Cebull, S.E. (1987). Tectonic interpretation of the westernmost part of the Ouachita-Marathon (Hercynian) orogenic belt, west Texas-Mexico. *Geology*, 15 (5): 458–461.

*Shvol'man, V.A. (1980a). Present problems in the tectonics of the Himalaya, *Izv. AN SSSR, Ser. Geol.*, no. 1, pp. 125–134.

*Shvolman, V.A. (1980b). Mesozoic ophiolite complex in the Pamirs. *Geotektonika*, no. 6, pp. 72–81.

Siedlecka, A. (1973). The Late Precambrian Ost-Finnmark supergroup—a new lithostratigraphic unit of high rank. *Norges Geol. Unders*, no. 289, pp. 55–60.

*Simonov, A.P. (1978). History of world oceans. *Sov. Geologiya*, no. 4, pp. 77–85.

Skipp, B., Sando, W.J. and Hall, W.E. (1979). The Mississippian and Pennsylvanian (Carboniferous) systems in the United States—Idaho. *Geol. Surv. Profess. Pap.*, no. 1110 M-DD, pp. 1–42.

Smedley, P.L. (1986). Relationship between calc-alkaline volcanism and within-plate continental rift volcanism: Evidence from the Scottish Palaeozoic lavas. *Earth Planet Sci. Lett.*, vol. 76, pp. 113–128.

Smith, D.L. and Gilmour, E.H. (1979). The Mississippian and Pennsylvanian (Carboniferous) systems in the United States—Montana. *Geol. Surv. Profess. Pap.*, no. 1110 M-DD, pp. 1–32.

Soper, W.J. (1986). The Newer Granite problem: A geotectonic view. *Geol. Mag.*, 123 (3): 227–236.

*Spassov, Khr. (1973). Devonian stratigraphy of south-western Bulgaria. *Izv. Geol. In-t B"lg. AN, Ser. Stratigr. i Litol.*, no. 22, pp. 5–38.

Spassov, Chr., Tenčov, J. and Janev, S. (1978). Die paläeozoischen Ablagerungen in Bulgarien. *Schriftenr. Erdwiss. Kommis. Österr. Akad. Wiss.*, vol. 3, pp. 279–296.

Splettstoesser, J.F. and Webers, G.F. (1980). Geological investigations and logistics in the Ellsworth Mountains, 1979–80. *Antarct. J.U.S.*, 15 (5): 36–39.

Stevanovic, P. and Veselinovic, H. (1978). Das Paläozoikum von Serbien. *Schriftenr. Erdwiss. Kommis. Österr. Akad. Wiss.*, vol. 3, pp. 297–311.

Stewart, J.H. (1988). Latest Proterozoic and Palaeozoic southern margin of North America and the accretion of Mexico. *Geology*, 16 (2): 186–189.

Stewart, J.H. and Poole, F.Y. (1974). Lower Palaeozoic and Uppermost Precambrian Cordilleran miogeosyncline, Great Basin, Western United States. *Soc. Econ. Paleont. and Miner. Spec. Publ.*, no. 22, pp. 28–57.

Stille, H. (1948). Ur- und Neuozeane. *Abh. Deutsch. Akad. Wiss. Math.-Nat. Kl.*, Jahrg. 1945/46, no. 6, Berlin.

Stöcklin, J. (1980). Geology of Nepal and its regional frame. *J. Geol. Soc. Lond.*, 137 (1): 1–34.

Stratigraphic Atlas of North and Central America. (T.D. Cook and A.W. Bally, eds.). Princeton Univ. Press, Princeton, New Jersey (1975).

Strong, P.G. and Walker, R.G. (1981). Deposition of the Cambrian continental rise: the St. Roch formation near St. Jean-Port-Joli, Quebec. *Can. J. Earth Sci.*, 19 (8): 1320–1335.

Stump, E., Laird, M.G. and Bradshaw, J.D. (1983). Bowers graben and associated tectonic features across northern Victoria Land, Antarctica. *Nature*, vol. 304, pp. 334–336.

Stump, E., Smit, J.H. and Self, S. (1986). Timing of events during the Late Proterozoic Beardmore orogeny, Antarctica: Geological evidence from the La Gorce Mountains. *Geol. Soc. Amer. Bull.*, 97 (8): 953–965.

Sturt, B.A. (1984). The accretion of ophiolitic terranes in the Scandinavian Caledonides, *Geol. en Mijnbouw*, 63 (2): 201–212.

Sturt, B.A., Soper, N.J. and Brück, P.M. (1980). Caledonian Europe, *Mém. BRGM*, no. 108, pp. 57–66.

Sturt, B.A., Roberts, D. and Furnes, H. (1984). A conspectus of Scandinavian Caledonian ophiolites. In: *Ophiolites and Oceanic Lithosphere*, pp. 381–391. Geol. Soc. Spec. Publ., no. 13.

*Suetenko, O.D. (1984). Stratigraphy of Lower and Middle Palaeozoic of eastern part of Southern Gobi region (Mongolia). *Tr. Mezhdunar. Geol. Eksped. v MNR*, no. 2, pp. 142–153.

Suggate, R.P. (Ed.) (1978a). *The Geology of New Zealand.* Vol. 1. E.C. Keating, Wellington, 343 pp.

Suggate, R.P. (Ed.) (1978b). *The Geology of New Zealand.* Vol. 2. E.C. Keating, Wellington, pp. v–xx and 345–820.

Summerhayes, C.P. (1986). Sea level curves based on seismic stratigraphy: Their chronostratigraphic significance. *Palaeogeogr., Palaeoclimat., Palaeoecol.*, 57 (1): 27–42.

Surlyk, F. and Hurst, J.M. (1983). Evolution of the Early Palaeozoic deepwater basin of North Greenland—aulacogen or narrow ocean? *Geology*, 11 (2): 77–81.

Surlyk, F. and Hurst, J.M. (1984). The evolution of the Early Palaeozoic deep-water basin of North Greenland, *Geol. Soc. Amer. Bull.*, 95 (2): 131–154.

Surlyk, F., Hurst, J.M. and Bjerreskov, M. (1980). First age-diagnostic fossils from the central part of the North Greenland fold belt. *Nature*, vol. 286, pp. 800–803.

*Surmilova, E.P. (1980). Devonian sedimentation characteristics in the Kolyma region. *Izv. Vuzov. Geol. i Razvedka*, no. 4, pp. 42–49.

Szatmari, P., Carvalho, R.S. and Simões, I.A. (1979). A comparison of evaporite facies in the Late Palaeozoic Amazon and the Middle Cretaceous South Atlantic salt basins. *Economic Geology*, vol. 74, pp. 432–447.

Tectonic Map of Eurasia. (A.L. Yanshin, ed.). Geol. Inst. AN SSSR and Main Direction of Geodesy and Cartography, Moscow (1966).

*Terekhova, G.P., Epshtein, O.G. and Solovyeva, M.N. (1979). Palaeozoic formations of the left bank of Khatyrka River (Koryak upland). *Byull. MOIP, Otd. Geol.*, 54 (5): 81–87.

Thery, J.M. (1985). Nouvelles données de l'Ordovicien colombien. Implications régionales. *Géodyn. Carraïbes Symp.* Paris, vol. 1, pp. 495–503.

Thomas, W.A. (1977). Evolution of Appalachian-Ouachita salients and recesses from re-entrants and promontories in the continental margin. *Amer. J. Sci.*, vol. 277, pp. 1233–1278.

Thompson, T.L. (1979). The Mississippian and Pennsylvanian (Carboniferous) systems in the United States—Missouri. *Geol. Surv. Profess. Pap.*, no. 1110 M-DD, pp. 1–22.

Thorpe, R.S., Beckinsale, R.D. and Patchett, P.J. (1984). Crustal growth and Late Precambrian-Early Palaeozoic plate tectonic evolution of England and Wales. *J. Geol. Soc. Lond.*, 141 (3): 521–536.

*Timofeev, P.P., Kholodov, V.N. and Zverev, V.P. (1986). Evolution of natural water masses on the Earth and sedimentary processes. *Dokl. AN SSSR*, 288 (2): 444–447.

Torsvik, T., Smethrust, M.A., Briden, J.C. and Sturt, B.A. (1990). A review of Palaeozoic palaeomagnetic data from Europe and their palaeogeographical implications. *Geol. Soc. Am. Mem.*, no. 12, pp. 25–41.

Trettin, H.P. and Balkwill, H.R. (1979). Contributions to the tectonic history of the Innuitian province, Arctic Canada. *Can. J. Earth Sci.*, vol. 16, pp. 748–769.

Trettin, H.P., Barnes, C.R. and Kerr, J.W. (1979). Progress in Lower Palaeozoic stratigraphy, northern Ellesmere island, district of Franklin. *Pap. Geol. Surv. Can.*, no. 18, pp. 269–279.

Truyols, J. and Julivert, M. (1976). La Sucesion paleozoica entre Cabo Peñas y Antromero (Cordillera Cantabrica). *Trab. Geol.*, no. 8, pp. 5–30.

*Tsukernik, A.B., Sharkova, T.T. and Kravtsov, A.V. (1986). Geological structure of the eastern part of the Mongolia-Altay range. *Izv. Vuzov Geol. i Razvedka*, no. 12, pp. 10–21.

Tucker, M.E. and Reid, P.C. (1973). The sedimentology and context of Late Ordovician glacial marine deposits from Sierra Leone, West Africa. *Palaeogeogr., Palaeoclimat., Palaeoecol.*, vol. 23, pp. 289–307.

Unified and Correlated Stratigraphic Sequence of the Urals. Explanatory Note to Unified and Correlated Stratigraphic Sequences of the Urals, pt. 1. Sverdlovsk (1980).

*Upper Ordovician formations of Mongolia. *Tr. GIN AN SSSR*, no. 354, pp. 26–63 (1981).

*Ustritsky, V.I. (1985). Relationships between the Urals, Pay-Khoy, Novaya Zemlya and Taimyr. *Geotektonika*, no. 1, 51–61.

*Ustritsky, V.I. (1990). Tectonic nature of Lomonosov threshold. *Geotektonika*, no. 1, pp. 77–89.

Vail, P.R., Mitchum, R.M. and Thompson, S. (1977). Seismic stratigraphy and global changes of sea level, Part 4: Global cycles of relative changes of sea level. In: *Seismic Stratigraphy—Applications to Hydrocarbon Deposition*, pp. 83–97. C.E. Payton (ed.). Mem. Am. Assoc. Petrol. Geol. vol. 26.

Vandenberg, A.H.M. (1978). The Tasman fold belt system in Victoria. *Tectonophys.*, 48 (3–4): 267–298.

Van der Voo, R. (1988). Palaeozoic palaeogeography of North America and intervening displaced terranes: Comparison of palaeomagnetism with palaeoclimatology and biogeographical patterns. *Geol. Soc. Amer. Bull.*, 100 (3): 311–324.

Veevers, J.J. (ed.) (1986). *Phanerozoic Earth History of Australia*. Oxford Sci. Publ., Clarendon Press, Oxford.

Veevers, J.J. and Powell, C. McA. (1987). Late Palaeozoic glacial episodes in Gondwanaland reflected in transgressive-regressive depositional sequences in Euramerica. *Geol. Soc. Amer. Bull.*, 98 (4): 475–487.

*Veimarn, A.B. and Milanovsky, E.E. (1990). Famennian rifting on the example of Kazakhstan and some other regions of Eurasia. Paper 1. *Bull. MOIP, Otd. Geol.*, no. 4, pp. 34–47.

*Vergunov, G.P. and Pryalukhina, A.F. (1980). Permo-Triassic stage of geological development of foreign Alpine Asia. *Izv. Vuzov Geologiya i Razvedka*, no. 7, pp. 14–27.

Vicente, J.-C. (1975). Essai d'organisation paléogeographique et structurale du Paléozoïque des Andes Méridiónales. *Geol. Rdsch.*, vol. 64, no. 2.

Visser, J.N.J. (1982). Upper Carboniferous glacial sedimentation in the Karroo Basin near Prieska, South Africa. *Palaeogeogr., Palaeoclimat., Palaeoecol.*, 38 (1): 63–92.

Walker, J.D., Klepacki, D.W. and Burchfiel, B.C. (1986). Late Precambrian tectonism in the Kingston Range, southern California. *Geology*, 14 (1): 15–18.

Wallin, E.T., Mattinson, J.M. and Potter, A.W. (1988). Early Palaeozoic magmatic events in the eastern Klamath Mountains, northern California. *Geology*, 16 (2): 144–148.

Wang Quan and Liu Xueya (1986). Paleoplate tectonics between Cathaysia and Angaraland in Inner Mongolia of China. *Tectonics*, 5 (7): 1073–1088.

Waterhouse, J.B. and Sivell, W.J. (1987). Permian evidence for Trans-Tasman relationships between East Australia, New Caledonia and New Zealand. *Tectonophys.*, vol. 142, pp. 227–240.

Watkins, R. (1986). Late Devonian to Early Carboniferous turbidite facies and basinal development of the eastern Klamath Mountains, California. *Sediment. Geol.*, 49 (1–2): 51–71.

Watkins, R. and Flory, R.A. (1986). Island arc sedimentation in the Middle Devonian Kennet formation, eastern Klamath Mountains, California. *J. Geol.*, 94 (5): 753–761.

Watts, A.B. (1982). Tectonic subsidence, flexure and global changes of sea level. *Nature*, vol. 297, pp. 470–474.

Webby, B.D. (1978). History of the Ordovician continental platform shelf margin of Australia. *J. Geol. Soc. Austral.*, 25 (1–2): 41–63.

Wehr, F. and Glover, L. (1985). Stratigraphy and tectonics of the Virginia-North Carolina Blue Ridge: Evolution of a Late Proterozoic-Early Palaeozoic hinge zone, *Geol. Soc. Amer. Bull.*, 96 (3): 285–295.

Weyl, R. (1980). *Geology of Central America*. Beitr. Reg. Geol. Erde, vol. 15. Gebrüder Borntraeger, Berlin-Stuttgart.

Williams, D.M. (1980). Evidence for glaciation in the Ordovician rocks of Western Ireland. *Geol. Mag.*, 117 (1): 81–86.

Williams, H. (1979). Appalachian orogen in Canada. *Can. J. Earth Sci.*, pt. 2, 16 (3): 792–807.

Williams, H. and Hatcher, R.D., Jr. (1982). Suspect terranes and accretionary history of the Appalachian orogen. *Geology*, vol. 10, pp. 530–536.

Wolfart, R. and Wittekindt, H. (1980). *Geologie von Afghanistan*. Beitr. Reg. Geol. Erde, vol. 14. Gebrüder Borntraeger, Berlin, Stuttgart, 500 pp.

Wright, A.E. (1977). The evolution of the British Isles in the Late Precambrian. *Estud. Geol.*, 33 (4): 303–313.

Wright, T.O., Ross, R.J., Jr. and Repetski, J.E. (1984). Newly discovered youngest Cambrian or oldest Ordovician fossils from the Robertson Bay terrane (formerly Precambrian), northern Victoria Land, Antarctica. *Geology*, 12 (5): 301–305.

Wrucke, Ch.T., Churkin, M. and Heropoulos, J.C. (1978). Deep-sea origin of Ordovician pillow basalts and associated sedimentary rocks, northern Nevada. *Geol. Soc. Amer. Bull.*, 89 (8): 1272–1280.

Xiang Dingpu (1982). Characteristics of the geological structures in the Qilianshan region, China. *Sci. Geol. Sin.*, no. 4, pp. 364–370.

Yanez, P. et al. (1991). Isotopic studies of the Acatlan complex, southern Mexico: Implications for Palaeozoic North American tectonics. *Geol. Soc. Amer. Bull.*, vol. 103, pp. 817–828.

Yang Shin-Pu, P'an Kiang, and Hou Hung-Fei (1981). The Devonian system in China. *Geol. Mag.*, 118 (2): 113–138.

*Yanshin, A.L. (1973). About the so-called world's transgressions and regressions. *Byull. MOIP, Otd. Geol.*, 48 (2): 9–44.

Young, G.M. (1977). Stratigraphic correlation of Upper Proterozoic rocks of northwestern Canada, *Can. J. Earth Sci.*, vol. 14, pp. 1771–1787.

Young, G.M. (1982). The Late Proterozoic Tinder group, east-central Alaska. Evolution of a continental margin. *Geol. Soc. Amer. Bull.*, vol. 93, pp. 759–783.

Young, L.M. (1970). Early Ordovician sedimentary history of Marathon geosyncline, Trans-Pecos, Texas. *Bull. Amer. Assoc. Petr. Geol.*, 54 (12): 2303–2316.

*Zaitsev, Yu.A. (1984). *Evolution of Geosynclines (Oval Concentric Zonal Type)*. Nedra, Moscow.

*Zamaletdinov, T.S. and Osmonbetov, K.O. (1988). Geodynamic model of development of the Earth's crust of Kirghizia in the Phanerozoic. *Sov. Geologiya*, no. 1, pp. 66–75.

Zeil, W. (1979). *The Andes. A Geological Review*. Beitr. Reg. Geol. Erde, vol. 13, VIII, 260 pp.

Zhang Zhijin (1984). Lower Palaeozoic volcanism of northern Qilianshan, NW China. In: *Marginal Basin Geology*, pp. 285–289. Geol. Soc. Spec. Publ. no. 16.

*Zhivkovich, A.E. and Chekhovich, P.A. (1985). *Palaeozoic Formations and Tectonics of Ufa Amphitheatre*. Nauka, Moscow.

Ziegler, P.A. (1982). *Geological Atlas. Western and Central Europe*. Shell Internationale Petroleum Maatschappij B.V., 130 pp.

Ziegler, P.A. (1986). Geodynamic model for the Palaeozoic crustal consolidation of Western and Central Europe. *Tectonophys.*, 126 (2–4): 303–328.

*Zonenshain, L.P., Kuz'min, M.I. and Kononov, M.V. (1987). Absolute reconstructions of positions of continents in the Palaeozoic and Early Mesozoic. *Geotektonika*, no. 3, pp. 16–27.

*Zonenshain, L.P. and Natapov, L.M. (1987). Tectonic History of the Arctic. In: *Actual Problems in the Tectonics of Oceans and Continents*, pp. 31–56. Nauka, Moscow.

Zu-qi Zhang (1984). The Permian system in South China. *Newslett. Stratigraph.*, B (3): 156–174.

Zwart, H.J. and Dornsiepen, U.P. (1980). The Variscan and Pre-Variscan tectonic evolution of Central and Western Europe: A tentative model. *Mém. BRGM*, no. 108, pp. 226–232.

INDEX

414

Vistula-Dnestr pericratonic trough, zone 4, 232
Volga-Uralian anteclise, basin, region 159, 227, 228, 230, 231, 232, 233, 318
Volta syneclise 85
Volyn-Podol basin 106
Voronezh arch, uplift 159, 228

Wagga basin 80, 101, 131, 132, 151
Wagga-Omeo metamorphic belt, plutonic belt, zone 130, 131
Wales basin, borderland, trough, uplift 144, 173
West African Craton, segment 11, 63
West Antarctic region 16, 31, 81, 133, 200, 221, 281, 351, 354-b, 359
West Asturia zone 21, 22, 72, 73, 74, 120, 121, 175, 177, 178
West Canadian basin 105, 153, 155, 156, 223
West European orogen, region 20–22, 72–74, 103, 119–121, 175–180, 181, 232, 253–256, 257, 258, 263, 298–302, 354-a, 358, 359, 369
West European Platform 233
West Pacific belt 28–31, 43–49, 79–81, 99–101, 137, 150–151, 194–200, 219–221, 270, 273–281, 305–309, 337, 350, 351, 354-b, 355, 356, 357, 361, 362
West Siberian Platform, system 43, 57, 92, 150, 268, 288, 304, 305, 319, 331
Western Congolides belt 12, 314
Western Kalba zone 194
Western Sayan basin, massif, region, trough 27, 43, 78, 96, 190, 191, 192, 350, 361
Western Texas aulacogen, basin 224, 225, 227, 290
Western Urals zone (*also see*, Uralian zone) 262, 263

Wichita aulacogen, mountains, trough 2, 224, 289, 297, 319
Wichita-Marathon system 297
Williston basin, syneclise 2, 53, 54, 153, 154, 155, 156, 209, 223, 224, 227
Wisconsin uplift 154
Wollomin trough 133
Wonnaminta microcontinent 47

Yana-Kolyma basin 59, 108, 109
Yangtze Craton, Platform 45, 60, 111, 130, 142, 163–164, 210, 220, 237–238, 315, 361
Yarrol thrust, trough 279–280, 309
Yarrol-Tamworth trough 133, 198, 199
Yenisei-Baikal orogen 234
Yenisei-Khatanga aulacogen 209, 235
Yenisei-Patom orogenic region, system 28
Yenisei-Sayan orogenic belt 8
Yerementau-Chu-Ili zone 24, 25, 76, 77
Ygyatty basin 161
Yucatan block 171
Yudoma belt, trough 7, 8
Yukon massif 282
Yulo uplift 197
Yunkai island arc, uplift 29, 79
Yunnan-Guangxi basin 60, 164, 237
Yunnan-Malaya basin, system 121, 122, 182, 216, 258, 259, 260, 261, 339, 351, 352, 354, 354-a, 356, 359, 361
Yustyd basin 190

Zambezi graben, trough 241, 243
Zechstein basin 255, 330
Zharma zone 194, 269
Zhe-Ming uplift 278, 309
Zilair series, trough, zone 185, 187, 217, 262, 335
Zimbabwe uplift 240